Micro and Nanoplastics in Aquatic Environment

Editors

Ashok Vaseashta
International Clean Water Institute,
Office of Strategic Research, Manassas, VA, USA
Ghitu Institute of Electronic Engineering and Nanotechnology,
Technical University of Moldova, and
Academy of Sciences of Moldova, Chisinau, Moldova
Institutul de Cercetare al Universităţii din Bucureşti, Strada
Atomiştilor, Magurele, Romania

Maris Klavins
Department of Environmental Science, University of Latvia
Riga, Latvia

Olena Stabnikova
Advanced Research Laboratory
National University of Food Technologies
Kyiv, Ukraine

CRC Press
Taylor & Francis Group
Boca Raton London New York

CRC Press is an imprint of the
Taylor & Francis Group, an **informa** business
A SCIENCE PUBLISHERS BOOK

First edition published 2026
by CRC Press
2385 NW Executive Center Drive, Suite 320, Boca Raton FL 33431

and by CRC Press
4 Park Square, Milton Park, Abingdon, Oxon, OX14 4RN

CRC Press is an imprint of Taylor & Francis Group, LLC

Library of Congress Cataloging-in-Publication Data (applied for)

ISBN: 978-1-032-48519-5 (hbk)
ISBN: 978-1-032-48520-1 (pbk)
ISBN: 978-1-003-38946-0 (ebk)

DOI: 10.1201/9781003389460

Typeset in Times New Roman
by Prime Publishing Services

Foreword

By Acad. Tiginyanu, Moldova

The pervasive challenge of plastic pollution has captured the attention of scientists, policymakers, and concerned citizens worldwide over the past several decades. Among the many forms of plastic waste, micro- and nanoplastics have emerged as particularly insidious pollutants, slipping through the cracks of conventional waste management systems and infiltrating the very fabric of our aquatic ecosystems. These microscopic particles are not only a testament to the durability and prevalence of plastics but also a stark reminder of the unforeseen consequences of our reliance on synthetic materials.

The timeliness and impact of the book titled *Micro- and Nanoplastics in Aquatic Environments*, cannot be overstated. It addresses a pressing and complex issue that affects ecosystems, economies, and public health on a global scale. By casting light on the sources, pathways, effects of micro- and nanoplastics on human health, and contamination mitigation strategies, this book serves as a vital resource for understanding micro- and nanoplastics that are often invisible yet have profound impacts on society.

As someone deeply invested in environmental science and policy, I am continually struck by the interconnectedness of the emerging challenges we face due to micro- and nanoplastics, which exemplify a catechism, viz., while plastics have revolutionized industries and improved lives, their improper disposal and overuse have created a global crisis. As described in the book, these particles originate from diverse sources such as degraded plastic debris, industrial processes, and everyday products and travel through atmospheric and aquatic systems, crossing geopolitical boundaries and accumulating in ecosystems far removed from their points of origin. Along the way, they interact with other environmental stressors, compounding their impacts and creating cascading effects throughout food webs.

The insights provided in this book are invaluable for several reasons. First, they underscore the complexity of the micro- and nanoplastics problem, integrating perspectives from chemistry, biology, ecology, and toxicology. This multidisciplinary approach is essential for grasping the full scope of the issue and for designing effective interventions. Second, the book goes beyond the academic realm to address practical solutions. By highlighting innovations in materials science, waste management, and policy development, it offers a roadmap for mitigating the impacts of micro- and nanoplastics while fostering sustainable practices.

Equally important is the book's emphasis on the human dimension of the problem. The presence of micro- and nanoplastics in our drinking water, seafood, and even the air we breathe raises profound questions about the long-term implications for human health. Addressing plastic pollution requires a multifaceted approach involving innovation, regulation, and public participation to transition toward a more sustainable future. The intersection of environmental and public health concerns highlights the need for urgent action, informed by robust science, and data-informed decisions guided by a shared commitment to the well-being of both people and the planet.

The book has eleven chapters written by various very well-known authors, including the editor, Prof. Ashok Vaseashta, an Academician and an honorary member of the Academy of Sciences of Moldova. The coverage of topics and solutions presented in this book can inspire many readers, as well as inform. The challenges posed by micro- and nanoplastics can be overwhelming, but the authors' dedication to exploring actionable solutions provides a source of hope. By engaging readers from diverse backgrounds—whether scientists, policymakers, educators, or environmentally conscious individuals—this book fosters a sense of collective responsibility and empowerment.

As the readers delve into the chapters that follow, I encourage you to approach this subject with both curiosity and resolve. The issue of micro- and nanoplastics in aquatic environments is emblematic of the broader environmental challenges we face in the 21st century. It requires not only scientific understanding but also a willingness to rethink our relationship with plastics and the natural world. Let this book be a catalyst for informed dialogue, innovative research, and meaningful action.

I highly commend Prof. Vaseashta as the corresponding editor along with co-editors and several authors for their dedication to explaining this critical issue and their overall commitment to fostering a more sustainable future. I hope that this book will not only deepen our understanding of micro- and nanoplastics pollution issues but also stimulate the collective effort needed to protect and restore our aquatic environments. I also hope that students will gain very useful insight into this subject to expand on this research and explore yet unexplored research topics.

Date: 6 January 2025

Prof. Dr. Ion Tiginyanu, Academician
President, Academy of Sciences of Moldova
Chisinau, Moldova

Foreword

By Prof. Grigoryan and Prof. Avetisyan, Armenia

The growing problem of microplastics in the environment represents one of the most pressing ecological challenges of our time. These tiny plastic particles, less than 5 millimeters in size, are pervasive, infiltrating ecosystems across the globe—from the depths of the oceans to the peaks of remote mountains. Originating from degraded larger plastics and synthetic materials, microplastics contaminate water, soil, and even the air we breathe. Their presence threatens wildlife, disrupts food chains, and poses potential risks to human health. Addressing this issue is crucial to safeguarding environmental sustainability and protecting future generations.

Addressing the growing problem of microplastics in the environment requires a multifaceted approach that emphasizes prevention, management, and innovation. Improving waste management systems is essential to prevent plastic discharge into natural ecosystems, including efficient recycling practices and reducing single-use plastics. Mitigation strategies, such as enhanced filtration technologies in wastewater treatment plants, can help capture microplastics before they reach waterways. Furthermore, investing in upcycling initiatives—transforming discarded plastics into valuable products—can reduce waste and promote a circular economy. Public awareness campaigns and policy regulations also play a pivotal role in driving collective action toward minimizing microplastic pollution.

The edited book *Micro and Nanoplastics in Aquatic Environment* by Prof. Vaseashta encapsulates the urgent nature of this growing concern from a holistic standpoint. The book consists of eleven chapters written by experts in the field, including the editor himself and several colleagues from Armenian institutions, such as the National Polytechnic University of Armenia, the National Academy of Sciences of Armenia, and Russian-Armenian University. This not only represents a collaborative effort to address the global challenge to mitigate microplastics but also a national concern about microplastics within Armenia. The topics presented include plastic waste management, mitigation strategies, and innovative ways to upcycle plastics into useful products.

Microplastics pose significant risks to both human and animal health, as their small size allows them to infiltrate ecosystems, food chains, and even the human body. In animals, ingestion of microplastics can lead to physical harm, such as internal blockages, reduced nutrient absorption, and exposure to toxic chemicals.

These plastics can accumulate in tissues, causing long-term health impacts and potentially affecting reproduction, growth, and survival rates in various species. For humans, microplastics have been detected in water, food, and even the air we breathe. Consuming these particles may expose people to harmful chemicals like phthalates, bisphenol A (BPA), and other additives linked to endocrine disruption, cancer, and developmental issues. Additionally, microplastics may serve as vectors for pathogens or pollutants, further amplifying their risks. The long-term consequences of microplastic exposure remain under study, but their ubiquity underscores the urgent need for solutions to mitigate their impact on health and the environment. These effects are also discussed in this book in cooperation with colleagues from Russian-Armenian University.

The need for continued studies in the field of microplastics is critical to fully understand their environmental, ecological, and health impacts. While significant progress has been made in identifying their sources and distribution, many questions remain unanswered, such as the long-term effects of microplastic exposure on human health and ecosystems. Research is essential to uncover the mechanisms by which microplastics interact with organisms, including their potential to accumulate and transfer toxic substances through food chains.

Additionally, more research is needed to develop innovative solutions for detection, removal, and prevention, as well as to assess the effectiveness of mitigation strategies. Understanding regional variations, degradation processes, and the role of microplastics in global pollution dynamics will also help shape more effective solutions, policies, and interventions. Continued research in this field is not only vital for advancing scientific knowledge but also for informing sustainable practices and safeguarding public health and environmental integrity. It is also imperative that the ever-expanding workforce needs books, reference manuals, and a comprehensive source of information for this specialized field. Typically, such information is available in several books, reference guides, and literature. It is quintessential to have the knowledge base in one consolidated book, such as this. We highly commend the work done by Prof. Vaseashta to put together this book and also contribute to several chapters along with colleagues in the Republic of Armenia and many other countries, demonstrating collaboration on topics that are essential to our future sustainability.

We wish the editor much success with this book and are sure that many students and researchers will be reading this book with great interest.

Prof. Dr. Areg Grigoryan
Head of the Scientific Laboratory,
National Polytechnic University of Armenia, Armenia

Prof. Dr. Armine Avetisyan
National Polytechnic University of Armenia, Armenia

Preface

In recent decades, the proliferation of plastic waste has emerged as one of our time's most critical environmental challenges, and it may be ranked as a grand challenge of the 21st century. The ubiquity of plastics in everyday life, coupled with their persistence in the environment, has proved it to be a critical and complex issue, viz., maintaining the valuable uses of plastics while minimizing the accumulation of micro- and nanoplastics in aquatic ecosystems. In fact, the accumulation of micro- and nanoplastics in our aquatic ecosystems has emerged as a particularly insidious pollutant, slipping through the cracks of conventional waste management systems and infiltration. This pervasive challenge has captured the attention of scientists, policymakers, and concerned citizens alike around the globe, and has caught everyone by surprise by the sheer magnitude of the challenge.

The challenge of micro- and nanoplastics pollution is multifaceted ranging from discarded plastic bottles, fishing gear, rigid plastic 6-beverage pack holders, plastic bags, etc., to the safety of marine life and negative health impacts on humans as well as marine animals. These tiny plastic particles, after environmental degradation, are invisible to the naked eye and pose significant risks not only to aquatic organisms but also to human health and the integrity of entire ecosystems. The intersection of environmental and public health concerns underscores the need for urgent action, informed by robust science and guided by a shared commitment to the well-being of both people and the planet. Hence, these micro-nanoscopic particles are not only a testament to the durability and prevalence of plastics but also a stark reminder of the unforeseen environmental consequences of our reliance on synthetic materials.

This edited book on *Micro- and Nanoplastics in Aquatic Environments* aims to provide a comprehensive overview of the origin, behavior, and consequences of these pollutants. The idea for this book was born out of an urgent need to bridge the gap between scientific research, public awareness, and actionable solutions—both technological and policy implementation. While there has been a growing body of research on micro- and nanoplastics, much of this knowledge remains confined to academic circles, leaving policymakers, environmental managers, and the general public struggling to grasp the full scope of the problem and the potential pathways for mitigation. This book provides an overview of sources of plastics, their impact on human and animal health, mitigation strategies, and recommendations for policies and regulations. Since the problem has far exceeded the technological scope of mitigation and the world has passed the "overshoot day", i.e., the day when the amount of plastic waste outweighs the world's ability to manage it, with environmental pollution occurring as a result. It calls for simultaneous multifaceted approaches, as are presented in this book by several authors. By casting light on the origins, pathways, and effects of micro- and nanoplastics, this book serves as a vital resource for understanding a pollutant.

The aquatic environment, encompassing oceans, rivers, lakes, and even groundwater systems, serves as both a recipient and a conduit for plastic pollution. Micro- and nanoplastics originate from a variety of sources—fragmentation of larger plastic debris, release from personal care products, synthetic textiles, tires, paints, and industrial processes, among others. Once released, these particles can persist for years, traveling across vast distances and interacting with a wide range of physical, chemical, and biological factors. Their small size allows them to infiltrate food webs, starting with plankton and other microscopic organisms and eventually reaching apex predators, including humans. One of the commonly ignored aspects of these particles is security, as neutronic particles can attract pathogens and other viral vectors through biofilms formed across them which is presented in this book.

This book is structured to provide a holistic understanding of the issue. An introductory chapter by Vaseashta et al. provides an extensive overview of statistics on plastic production and waste management. The stark difference between the two highlights the pollution problem that we are facing today. The chapter highlights several other sources of plastics followed by their environmental and health impacts. A section also addresses the potential implications for human health, highlighting the urgent need for interdisciplinary research to fully understand these risks. The chapter delves into waste management and mitigation strategies and concludes by making some policy recommendations and regulatory controls. A chapter by Stabnikova et al. describes that microplastics constitute a widespread and ever-increasing pollution globally in water systems and how the natural processes can mitigate microplastics from the water environment. In a chapter by Khachatryan et al., they also present sources of microplastics in aquatic environments from some sources that are commonly overlooked yet serve as potential sources of these particles. It begins with an exploration of the sources and pathways of micro- and nanoplastics, delving into the complex processes that govern their distribution and fate in aquatic environments. The chapter also presents some interesting remediation strategies. A chapter by Laforsch et al. explores the wide range of microplastic physicochemical properties and how these might influence interactions with and effects on organisms and cells. A chapter by Zoppas et al. presents a review of the intricate and dynamic relationship of microplastics by exploring how biofilms influence the transport, degradation, and potential toxicity of microplastics in aquatic environments. By comprehensively unraveling the tangled web of microplastic-biofilm interactions, the chapter of Zoppas et al. aims to pave the way for informed mitigation strategies and environmental protection against microplastic pollution. Using advanced diagnostics, it is interesting to see the evidence of microplastic dispersion in the sea and how plastic is reduced to micro-nano-sized dimensions. A chapter by Gatti et al. uses a Field Emission Gun Environmental Scanning Electron Microscope coupled with an X-ray microsensor of an Energy Dispersive System to assess their elemental chemical composition. A chapter by Hovhannisyan et al. presents toxicokinetics effects of micro-nano particles on human health and serves as an important reminder that it is critical that we find mitigation strategies to counter such effects. The subsequent sections examine the ecological and physiological impacts on aquatic organisms, including the bioaccumulation and biomagnification of toxic substances associated with plastic particles. Chapters by Baruah et al. and Singh et al. discuss micro-nano particles in soils in terms of its effects on management and policies for sustainable agriculture. Lastly, as a case study, a chapter by Hambaryan et al. studies microplastics in lakes in Armenia

Importantly, this book does not stop at outlining the problems. It also explores innovative solutions and strategies to combat micro- and nanoplastics pollution. From advances in waste management and plastic recycling to the development of biodegradable alternatives and policy interventions, the book seeks to inspire a collective response to this global challenge. Implementation of policies for waste management and regulatory control for recycling and controlled waste disposal will be critical in addressing this global challenge. Furthermore, input from concerned citizens will be necessary to collectively address this challenge. To this point, Vaseashta in his initial chapter has proposed a resolution to barcode plastic nets, enabling attribution for abandoned equipment, especially if a sea animal is found trapped and suffering. Furthermore, the role of education, public engagement, and international collaboration is also emphasized as crucial components of a sustainable solution.

Selecting articles for this book has been both a challenge and a privilege. The complexity of the subject reflects the interconnectedness of environmental, social, and economic systems, reminding us that addressing plastic pollution requires a multidisciplinary and multi-stakeholder approach. As you journey through these pages, we hope you will not only gain a deeper understanding of the science behind micro- and nanoplastics but also feel empowered to contribute to the solutions—whether through research, advocacy, or changes in personal and organizational practices. Let this book serve as a call to action. The health of our aquatic environments, and by extension the well-being of all life on Earth, depends on our ability to tackle the micro- and nanoplastics crisis with urgency and determination. Together, we can move toward a future where the resilience and beauty of our water bodies are restored and preserved for generations to come.

The editors acknowledge the contributions of many fellow scientists, and reviewers who assisted in providing their input, remarks, and critical assessment of the subject matter. Such feedback was critical to the quality of this book. The editors, in no way, claim that the contents of this book completely address the challenge of the scientific issue presented in this book. In fact, in subsequent editions, the authors plan to address additional ongoing and upcoming challenges and opportunities.

Disclaimers: The opinions, recommendations, and propositions presented in various chapters represent the findings, views, and opinions of the editors and may not represent the position(s), views(s), and/or opinion(s) of their institutions. Also, the editors do not promote, endorse, or express their opposition against any product(s) that may have been mentioned in the chapters. Although due precautions are exercised before publication, the authors, however, do not take any responsibility for unintentional infringement of any material of either their own or from elsewhere. For all suggestions, corrections, or exchanges of information, the readers are advised to contact the authors directly.

Ashok Vaseashta
Maris Klavins
07 January 2025 **Olena Stabnikova**

Abbreviations

AAS	Atomic absorption spectroscopy
AFM	Atomic force microscopy
AI	Artificial intelligence
AML	Amlodipine
AMP	Antimicrobial peptides
ANT	Adenine nucleotide translocator
ARB	Antibiotic-resistant bacteria
ARG	Antibiotic-resistant genes
ATV	Atorvastatin
BCF	Bioconcentration factors
BET	Brunauer–Emmett–Teller
BPA	Bisphenol A
CAGR	Compound annual growth rate
CAT	Catalase
CEP	Cephalosporin
CNS	Central nervous system
COPD	Chronic obstructive pulmonary disease
CTD	Conductivity, Temperature, Depth
CyD	Cyclophilin D
DCF	Diclofenac sodium
DDT	Dichlorodiphenyltrichloroethane
DHS	Department of Homeland Security
DIFE	Difenoconazole
DLS	Dynamic light scattering
DNA	Deoxyribonucleic acid
DSC	Differential scanning calorimetry
DTA	Differential thermal analysis
ECHA	European Chemical Agency
EDC	Endocrine disruption
EDS	Energy dispersive spectroscopy
EDX	Energy dispersive X-ray photoelectron spectroscopy
EEF	Estradiol equivalency factor

ELISA	Enzyme-linked immunosorbent assay
ELS	Electrophoretic light scattering
EPA	Environmental protection agency
EPR	Extended producer responsibility
EPS	Exopolymeric substances
EPS	Exopolysaccharides
EPS	Extracellular polymeric substances
ESEM	Environmental scanning electron microscope
FACS	Fluorescence-activated cell sorting
FDA	Fluorescein diacetate hydrolysis
FDA	Food and drug administration
FFF	Field-flow fractionation
FSC	Forward scatter
FTIR	Fourier-transform infrared analysis
GC	Gas chromatography
GIT	Gastrointestinal tract
HEX	Hexaconazole
HM	Heavy metals
HMWPE	High molecular weight polyethylene
HOC	Hydrophobic organic chemicals
HTC	Hyper thermophilic composting
ICPMS	Inductively coupled plasma mass spectrometry
IoT	Internet of Things
IR	Infrared
LCA	Life cycle analysis
LDPE	Low-density polyethylene
LIBS	Laser-induced beam spectroscopy
LIDAR	Light detection and ranging
MDA	Malondialdehyde
MFC	Microbial fuel cell
MFWA	Microbead-free Waters Act
MGE	Mobile genetic elements
MNP	Micro-nano plastic
MP	Microplastic
MS	Mass spectrometry
MYCN	Myclobutanil
NGO	Non-governmental organizations
NMR	Nuclear magnetic resonance
NP	Nanoparticles
NP	Nanoplastics

NPX	Naproxen
NR	Nile Red
NTA	Nanoparticle tracking analysis
OPFR	Organophosphorus flame retardants
PA	Polyamide
PAH	Polycyclic aromatic hydrocarbons
PBDE	Polybrominated diphenyl ethers
PCB	Polychlorinated biphenyls
PCP	Personal care products
PE	Polyethylene
PET	Polyethylene terephthalate
PFAS	Poly-fluoroalkyl substances
PFR	Phosphorus flame retardants
PLA	Polylactic acid
PMMA	Polymethylmethacrylate
POP	Persistent organic pollutants
PP	Polypropylene
PPE	Personal protective equipment
PS	Polystyrene
PU	Polyurethane
PVC	Polyvinyl chloride
QoL	Quality of life
RCRA	Resource Conservation and Recovery Act
REACH	Registration, Evaluation, Authorization, and Restriction of Chemicals Committee
ROS	Reactive oxygen species
SDG	Sustainable development goals
SEM	Scanning electron microscope
SMT	Sulfamethazine
SMX	Sulfamethoxazole
SOC	Soil organic carbon
SOD	Superoxide dismutase
SOM	Soil organic matter
SPME	Solid phase microextraction
SSC	Side scatter
TCS	Triclosan
TEM	Transmission electron microscope
TIC	Toxic industrial chemicals
TIM	Toxic industrial metals
TIR	Total internal reflection

TMO	Tire microplastics
TRIP	Toxics release inventory program
TSCA	Toxic Substances Control Act
TWP	Tire wear plastics
UAV	Unmanned aerial vehicles
UHMWPE	Ultra-high molecular weight polyethylene
UNEA	United Nations Environmental Assembly
UNEP	United Nations Environmental Program
UV	Ultraviolet
VDAC	Voltage-dependent anion channels
VOC	Volatile organic compounds
WWPT	Wastewater treatment plant
XPS	X-ray photoelectron spectroscopy

Contents

1 | Micro/Nano Plastics in Aquatic Biota: Concerns, Risks, Mitigation, and Policies

Ashok Vaseashta[1,2,3(0000-0002-5649-0067)], **Ioan Stamatin**[3] **and Surik Khudaverdyan**[4]

[1] International Clean Water Institute, Office of Strategic Research, Manassas, VA, 20112 USA

[2] Ghitu Institute of Electronic Engineering and Nanotechnology, Technical University of Moldova, and Academy of Sciences of Moldova, Chisinau, Moldova

[3] Institutul de Cercetare al Universităţii din Bucureşti, Strada Atomiştilor, Magurele, Romania

[4] National Polytechnic University of Armenia, 104 Teryan St. Yerevan, Armenia

Abstract: A large variety of polymers and the versatility of their properties have transformed almost everything around us. While their widespread use for plastics, including disposable aspects, was anticipated, some of the recent reports associated with the extent of environmental pollution created by nano/microplastics were not articulated. The presence of microplastics in treated water and food products has presented serious questions and concerns about the impact the fragmented plastics might have on human health and aquatic well-being. However, to reap the benefits of plastics while keeping pollution to a minimum, it is critical to reimagine the topic of plastics from a holistic standpoint. It is important to identify specific uses that offer clear advantages and to refine national and international standards and associated product labeling to specify suitable usage and appropriate disposal methods. Since a large fraction of plastic waste ends up in the aquatic environment, it is critical to consider the ecological risks presented by micro- and nano plastics in aquatic biota through trophic transfer, that eventually impacts human health through the food chain supply. Aquatic ecosystems have already been affected by a multitude of environmental stressors, and now the presence of micro/nano plastics represents a significant additional risk to human health. In this context, the chapter provides an overview of various aspects of nano/microplastics in the environment using

*Corresponding author: prof.vaseashta@ieee.org

temporal trends in MNPs in aquatic environments to better assess their risks to human health to develop appropriate mitigation and management strategies, such as identification of key knowledge gaps, strategies to mitigate environmental pollution, and policies to minimize their impact.

1. Introduction: Framing the Global Concerns

Historically, humans have been using natural polymers, such as rubber derived from the sap of rubber trees (*Hevea brasiliensis*), Shellac—a resin secreted by the lac bug, and Gutta-percha—a rubber-like material derived from trees native to Southeast Asia for a long time. The first synthetic plastic dates back to the mid-19th century, when the first synthetic plastic, Parkesine, was developed from cellulose by Alexander Parkes in 1862, marking the beginning of man-made polymers. In 1907, Belgian chemist Leo Baekeland created Bakelite, the first fully synthetic plastic, which revolutionized industries due to its heat resistance and electrical insulating properties. The 20th century experienced rapid advancements, with the invention of polyethylene in the 1930s and nylon in 1935, which brought plastics into our daily lives. During World War II, plastic production was ramped up to meet demands from the defense industry. In the post-war consumer culture, using technological innovations and advanced synthesis methods to create and manipulate isomers, synthetic polymers became an existential part of our daily existence due to several useful applications supporting our daily functionalities. Since then, Global plastics production has increased exponentially, especially in the past two decades. At present, the production is over 502.5 million tons (MT) worldwide. At the current trajectory and barring any binding treaty to limit plastic production, the number is on track to more than double by 2050. During the same period, mismanaged plastic waste is expected to double, reaching over 120 MT.

With increasing applications of plastics, there has been growing global concern with its use, reuse, recycling, and overall life cycle. According to recent reports, approximately 52 MT of plastic waste enters the environment annually. As per recent sets of data, for the U.S. about 25% of used plastics are recycled, 8–12% are used for energy conversion (with a significant environmental pollution footprint), and slightly over 53% of plastic waste ends up in landfills, rivers and estuaries, and oceans (Ritchie et al., 2023) (Figure 1). A 2024 analysis by the United Nations Environment Program (UNEP) warns that at the current trajectory, global municipal solid waste could increase from 2.1 BT in 2023 to 3.8 BT by 2050 (Andersen, 2024). The economic cost of waste mismanagement, including hidden costs like pollution and health impacts, could rise dramatically to $640B annually by 2050 without intervention. The report further emphasizes the need for enhanced waste management strategies, promoting circular economies to decouple waste generation from economic growth. September 5th, 2024 marked the "Plastic Overshoot Day" when plastic waste generation surpassed global management capacity (Perreard et al., 2024).

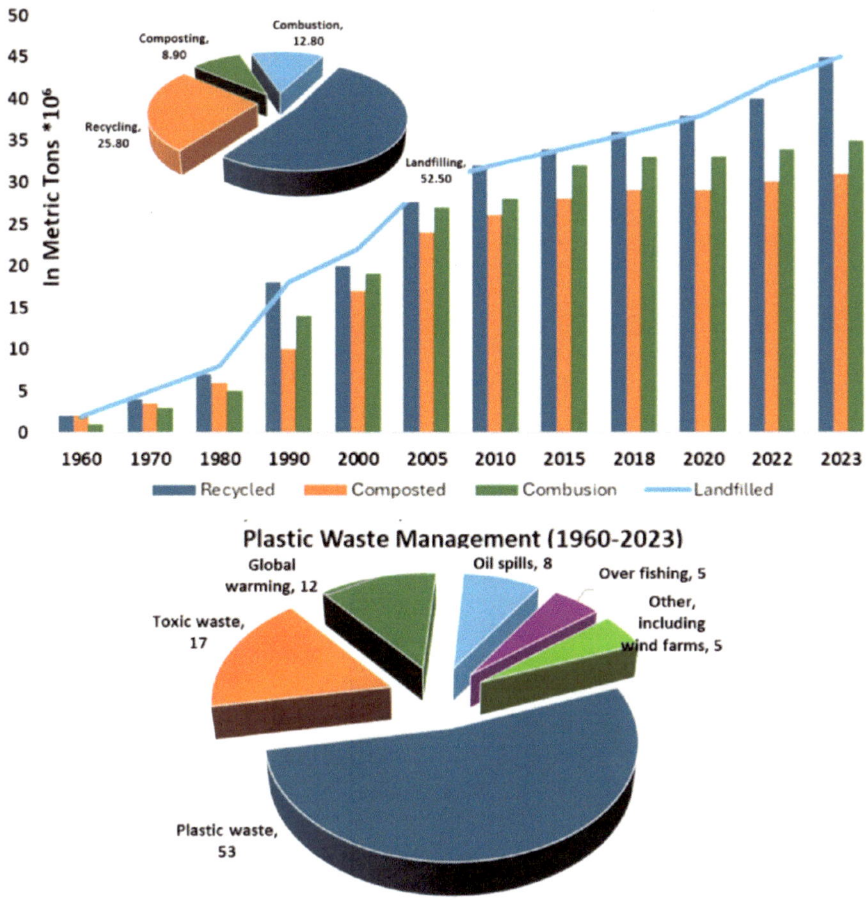

Figure 1: (a) Plastic waste management data from 1960–2023 (Data extracted from various sources). (b) Threat to Aquatic Environment.

Microplastics are generally produced due to physical disintegration and chemical, biological, and UV-induced degradation of plastic materials, arising mainly from tableware, single-use beverage bottles, grocery bags, package wrappings, and additives in cosmetics – which fragment into microbeads, microspheres, and fibers (Stabnikova et al., 2021; Vaseashta, Ivanov, et al., 2021). Additional processes include industrial abrasives and pallets used in plastic production, generally called nurdles. These byproducts enter our environment through various mechanisms, primarily by discarding plastic waste on land or in water. Micro-nanoplastics (MNPs) are found across our ecosystems, from oceans to freshwater systems and even in the air. Figure 2, shows various sources of plastic waste, the largest being due to the packaging of goods (data extracted from various open-source databases).

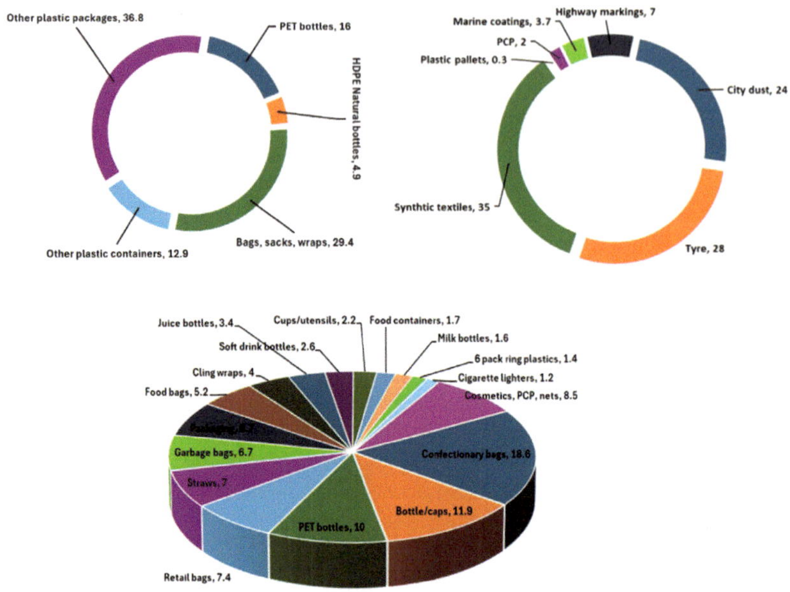

Figure 2: Sources of plastics by percentages in aquatic environment: (a) packaging-related waste, (b) incidental contaminants (viz. paints, coatings, synthetic textiles, nurdles, etc.), and (c) various household sources.

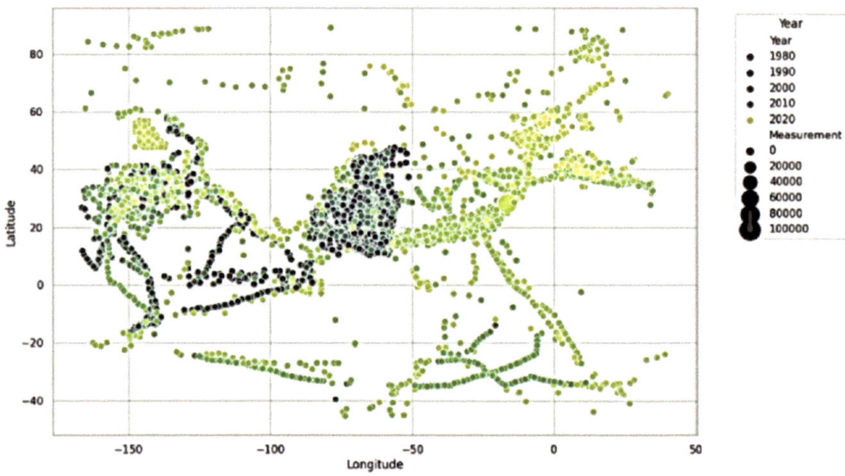

Figure 3: Scatter plot of microplastic density by latitude and longitude over time (1980–2020s).

Using temporal trends in marine microplastic density in open-source datasets, we see that a vast amount of plastic trash exists in our aquatic environments. Several databases exist that provide data on the MNP debris on the ground and in the oceans, depending upon sampling and processing based on factors such as budget, equipment availability, labor, and the specific research questions, however qualitative data on microplastic density distribution since 1980, is quite revealing, as shown in Figure 3. Albeit the data shown is not quantitative, it projects a rather bright light on the growing trend of MNPs in the aquatic environment, with the hope and vision to create a cleaner and more sustainable planet by developing zero-sum ways to manage the plastic waste and ways to mitigate environmental plastic in environmentally friendly by products.

Microplastics (MPs) are often defined as plastics either specifically manufactured or environmentally fragmented particles <5 mm in size (Masura et al., 2015; Vaseashta, Ivanov, et al., 2021). MPs occur in several forms: (a) primary MPs, i.e., plastic particles specifically manufactured for their abrasive qualities (e.g., microbeads and industrial scrubbers), (b) secondary MPs originating from parent materials such as discarded plastic items and synthetic textiles, and (c) tertiary MPs which includes any pre-production pellets used for molding plastic goods (Song et al., 2024). In the context of plastics recycling and circular economy, it is critical to study various aspects of its interaction with the environment due to growing concern resulting from the presence of MNPs in the environment, and other forms of plastic debris in municipal and industrial wastewater, rivers, lakes, and seawater. Conventional classes of plastic pollutants consist of nanoplastics, microplastics, mesoplastics, and macroplastics (Song et al., 2024; Stabnikova et al., 2021; Vaseashta, Ivanov, et al., 2021). As discussed in a later section, using various routes, such as ingested by marine life, consumption through soil, air, and water, and indirect consumption, the plastics enter the food chain and have been detected in human food, seafood, and other regular nutritional pathways, including water supplies, as shown in a conceptual schematic in Figure 4.

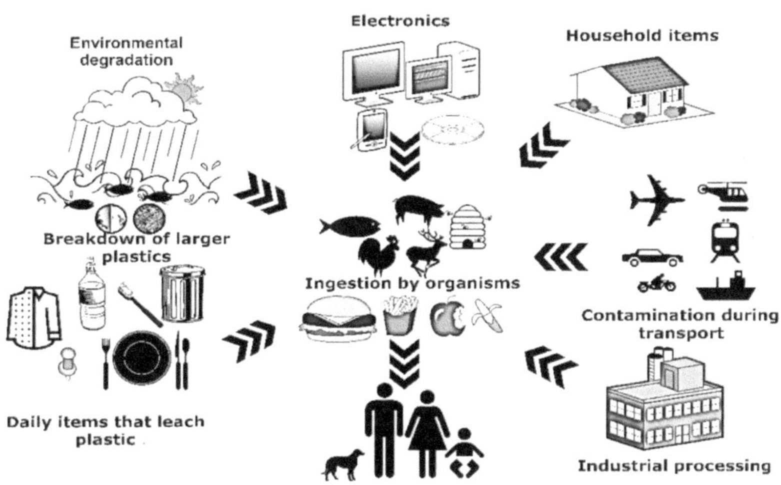

Figure 4: Potential pathways for micro/nano plastics to contaminate the food chain supply.

It is estimated that over 5 trillion plastic pieces float in the ocean and MPs far outweigh MPs by mass, as plastic concentration in the whole ocean is ~ 2×10^{-9} g/L^2 compared to 8×10^{-2} g/L^2 in highly contaminated rivers (Eriksen et al., 2014). This indicates that our municipality's water intake is at greater risk of pollution, leading to higher plastic pollution through the intake of municipality water. The pollution due to MNPs significantly impacts our ecosystems, leading to reduced biodiversity, weakened fisheries sustainability, disruption of phytoplankton and zooplankton life cycles, and deterioration of human and animal health, while also threatening economic stability, posing an ever-increasing ecological threat. Recent studies show MPs are accumulating in human tissues, thus raising concerns about potential health effects. Studies have shown that MNPs can penetrate biological membranes, enter cells, and disrupt cellular functions, potentially leading to toxicity. Recent research has highlighted the relationship between environmental chemical exposures and cardiovascular diseases (CVD), viz., cardiovascular and metabolic disturbances that could ultimately lead to atherosclerosis, hypertension, and even premature death (Zheng et al., 2024). Recent studies have identified MNPs in a variety of tissues, including the heart, blood vessels, and atherosclerotic plaques (Prattichizzo et al., 2024). There is a conclusive relationship between the size of MNPs and their toxicological effects on the body. Since MNPs can bind pollutants and heavy metals, they pass through biological barriers or are adsorbed on various microbial surfaces (Kokilathasan & Dittrich, 2022) and physically obstruct light transmission, and disrupt photosynthesis or respiration, thereby leading to oxidative stress (Dissanayake et al., 2022; Gao et al., 2021), causing the generation of reactive oxygen species (ROS) (Das, 2023; Shi et al., 2023; Singh et al., 2022), damaging DNA, lipids, and other biomacromolecules (You et al., 2022). Plastic-related health risks are also linked to additives used with plastics, such as plasticizers–Bisphenol A (BPA) and phthalates, which are known for their endocrine-disrupting characteristics (Kahn et al., 2020). MNPs can enter the soil through a variety of routes, such as mulching with plastic, compost, and sewage sludge, as well as precipitation, and have negative impacts on the soil, food safety, and living organisms. Thus, through seepage, the groundwater is also contaminated with MNP pollution and poses a risk to drinking water supplies (Vaseashta, Gevorgyan, et al., 2021; Vaseashta, Duca, et al., 2022). Accumulated MNPs in water detrimentally affect the structural and functional traits of aquatic biota and the related ecosystem functions (Adegoke et al., 2023; Di Pippo et al., 2023). MNPs directly affect living beings, either through ingestion or toxicity, and could act as a vehicle for invasive species by adsorption on their surface, or via other synthetic chemical pollutants, thus potentiating or synergizing their toxic capabilities due to their ingredients (Vethaak & Legler, 2021). Additionally, negatively charged particles are more likely to accumulate in vascularized organs due to their ability to cross cellular membranes. However, so far only limited toxicokinetic data exists and a thorough and systematic study is required to evaluate MNP entry routes, e.g., through inhalation, ingestion, absorption, or injection, and related health impacts. This is further elaborated in a later section, highlighting specific health-related risks, with examples.

Plastic waste is littering aquatic ecosystems, contaminating the air of communities near production sites, and threatening public health with MNPs, notwithstanding the beneficial uses of plastics in most aspects of our daily lives. Hence, it is vital to study the ways to detect, sample, and filter these MNPs in marine environments, soil, and air. In addition to their capture, it is necessary to characterize and identify the source. There are limited capabilities to characterize fragmented plastics, however, a set of capabilities must be developed to quickly sample a cocktail of MNPs and additives that are used for the custom design of plastic-based products. We describe below the various sources of plastics, methods to sample MNPs in different environments, and ways to characterize the MNPs, especially, in water, soil, and air. The identification of MNPs is necessary not only for source attribution, especially for regulatory controls, but also to study health impacts on marine and human health and develop strategies for mitigation of MNPs from our immediate environments.

2. Ubiquitous Nature of the Plastics – Type, Origin, and Sources

The pervasive presence of plastic in nearly every facet of human life seems to be inextricably essential, which underscores the necessity of a thorough analysis, such as reported in this chapter. One of the principal uses of plastics is for consumer packaging products, such as single-use plastic containers for food and beverages. Plastic is now used in cosmetics and personal care products (PCP) for hygiene and longevity. Plastic is also extensively used in medical devices and tools, pharmaceutical packaging, and personal protective equipment (PPEs), as was the case during COVID-19 (Baldwin et al., 2023). Due to plastics being lightweight, flexible, and having a high dielectric constant, they are used as insulation in cables and wires, consumer electronics, and wearable electronics. The use of plastics also dominates the automotive and aviation industries as these materials are lightweight, flexible, and can absorb vibration and kinetic energy through fiber reinforcements. The co-existence of the convenience of plastics in our daily lives and their environmental impact makes it quite complex to balance the two and hence necessitates strategic management through technological innovations, policy regulations, and the development of sustainable alternatives. Table 1 lists below commonly used plastics and their common use. The table is not exhaustive and serves only illustrative purposes.

Besides its intrinsic characteristics, additives are introduced during the fabrication of plastics products, since they enhance certain desired features and improve performance. Without these additives, polymers would be limited in their use and durability, However, as a part of life cycle management, it is critical to study their environmental impact, health risks, and compatibility which is crucial

Table 1: List of common plastics, type, properties, and their common uses

Name of Plastic	Plastic Type	General Properties	Common Uses
Acrylonitrile Butadiene Styrene (ABS)	Acrylonitrile C_3H_3N, Butadiene, CH_2=CH-CH=CH_2 and Styrene C_8H_8	Strong, impact-resistant, rigid and glossy	LEGO bricks, electronic casings (computers, game consoles)
High-density Polyethylene	HDPE, ethylene (C_2H_4) monomers	Durable, resistant to chemicals, lightweight	Milk bottles, detergent containers, pipes, grocery bags
Low-density Polyethylene	LDPE, ethylene (C_2H_4) monomers	Flexible, moisture-resistant	Plastic bags, film wraps, squeezable bottles, food packaging
Polyamide	Polymer with repeating units linked by amide bonds	Strong, abrasion-resistant	Clothing, ropes, machine parts, toothbrush bristles
Polybutylene Terephthalate	PBT – $C_{12}H_{12}O_4)_n$	Tough, heat-resistant	Electrical components, automotive parts, fibers
Polycarbonate	PC – functional groups linked by carbonate groups –O–(C=O)–O–.	Transparent, shatter-resistant	Eyeglass lenses, water bottles, CDs, bulletproof glass
Polyethylene Terephthalate	Repeating ($C_{10}H_8O_4$) units	Lightweight, transparent, strong, recyclable	Bottles (water/soft drinks), food containers, clothing (polyester)
Polylactic Acid	Lactic acid with cyclic di-ester lactide	Biodegradable, compostable	3D printing, disposable cups, cutlery, food packaging
Polymethyl Methacrylate	Poly(methyl 2-methyl propanoate)	Transparent, scratch-resistant	Acrylic glass (Plexiglass), aquariums, signage
Polypropylene	Chain-growth polymerization from the monomer propylene	Heat-resistant, tough, lightweight	Bottle caps, straws, yogurt containers, automotive parts
Polystyrene	PS – $(C_8H_8)_n$	Lightweight, insulating, brittle	Disposable cutlery, foam packaging, insulation panels
Polytetrafluoroethylene	Teflon (PTFE) – $(C_2F_4)_n$	Non-stick, heat-resistant, chemically inert	Non-stick cookware, gaskets, plumbing tape
Polyvinyl Chloride	PVC – $(C_2H_3Cl)_n$	Rigid or flexible, resistant to weathering	Pipes, vinyl flooring, electrical cables, medical tubing

Table 2: Additives in polymers – types, purpose, and common uses

Additive Type	Example	Purpose	Common Uses
Antimicrobials	Ag, Cu, Z, QACs, Organics	Inhibit bacterial or fungal growth	Medical devices, food containers, textiles (Shrivastava, 2018)
Antioxidants	Primary, secondary, synergistic	Prevent oxidation during processing, aging	Food packaging, cables, automotive plastics (Almeida et al., 2022)
Antistatic Agents	Ionic, non-ionic, conductive, polymeric. Alkylamines, ethoxylated surfactants, Cabon, CNT	Reduce static electricity buildup	Electronic packaging, conveyor belts (Almeida et al., 2022)
Colorants (Pigments/Dyes)	Pigments and dyes: phthalocyanines (blue/green) and azo pigments, TiO_2, iron oxides, carbon black	Provide color or opacity	Toys, bottles, packaging, textiles (Hahladakis et al., 2018)
Coupling Agents	Silane, maleic anhydride-modified polymers, titanate and zirconate, and isocyanate	Enhance bonding between plastic fillers or fibers	Composite materials, reinforced plastics (Y. Shi et al., 2021)
Fillers	$CaCO_3$, silica, clay, glass fibers, carbon black, organic fibers	Reduce cost or enhance mechanical properties	Calcium carbonate in packaging, talc in automotive plastics (Hahladakis et al., 2018)
Flame Retardants	Halogenated (TBBPA, Deca-BDE), non-halogenated (APP, ATH), and phosphorus based	Reduce flammability	Electronics casings, furniture, building materials (Tawfik, 2017)
Foaming Agents	Chemical (ADC, sodium bicarbonate, hydrazine), physical foaming – gases, volatile liquids	Create lightweight foam structures	Packaging materials, insulation, disposable cups (Hahladakis et al., 2018)
Impact Modifiers	ABS, MBS, PMMA, EPDM	Improve toughness and impact resistance	Car bumpers, helmets, appliance housings (Hahladakis et al., 2018)

(Contd.)

(*Contd.*)

Additive Type	Example	Purpose	Common Uses
Lubricants	Internal lubricants, fatty acid amides – erucamide, oleamide, and esters-glyceryl monostearate	Reduce friction during processing and use	Gears, automotive parts, household appliances (Gubała et al., 2024)
Nucleating Agents	Inorganic, organic, and polymeric	Promote faster crystallization, improving clarity and strength	Food packaging films, bottles (Murphy, 2018)
Plasticizers	Phthalates: DEHP, DINP. Non-phthalate plasticizers: DOTP, ATBC. Bio-based plasticizers-ESBO, Polymeric – Polyesters, poly adipates	Increase flexibility and softness	PVC pipes, vinyl flooring, medical tubing (Shrivastava, 2018)
Stabilizers	Heat, UV, antioxidants, flame, and filler – examples in this table	Prevent degradation from heat, light, or oxygen	Outdoor furniture, pipes, and automotive parts (Hahladakis et al., 2018)
UV Stabilizers	Benzotriazoles and hindered amine light stabilizers (HALS)	Protect from UV radiation to prevent cracking and discoloration	Outdoor products, garden furniture, car dashboards (Murphy, 2018)

for sustainable development, especially from a waste management viewpoint. Additionally, it is critical to examine the health impact of MNPs that display different characteristics than the host polymer, especially due to the presence of additives in the polymer matrix. Table 2 lists a series of commonly used additives with plastics, with examples of their purpose and common uses.

3. Sampling and Identification of Microplastics

Sampling and identifying global MNPs from debris are critical for assessing the extent of environmental contamination and potential impacts on ecosystems and human health. In this context, a brief overview of sampling methods is provided here depending on the sampling environment matrices, viz., marine, soil/sediment, atmospheric, or biota. By gaining more understanding of this emerging pollution through sampling, we can model its fate and transport (Vaseashta, 2015), and develop strategies to mitigate the broad impacts of plastic pollution to foster a healthier planet.

3.1. Sampling Methods for MNP

Sampling of macro- and nanoplastics is critical for addressing environmental, health, and technological challenges presented by this pollution. These particles range in size from visible fragments to macro, micro (and even nano) sizes. Sampling of these particles is vital for well-informed decision-making for attribution, regulatory controls, and also for correlating their impact on human health.

3.1.1. Surface Water Sampling

Typically, it consists of Manta trawls for collecting water samples either in the river or coastal waters using fine meshes or utilizing a neuston net to drag it through the water. The neuston net was originally utilized to capture plankton, however, presently using such nets enables the sampling of MPs through a large volume of water. The method utilizes a net with a mesh of sizes between 300–500 μm. The net can capture small plastic balls, nurdles, and larger pieces of plastic that have degraded over time. The Plankton nets typically capture microplastics at various depths with nets that have smaller mesh sizes (100–300 μm). Additionally, nets with smaller mesh sizes (~20-100 μm), are also used for finer plastic particles. To filter yet smaller particles (~0.45–5 μm), pump sampling is utilized to transport through large volumes of water. The methods described above are sampling, however, *in situ* filtration systems are installed directly in rivers, or water pipelines to capture these particles for monitoring. Furthermore, centrifugation and membrane filters are used to concentrate these particles. For bulk water sampling Niskin bottles, obtained from Aquatic Biotechnology companies, are used to collect water at specific depths. The samples are further filtered in laboratories for analysis. In a study involving the widespread distribution of microplastics in subsurface seawater (Desforges et al., 2014), the water samples were collected and the large plastic debris was filtered utilizing a coarse 5 mm filter to remove large debris and biomass, followed by 250 μm and 125 μm, respectively, aperture size nets to filter the microplastics. Conductivity, Temperature, Depth (CTD) Rosettes were equipped with multiple bottles to sample MPs at various depths. One of the important aspects of research related to water sampling of MPs is its separation from the biomass. Most studies have implemented the flotation separation method to extract MP from water samples, either as a standalone method or mixed with a hypersaline solution (Prakash et al., 2018). The most widely used method is the hypersaline solution with sodium chloride. The water sample is mixed with the sodium chloride solution agitated for a few minutes and left overnight for the separation process (Rani et al., 2023). A study utilized two-part digestion using 30% hydrogen peroxide (H_2O_2) and 40% hydrofluoric acid (HF) for suspended MPs and carbon black particles (Dubaish & Liebezeit, 2013). Even though chemical digestion is very effective in reducing organic matter, MPs may be susceptible to certain acids. Yet, in another study, the organic matters within the water samples collected were digested with hydrochloric (HCl) acid while microplastics were left uncompromised (Lavoy & Crossman, 2021).

3.1.2. Separation and Preconcentration Techniques

Membrane filtration is generally used to isolate particles based on size, while centrifugation and ultracentrifugation separate particles by density. The Field-Flow Fractionation (FFF) separates particles based on their size and molecular weight in the liquid media. The separation studies need additional methods for MNPs pre-identification as the techniques may vary from one particle to the other. Table 3 lists several separation, sampling, and methods that are used to separate MPs in the river and marine environments globally. These techniques are conventional and are widely used. With increasing MNPs in aquatic environments, more sophisticated and *in situ* separation methods must be used to isolate these particles.

3.1.3. Sediment Sampling

The sediment sampling is used to investigate MP accumulation of the presence of meso and MPs from benthic items collected from beaches, estuaries, and the seabed (Van Cauwenberghe et al., 2015) over time. There are various sediment sampling types, which include deep-sea sand, beach sand, river sand, municipal soil, and mangrove mud with the maximum depth of sampling varying from a range of 2–5 cm based on the sampling types conducted (Van Cauwenberghe et al., 2015) using grab samplers. Several methods are used to separate MPs from the benthic material including saline water, or mineral salts. Furthermore, the sieving and density separation methods involve the sieving of sediment to isolate particles followed by density separation using salt solutions (e.g., NaCl

Table 3: Water sampling techniques for microplastic identification.

Sampling Method	Mesh size (μm)	Sampling Depth (m)	Separation Methods	Separation Chemicals
Continuous intake	0.5–5	5	Visual	Manual
Manta trawl	300–500	0.05	Visual	Manual
Neuston net	100–300	Surface	Flotation, visual	NaCl
Sieves or mesh	20–100	Surface	Flotation, visual	Milli-Q water filtered using GF/F 0.7 μm (47 mm φ)
Fine sieves or fine mesh	<10	Sub-surface	Density, flotation, oxidation	ZnCl and 30% H_2O_2

or $ZnCl_2$). Larger plastic fragments, as observed under visual inspection can be extracted using forceps under optical microscopes. The density flotation method normally utilizes sodium-based salts of chloride or iodide. There are several other alternative methods for characterizing the MPs from sediment samples, such as via staining the samples with a Nile Red (NR) acetone solution (Ho et al., 2024). While the method has shown improvement in enhancing the speed for visual monitoring and promises general particle categorization, the NR-stained MP particles rely on utilizing small amounts or concentrations of suspension and necessitate the use of more accurate methods, such as Fourier Transform Infrared (FTIR) spectroscopy. Other techniques, such as chemical characterization by Raman spectroscopy or physical characteristics, are also used to characterize the MPs. For soil samples, columns are used to investigate plastic contamination at different depths by using systematic grids, which ensure unbiased sampling from agricultural fields or urban environments. Density separation is very similar to sediments, where lighter plastic particles are separated from soil using density gradients. Table 4 shows the separation methods, chemicals used, and sampling and characterization techniques applied to acquire and characterize MPs in rivers, estuaries, and marine sediments.

3.1.4. Sampling in Air and Biota

These methods, although not directly related to the topic of the aquatic environment, however, it is vital to mention that MNPs, being lightweight, remain suspended in the air. Hence, air sampling is conducted by pumping air through special filters, such as glass fiber or quartz filters to collect airborne particles. The samples are collected through falling particles over a specific period. For urban or industrial environments, high-volume air samplers are employed to collect airborne MNPs, including fibers over a period of time. Through air or water, the MNPs enter the biota and are normally collected from the digestive

Table 4: Sediment sampling techniques to identify microplastics.

Depth (cm)	Sediment Type	Separation Method	Separation Chemicals	Characterization Methods
2–5	River	Density flotation	NaCl	Physical characterization
2	Beach	Density flotation	NaCl	FTIR spectroscopy
5	Beach	Oil extraction protocol	Canola oil, NaI, $CaCl_2$	FTIR spectroscopy
10	River	Density flotation	$ZnCl_2$	Raman spectroscopy
>10	Deep-sea	Density flotation	NaI	Raman spectroscopy

tracts of fish, birds, or invertebrates. The gut contents are filtered to identify MNPs, using chemical methods, such as enzymatic or chemical digestion (e.g., using KOH or H_2O_2) to break down the organic material, leaving behind plastic fragments. This will be discussed in a bit more detail later while dealing with the health effects and toxicity of MNPs.

3.2. MNP Characterization

It is critical to characterize MNPs in aquatic environments to understand their sources, composition, size distribution, behavior, and potential health and environmental impacts. This is also important from a policy implementation, enforcement, and regulatory control standpoint. These methods often require a combination of several existing, new, and innovative techniques to analyze physical, chemical, and biological characteristics of MNPs of various sizes, shapes, forms, and in different media. Due to the limited scope of this chapter, an overview of some of the key methods is provided here, followed by a listing of these methods in Table 5. The description and list may not be exhaustive, however, it provides an overview of such characterization methods.

3.2.1. Physical Characterization

These methods focus on determining the size, shape, surface area, and morphology of particles. The methods include Dynamic Light Scattering (DLS) for particle size distribution in colloidal suspensions (size ranging from a few nm to microns), Nanoparticle Tracking Analysis (NTA) (Malloy, 2011)—for size distribution and concentration (size 10 nm to 1 μm), Scanning and Transmission Electron Microscopy (SEM and TEM) for observing surface morphology, shape, and internal structure, including high-resolution images of nanoscale particles, and Atomic Force Microscopy (AFM) for 3D topography of particles and surface interactions at nanometer resolution (Vaseashta & Irudayaraj, 2005).

Table 5: Measurement techniques of MNPs and their advantages and disadvantages

Techniques	Particle Size – Range	Advantages	Limitations
Optical microscopes	Large sample size	Easy, measures size, color, morphology, surface topography	Only visual, no polymer characterization
Stereo Microscopes	~100 μm and above	Easy, fast, color, morphology and surface topography	Only visual, no polymer characterization
Scanning Electron Microscope	Few mm	High res.images, morphology, topography	Slow, sample prep is required.
Transmission Electron Microscope	~ Few nm	High res. images, morphology	Very slow, sample prep required
Atomic Force Microscopy	~ A few nm	High res, surface analysis, blended polymer characterization	Sample prep, may damage sample
Fluorescence Microscopy	~ A few μm	Characterizes transparent MNPs	Sample prep, sample damage
TGA-DCS	~ A few μm	Easy, polymers and additives characterization	Destructive
GC-MS	~ A few μm	Polymers and additives characterization	Destructive
HPLC	~ A few μm (MNPs)	Highly sensitive, polymers and additives characterization	Long sample prep
Cytometry	~ A few hundred nm	Fast, liquid samples, sample characterization	Sample contamination due to dye, interference (Salvia et al., 2023)
FTIR	10 μm and above	Non-destructive, identifies polymer, no sample prep.	Slow characterization
FTIR variant (1)	ATR – 0.5 mm	Non-destructive, identifies polymer, no sample prep.	Slow characterization

(Contd.)

(*Contd.*)

Techniques	Particle Size – Range	Advantages	Limitations
FTIR variant (2)	Micro – 20 μm or less	Nondestructive, identifies polymer, no sample prep.	Slow characterization
Raman	10 μm and above	Nondestructive, identifies polymer	Slow characterization
Raman variant	Micro ~1 μm and above	High res, polymer identification	Signal interference
NMR (IR)	Micro ~1 μm and above	Function group identification	Expensive, sample prep, Frequency selection req.
EDS/XPS	Same as SEM	Chemical composition, identifies additives	Sample prep. req.
ELISA	~ A few hundred nm	Measure the growth of bacterial communities	Sample prep req. (Vaseashta & Dimova-Malinovska, 2005; Vaseashta & Irudayaraj, 2005)
ELS	~ A few hundred nm	Measures particle aggregation	Sample prep. req.
Zeta potential	~ A few hundred nm	Measures surface charge	Sample prep. req.
BET	~ A few hundred nm	Measures adsorptive behavior and reactivity of MNPs	Sample contamination
LIBS	~ Size of laser beam – a few μm	Remote sensing	Destructive (V. K. Singh et al., 2018)

3.2.2. Chemical Composition

The chemical analysis focuses on the characterization of the elemental or molecular composition of particles (Brčeski & Vaseashta, 2021). The most commonly used method is FTIR spectroscopy (Gong et al., 2024) to identify organic compounds, especially polymers, such as microplastics. Raman spectroscopy is another commonly used method to detect molecular vibrations encompassing vibrational and rotational coupling to identify plastics (Lv et al., 2020), and provides a higher spatial resolution than FTIR. In Raman spectroscopy, the samples exhibit fluorescence when stimulated by the laser and hence need preliminary filtration to minimize fluorescence effects (Adhikari et al., 2022). FTIR involves infrared absorption patterns as compared to photonic dispersion. IR can furnish additional details based on carbonyl moieties linked to the oxidation of materials based on total internal reflection (TIR) (Chen et al., 2020). Most SEM equipment is outfitted with energy-dispersive X-ray spectroscopy (EDS or EDX) to analyze the elemental composition of samples under test. X-ray photoelectron spectroscopy (XPS) is yet another method that is used to measure surface chemistry and oxidation states of elements. Matrix-assisted laser desorption/ionization time-of-flight mass spectrometry is yet another method for examining plastics. It can be used to differentiate polymer samples, as they are ionized using a laser, depending upon their molecular weights (Lin et al., 2020).

3.2.3. Electrophoretic Light Scattering, Zeta Potential, and Surface Charge Measurements

These methods are used to determine the stability of particles in suspension. This is generally accomplished by Electrophoretic Light Scattering (ELS) (McNeil-Watson, 2013), which is based on DLS, in which the suspension is illuminated by a light source, using a laser of wavelength (~400–800 nm). The scattered light from the suspension is sampled by an optical system connected to a photodetector and signal processing system that allows the measurement of particle motion utilizing the Doppler effect. ELS differs from DLS in that the random Brownian motion of the suspended entities is superimposed on a directed motion which is dependent on the electric charges carried by the suspended particles due to the presence of the suspending liquid. The method is used for measuring electrophoretic mobility, from which zeta potential can then be calculated. The zeta potential is used to assess the electrostatic attraction/repulsion between particles and is a predictor of emulsion formation and dispersion stability. Zeta potential accounts for both the particle surface and the dispersant and hence can be used to optimize the formulations of suspensions, emulsions, and proteins to predict interactions with surfaces.

3.2.4. Particle Size and Concentration in Fluids

Such measurements are used to estimate particle count and size ranges. This is usually measured utilizing optical microscopy to provide qualitative observations and counts of micro-sized particles. For biological samples, flow cytometry is a powerful and versatile laboratory technique used to analyze the physical and chemical characteristics of cells or particles suspended in a fluid. The sample is injected into a flow cytometer, where cells are aligned in a single file by a process called hydrodynamic focusing, while the cells or particles are prepared in a suspension, often labeled with specific fluorescent markers (fluorophores) that bind to particular proteins, antigens, or other molecular targets of interest. A laser beam illuminates the cells as they flow through the interrogation point. The cells scatter light and emit fluorescence (if labeled) and are measured using forward scatter (FSC) for cell size and side scatter (SSC) for internal complexity or granularity. The signal is digitized and analyzed to quantify and categorize cells based on their physical and chemical properties. Specific fluorophores are used for biological and microplastic particle detection using fluorescence techniques and nano cytometry, in conjunction with the staining of the lipophilic dye NR, to demonstrate that NPs can be accurately detected using flow cytometry (Salvia et al., 2023). A specialized form of flow cytometry, fluorescence-activated cell sorting (FACS), physically separates specific cell populations for downstream applications.

3.2.5. Surface Area and Porosity Analysis

This is usually done to determine the surface properties of materials, which influence particle behavior, especially in water. The Brunauer–Emmett–Teller (BET) method is used to determine the specific surface area of materials by analyzing gas adsorption. It has potential applications in studying MPs, particularly for assessing their surface characteristics, which play a crucial role in environmental interactions, pollutant adsorption, and degradation behavior. The BET method quantifies surface area by measuring the physical adsorption of a gas (commonly nitrogen) onto a material's surface at a specific temperature to study the adsorptive behavior and reactivity of MNPs, especially to analyze membrane fouling (Virtanen et al., 2020).

3.2.6. Mass Spectrometry

Inductively coupled plasma mass spectrometry (ICP-MS) is an elemental analysis technology capable of detecting most of the periodic table of elements at milligram to nanogram levels/liter based on their mass/charge ratio (m/e). Elements with a specific mass are ionized by Ar plasma and become monovalent ions. The technique reveals the elements present in the sample (qualitatively), and the amount of each element in the sample (quantitatively), based on the m/e of detected ions and the number of ions detected during a specified period of detection. Since liquid samples are introduced to the Ar plasma as an aerosol and undergo very efficient ionization, ICP-MS can measure most sample elements at parts per trillion (ppt) levels. Using the single particle (SP) mode of ICP-MS, the microplastics are detected using ultra-fast monitoring of transient signals, by using a quadrupole-based ICP-MS unit in single-event mode and registering the signal spikes produced by individual microparticles by monitoring the signal intensity at a (m/e) ratio of 13 ($^{13}C^{+}$). The results for microspheres in terms of size compare well with the size as determined using electron microscopy. ICP-MS operated in single-event mode thus allows information on both the size distribution and mass concentration of microplastics (Bolea-Fernandez et al., 2020). Laser Ablation ICP-MS allows spatially resolved analysis of elements in particles.

3.2.7. *Thermal Analysis*

Thermogravimetric Analysis (TGA) is generally performed to determine the thermal properties of plastic particles. The methodology is based on the identification of the polymer according to its degradation products, and it measures weight changes as a function of temperature, identifying fractions (Peñalver et al., 2020). Yet, another method for thermal analysis is Differential Scanning Calorimetry (DSC), which assesses phase transitions, such as melting or crystallization of polymers. Other heat-induced degradation methods include vapor phase chromatography using spectral analysis. Methods such as Pyrolysis gas chromatography/Mass spectrometry offer particle recognition using relatively small samples of PE, PET, PS, and PA (Enyoh et al., 2021; Picó & Barceló, 2020).

3.2.8. *Biological Characterization*

MNPs are known for their Neustonic nature (Vaseashta, Ivanov, et al., 2021) and it is necessary to study the surface coating for biofouling, accumulation of bacterial communities, or any other colonization of viral vectors. Fluorescence Microscopy visualizes biofilms, algae, or bacterial communities on particle surfaces. Another commonly used method is Enzyme-Linked Immunosorbent Assay (ELISA), which detects specific biomolecules or pathogens associated with these particles (Kadac-Czapska et al., 2024; Vaseashta & Dimova-Malinovska, 2005). ELISA can indirectly characterize MPs by detecting associated chemical additives, adsorbed environmental contaminants, or surface-bound biomolecules through a process called 'eco-corona'. The method extracts proteins or biomolecules from the surface of microplastics. By detecting specific biomolecules using antibodies, the method provides insights into biological interactions, viz., MP uptake by cells or biofilm formation.

3.2.9. *Remote Monitoring Platforms*

Several autonomous platforms are used and commissioned using unmanned aerial vehicles (UAV), or drones with hyperspectral imagers, Laser-Induced Beam Spectroscopy (LIBS), and Light Detection and Ranging (LIDAR), along with Internet of Things (IoTs) based sensors for remote monitoring to aquatic environment with MNPs. The nanomaterials-based sensors used nanophotonics to image various spectrums of materials present in the surface waters. The authors have used sample collection platforms by water-based vessels and using UAVs, in several locations for laboratory analysis as well (Vaseashta, 2018; Vaseashta & Khudaverdyan, 2013). The methods described here with several other advanced optical imaging technologies for in-situ detection, such as polarization imaging and digital holography are used to characterize MNPs from aquatic, soil, and air mediums to identify, possibly trace their origin for forensics, and will likely be a testbed for regulatory controls.

4. Trophic Transfer of MNPs and Associated Health Impacts

MNPs are observed to be almost everywhere and have recently been found, even in human bodies. The process of generation of MNPs takes complex stages and routes before they enter the environment and find their way to humans (and animals), through several entry points. Generally, an estimation of MNPs in the environment is done based on their source and size through various stages of deterioration and segmentation. The sources and distribution of MNPs are provided in Figure 2 through a variety of processes. Since water is a major part of our surroundings, one of the most common pathways is to find its way to sources of water, including wastewater as the particles amass in aquatic environments. Since our existence and actions are indistinguishable from aqueous media; the aquatic surroundings eventually become the most extensive repository of MNPs. It is estimated that over 80% of oceanic polymers result from terrestrial origins – such as waterways transport, releases from watercraft, fishing nets, exfoliants, personal care products, runoff from motorways consisting of paint, tire microplastics (TMPs), or tire wear plastics (TWPs), and pharmaceutical shells. Moreover, several farming practices play a significant role in the emission of MNPs. Using wastewater for plant nourishment represents a sizable source of MNPs, since as the liquid moves, so do these particles and are transported to any location wherever the liquid mass traverses. This observation is not only associated with the characteristics of MNPs but also relies on the physicochemical features of the marine surroundings, such as fluid dynamics, adhesion, and absorption by marine animals, which eventually influences the deposition, re-floating, and travel pathlength of plastic fragments, and finally influence their ecological distribution. MNPs from terrestrial environments are discharged directly into streams and ponds through downstream water management which consists of sewage management facilities or water runoff (Phuong et al., 2022). Typically, light plastic fragments float on the water's surface, while bulky plastic fragments tend to settle towards the bottom. Due to an abundance of organic matter and the presence of microbes in their natural surroundings, MNPs tend to conglomerate on the surface due to their neustonic nature. The MNPs settled at the bottom of the water may be stored for the short term or perpetually in the sediment, whereas the floating plastic will continue drifting toward the sea. When the river extends into the estuarial zone, the flow speed abruptly reduces due to an increase in surface area. Typically, an estuary shelters a higher level of MNPs in the transition zone. However, plastic particles in riverbed deposits far exceed estuaries. Hence, rivers contain more MNP contamination than oceans (Eriksen et al., 2014). In fact, plastic refuse from human actions is far more actively linked to shoreline waters. Nevertheless, the coastal deposits are inherently variable, dependent on the topography of the border. Depending upon the geologic dynamic margins, plateau, and seabed characteristics, it may be possible to model the characteristics of trophic transfer, as described in the following section.

4.1. Accumulation of Micro- Nanoplastics via Trophic Transfer

Due to the increased number of MNPs in our environment, there are concerns about their inclusion in the food supply web and water intake. Hence, assessing the trophic transfer from a life cycle analysis (LCA) perspective is critical, although the transfer conditions from plastic waste to aquatic environments vary significantly from one location to another. Consistent with a general trophic transfer rule, the cycle starts from the bottom level of the food web in the aquatic environment. The dynamics of micro, meso, and nano plastics reaching the lowest level are quite complex and vary significantly depending upon terrain and other environmental factors. In contrast to hydroecology, research on the MNPs ecosystem is extremely limited since such studies tend to miscue the inherent connection between hydrology and the transport of MNPs, especially due to the ubiquitous presence of plastic parts being pervasive in our daily lives, thus creating a technical knowledge gap. Furthermore, although not new, the growing concerns about MNPs in aquatic environments appear to be relatively recent, again due to the technical gap created by more focus on the positive aspects of plastic as compared to the negative health impacts—perhaps being a natural human tendency. We intend to frame this complex subject in a balanced perspective.

Since MNPs are of the same size as plankton and also the size of organic food grains, they are consumed by a variety of organisms through several feeding schemes. The variation in densities, shapes, and sizes of these polymers influences their characteristics and distribution into different trophic regions of the aquatic environment (Carbery et al., 2018), which impacts their availability to organisms (Saikumar et al., 2024). Additionally, biofouling may also play a major role in the erroneous identification of plastic as a nutritious food source by organisms (Kooi et al., 2017). Studies demonstrate that the formation of biofilms increases the likelihood of MNP ingestion by altering the vertical distribution, and they attract organisms relying on their chemoreceptors for odor associated with organic matter produced during the breakdown of biofilms on the plastic surface (Sooriyakumar et al., 2022). It was reported that Sea turtles ingested plastics for the same reason, with visual signs playing an important function (Schuyler et al., 2014). The ingestion of fouled plastic particles is not just restricted to higher trophic organisms, in fact, Daphnia magna and copepods – both planktonic and benthic, are exposed to clean as well as fouled particles and ingest such fragmented plastic particles at a higher rate in the presence of biofilm (Pikuda et al., 2023). In aquatic ecosystems, the availability of nutrient cycling is dependent on zooplankton, and the primary producers of higher trophic organisms are linked through them. As a primary consumer, they convert the organic matter into fecal pellets of high density and a sinking velocity (Wotton & Malmqvist, 2001), which then becomes available to sedimentary organisms and also contributes to the carbon sink in the aquatic ecosystem (Butman et al., 2016). Through the food intake, MNPs are ingested and egested with fecal pellets. The MNPs in the fecal pellet can potentially be transferred to other coprophagous (κόπρος "feces" and φαγεῖν "to eat") sea animals. The microplastics ingested by the zooplankton may not be completely egested, thus, the residual MNPs in the gut of zooplankton get directly transferred to the higher trophic levels by preying on them. Additionally, large-sized MPs can stay on the bottom of water bodies, altering the physical properties of sediments and, thus, affecting the behavior, habitats, and survival of benthic organisms.

Plastics, either from primary sources or degraded over time, enter the food chain through algae, bacteria, and/or other similar mechanisms and can have a significant impact on aquatic ecosystems. Algae and phytoplankton are credited for significant photosynthesis on earth and conversion of solar energy to biomass and encounter a significant challenge due to the MNPs. The effects are physical, biological, and toxicological with severe ecological implications. Phytoplankton and algae can inadvertently ingest MNPs, which may obstruct nutrient uptake, affect buoyancy, and/or damage cell membranes. Micro- nanoplastics in water can scatter light, reducing the amount of sunlight available for photosynthesis. Additionally, MNPs may adhere to algal surfaces, interfering with cellular processes, such as gas exchange or nutrient absorption. The biological impacts include reduced growth and reproduction, altered community composition, and reduced photosynthetic efficiency potentially due to decreased chlorophyll content or disrupted electron transport chains. The toxicological effects are due to disrupted cellular functions since the pollutants can bioaccumulate within algae and phytoplankton, potentially affecting growth and reproduction. MNPs can also induce oxidative stress in phytoplankton, leading to damage to proteins, lipids, and DNA. A study on marine phytoplankton, viz., *Thalassiosira weissflogii*, found that micro- nanoplastics diminished their growth rates and photosynthetic efficiency (Sjollema et al., 2016). MPs have been shown to accumulate in freshwater algae, e.g., *Chlamydomonas reinhardtii*, affecting nutrient uptake and causing oxidative stress. Bhattacharya demonstrated the physical influence of MNP beads on two types of algae species, viz. *Chlorella* sp. and *Scenedesmus* sp. (Bhattacharya et al., 2010). In this investigation, the adsorption of NPs appears to be due to the electrostatic attraction between the positively charged algal cellulose and negatively charged MNPs. This observed adsorption depends on the cell morphology and motility of the algae leading to reactive oxygen species (ROS) generation in the algae. In yet another study, a significant lessening of photosynthesis is observed in algae *Dunaliella tertiolecta, Thalassiosira pseudonana,* and *Chlorella vulgaris* upon 72-h exposure to micro polystyrene (PS) particles, but no changes in algal growth rate were found (Sjollema et al., 2016). The effects of food web transfer demonstrated that an algal species exposed to nano PS particles resulted in a reduction in chlorophyll and zooplankton *Daphnia* sp. displayed some alterations in reproduction and reduced body size (Besseling et al., 2013). A study on the next level of the food chain was conducted in which *Daphnia magna* exposed to PS MNPs were fed to fish (*Crucian carp*). The results not only show the direct uptake of PS MNPs from *Daphnia magna* to fish, but also the PS MNPs entered the brain of the fish, resulting in behavioral disorders and their survival rate (Mattsson et al., 2017). Through the food chain web, this kind of transfer presents serious consequences for humans, as discussed in the following section.

Benthic and pelagic food webs are also affected by MNPs via accumulation and transfer. The studies have shown the trophic transfer of MPs from blue mussels to crabs. In an experiment, the blue mussel *Mytilus edulis* was exposed to PS MPs and was subsequently fed to crabs *Carcinus maenas*. Following exposure for over 1 day, ~0.027% of the concentration of MPs fed to the mussels was observed in the crab hemolymph (Farrell & Nelson, 2013). It confirms the feeding type or habits that may influence the uptake of MPs in organisms. A study further reveals that omnivorous fish species have logged higher MPs content than carnivores and herbivores organisms. It was also presumed that a varied diet source of omnivorous organisms may be the justification for having higher MP content (Mizraji et al., 2017). These results, collectively, show the possible accumulation of MNPs through the food chain cycle via direct or indirect uptake. While direct uptake transpires when the organisms can't distinguish between natural prey and synthetic items, indirect uptake arises through feeding on previously exposed prey (Lusher, 2015). The food chain cycle is impacted by MNPs, which have cascading effects through the entire food network. For human consumption of fish or other aquatic organisms, MNPs are likely to accumulate when the parts with tissues are consumed. However, certain small pelagic fish species, such as *Engraulidae, Sardinops, Scomber,* and *Clupea,* and large species such as *Thunnus, Istiophoridae, Xiphias gladius,* and *Prionace glauca* pose a risk to human food security being the primary, secondary and even tertiary consumer of other organisms.

4.2. Trophic Uptake and Toxicity Bioaccumulation

Furthermore, as predators consume contaminated prey over time, the concentration of MPs and associated contaminants can increase through a process known as biomagnification. Marine mammals and seabirds are particularly vulnerable to the effects of biomagnification because they can accumulate high levels of MPs and associated toxins. MPs can easily bioaccumulate in fish organisms, which are an important food source for humans. As stated earlier most plastics contain additives to modify their characteristics, hence, in addition to MNPs, these additives and chemical contaminants also pose a greater risk to the overall health of the ecosystem. Since fish play a crucial role in the marine ecosystem, contributing to material circulation, energy flow, and information transfer, the health of fish populations directly impacts the stability of marine ecosystem structure and function. Figure 5 presents a generic trophic transfer pyramid for MNPs, as the exact transfer depends on many other geohydrological and environmental factors.

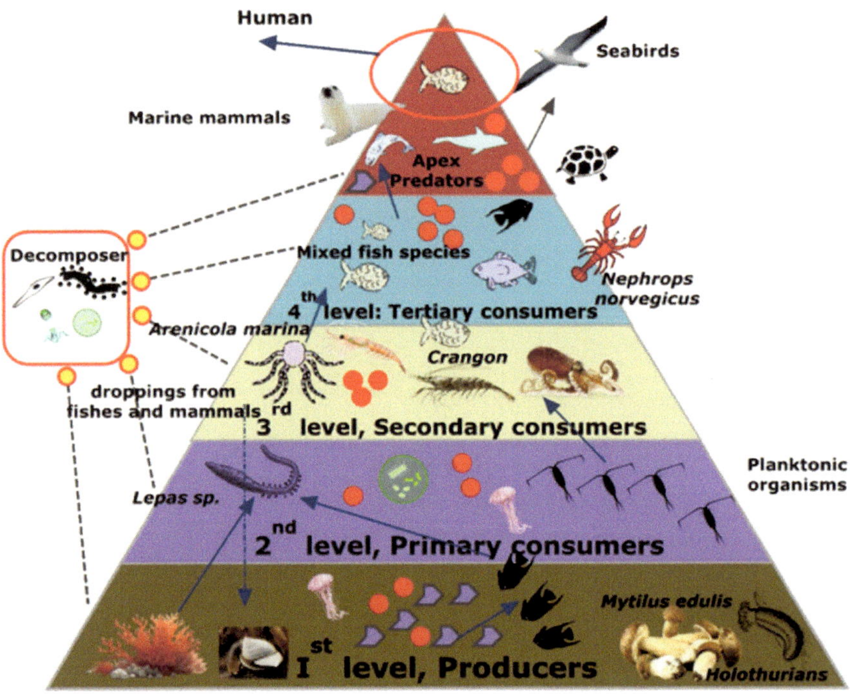

Figure 5: Trophic transfer pyramid for MNPs: A generalized overview.

The uptake of MNPs in the human body through trophic transfer involves their movement through the food chain, eventually entering human systems. As elaborated in the section above plankton, algae, or other small organisms ingest MNPs directly from water. As predators consume prey, MNPs accumulate in higher concentrations in higher organisms in the food chain hierarchy. Humans ingest MNPs by consuming contaminated seafood, drinking water, and even processed foods. The ingestion of MNPs causes several physical and biological impacts in humans. As stated earlier, in addition to MNPs, adverse health effects also occur from the additives, sorbed contaminants, and other chemicals that are known to have carcinogenic and endocrine disruption (EDC) characteristics. Although a simplistic and general model of trophic transfer is illustrated above, the ingestion could also be due to a failure to distinguish MNPs from the prey, or it could be due to the intake of lifeforms from lower trophic levels that contain these particles (Ivar Do Sul & Costa, 2014). The ecological impact of MPs extends beyond individual organisms to entire ecosystems. Changes in the abundance, distribution, and composition of aquatic populations can disrupt ecological equilibrium, trophic dynamics, and ecosystem functioning. MNPs may also adhere directly to organisms (Cole et al., 2013) and are currently listed as new and emerging contaminants and health threat vectors through aquatic-based products (Correia & Loeschner, 2018). Only limited data is available in the literature on the transport pathways, viz., gastrointestinal tract (GIT), respiratory tracts, and lymphatic and circulatory systems. Similarly, the accumulation of MNPs in various human organs and the toxicity of these particles are not fully understood, and several models are developed simply based on the extrapolation of laboratory studies on nanoparticles. To better understand the discussion concerning the transfer of MNPs within the human body, Figure 6 shows potential entry points of MNPs within the body and their size classification.

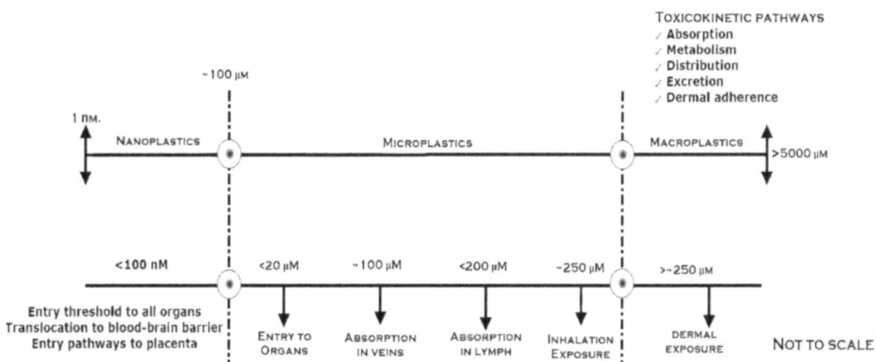

Figure 6: Toxicokinetic pathways and MNPs size classification (Vaseashta, Ivanov, et al., 2021).

Several investigations have reported the exposure and transfer of MNPs via different routes on the human body (Usman et al., 2020). Recently the presence of MNPs was observed even in human blood (Leslie et al., 2022) and in lung tissue (Amato-Lourenço et al., 2021). The literature regarding the trophic transfer of MNPs from aquatic food chains, particularly in freshwater ecosystems to humans, is very limited. Generally, the significant exposure pathways of MNPs to humans are inhalation and ingestion, while the latter mostly occurs through aquatic-based food consumption. Transfer of MNPs to humans in the freshwater ecosystem occurs through the consumption of aquatic organisms. As an example, the probability of MNP transfer is greater when some species of fish are consumed as a whole or through partial exclusion of the viscera. Studies further show that the presence of MNPs in the edible tissue of fish increases the probability of exposure to MPs (Daniel et al., 2021; Garcia et al., 2021). Although accurate data on the quantity and mass of MNPs ingested by humans through the aquatic food chain alone is incomplete; however, a recent study shows that it is estimated that the per capita MNP intake through the consumption of shellfish can be approximately 13 ± 58 microplastic particles/year, while an average of 3.2 ± 10 microplastic particles/kg was reported in the edible tissue of *Portunus pelagicus* (Daniel et al., 2021). In a study, the evidence seems to suggest the ingestion of MNPs by humans as demonstrated through their presence in human feces (Schwabl et al., 2019). A study on mice shows the accumulation of MNPs in the liver, kidney, and gut, with adverse effects on the liver, viz. difficulty in lipid and energy metabolism, oxidative stress, and neurotoxic responses (Deng et al., 2017). This causes concerns about the toxicity of MNPs in humans, as well (Wang et al., 2019), and is described in the following section.

4.3. Toxicity of MNPs: Impact Models, Public Health

Several reports suggest that the ingested MNPs disrupt the gut microbiome causing metabolic stress leading to inflammation and insulin resistance. Additionally, the accumulation of MNPs in the kidney and liver has damaging effects over time. Limited studies further show that the MNPs generate ROS that can damage DNA, and potentially initiate cancer. Certain MNPs, such as ultrafine diesel exhaust particles, can enter the brain through the olfactory nerve and can cause inflammation, potentially contributing to neurodegenerative precursors. MNPs accumulate or agglomerate in the lungs to cause lung damage, inflammation, oxidative stress, and translocation through the lung barrier into the bloodstream. Increased accumulation can cause vascular dysfunction, such as increased blood pressure and clotting. It can also cause atherosclerosis due to chronic exposure. In addition, however, the additives in polymers, when released upon division, are also known to have an adverse impact on human health depending on the properties of the additive. Within the limited scope of this chapter, we briefly provide limited studies showing the toxicity of MNPs and their impact model on human health.

The sorption mechanisms for MNPs generally consist of electrostatic interactions, Van der Waal forces, hydrogen bonding, π-π interactions, pore filling, and other mechanisms that are subject to sorbent-sorbate interaction (Agboola & Benson, 2021; Torres et al., 2021). Most commonly used polymers are hydrophobic, hence the sorption is due to hydrophobic interaction, when an aggregate or cluster is formed between two nonpolar substances. The pore-filling mechanism depends on the polymer, particularly its pore diameter and also the molecular size of the pollutant/chemical, and the studies show that MNPs can adsorb chemicals ranging from 1 to 10,000 ng/g (Bhattacharjee et al., 2017). The greatest concern is the role of MNPs as vectors for hydrophobic organic chemicals (HOCs) and heavy metals (HMs). The sorption of HOCs and HMs to MNPs is significant since this affects the mobility and bioavailability of these pollutants (Campanale et al., 2020). MNPs react with pollutants before, during, and following their release into natural environments (Hartmann et al., 2017). Adsorption kinetics vary due to several parameters, viz., polymer type, density, crystallinity, environment, and the type of contaminants (Wagner & Lambert, 2018). Additional parameters that affect bioaccessibility are photodegradation, which increases the absorption capacity for HOCs (Liu et al., 2019). In the freshwater, MNPs coexist with other emerging contaminants such as returned pharmaceuticals, PCPs, VOCs, flame retardants, fluorinated and sulfonated organic compounds, household degreasers, toxic industrial chemicals (TICs), and toxic industrial materials (TIMs). The polarization influences the sorption process, since polarized molecules adhere to MNPs via surface adhesion, while non-polarized molecules adsorb within the core structure. Since MNPs have a high surface area/diameter ratio, they show higher sorption affinities for toxic compounds relative to microplastics (Velzeboer et al., 2014). The large surface area of MNPs also results in the formation of microbial communities in the natural environment (Vaseashta, Ivanov, et al., 2021; Vaseashta, Stabnikov, et al., 2022), and changes in sorption processes due to different additives used in the polymer matrices (Munoz et al., 2021). Transport through the polymer matrix, following adsorption, is influenced by other factors such as the free volume within the polymer, the distance between the polymer chains, and the segmental mobility. If an organism ingests MNPs, the sorbed contaminants can be transferred to the organism's tissue via extracellular/interstitial fluids or direct contact between the MNPs and the organism's interior walls. The uptake mainly occurs through intra-organismal fluids and desorption of HOCs from the sorbent forms freely dissolved molecules, which depends on several factors, and it mainly decreases with an increase in partition ratios and increasing binding strength (Hartmann et al., 2017). Furthermore, it was shown that bacterial biofilms formed on MNPs include bacteria with antibiotic-resistant genes (ARGs) (Mayer et al., 2007) which may originate from human and animal populations exposed to antibiotics and subsequently transferred downstream through wastewater into aquatic ecosystems (Berkner et al., 2014). These antibiotics have shown higher adsorption to MNPs in freshwater than in marine ecosystems. The spread of these antibiotic-resistant bacteria (ARB) and/or ARGs renders water bodies as natural sinks of antibiotic resistance and is a cause for yet another major toxicological concern for society and the economy.

In humans, the MNPs traverse via microvilli to access the bloodstream to form protein corona, which is affirmed by *in vitro* studies. This is crucial to comprehend the malignancy of MNPs in humans, as the engagements between organs and tissues with protein-coated particles vs. uncoated nanoplastics. A study of living organism hepatocytes demonstrated that polymeric protein complex may induce elevated cell damage and genetic toxic impacts as compared to uncoated MNPs (Zhou et al., 2022), perhaps due to the formation of bio-coating on the outer layer of MNPs, which prevents detection by the immune network leading to their prolonged presence in the vascular system. Weak intermolecular interactions, H-bonding, and π electron delocalization also compete for a role in creating a polymer-protein composite. Laboratory toxicokinetic tests show that MNPs may harm the immune system and have adverse impacts on living organisms due to stimulated cytokine release linked to stomach disorders, interference with Fe transport, stimulation of programmed cell death, disruption of endoplasmic reticulum homeostasis, and oxidative damage (Chen et al., 2024; Gao et al., 2022; Zheng et al., 2024). In addition to many additives that display severe toxic responses, BPA is known for its hormone-mimicking attributes, contributing to various metabolic disorders and affecting fertility and maturation processes. Laboratory studies to calculate Estradiol (17-beta-estradiol, 17β-estradiol, E2) Equivalency Factor (EEF) is a measure of estrogen, a steroid hormone having a molecular weight of 272 Daltons, show that BPA produces EEF of $2\times10^{-}4$, which is higher level as compared to other additives, such as 4-*tert*-octylphenol ($\sim2\times10^{-5}$) and 4-n-nonylphenol ($\sim1\times10^{-5}$) (Koelmans et al., 2015; Yang et al., 2011). More specifically, plastic mineral water bottles utilized for infants exhibited substantial seepage of BPA, since infants have a higher vulnerability compared to grown-ups, due to heightened absorption. An example of the toxicity of PS-MNPs in the human lung epithelial BEAS-2B cells shows that it can cause cytotoxic and inflammatory effects in BEAS-2B cells by inducing ROS formation thereby posing a risk to human respiratory health (Dong et al., 2020).

There is some uncertainty regarding whether plastic particles can permeate the epidermis. One of the concerns is due to the increased use of nurdles in the cosmetic industry. The plausible interaction of MNPs with dermal tissue occurs via contact by the application of hygiene products. The epidermal layer functions as a barrier to the skin and its water-repellent nature makes it challenging for MNPs to infiltrate the skin when exposed to an aqueous medium, but the hygiene components could also support the infiltration. Broadly, the permeation of the skin could play a key role in the absorption of MNPs, but additional experimentation is required to ascertain.

One of the principal repositories for MNPs is the digestive system, boasting an average area of about 32 m^2. After entering the gastrointestinal tract (GIT), MNPs gather in the stomach and intestine. Plastic fragments measuring <150 μm are absorbed by lymphoid cells interrupting key physiological processes. This can result in oxidative stress, including triggering ROS generation and inflammatory response, weakening gut immunity, and damaging lipid metabolism and gut microbiota. In an investigation on Benthic Clams contaminated with polystyrene (PS), disruption of the intestinal lining was observed with an escalation in harmful microorganisms and apoptosis (Gagné et al., 2023). Another research showed the adverse effects of MNPs in the guts of rodents and hepatic systems (Liang et al., 2021). Yet another study investigated the impacts of three plastics of different sizes on the mucosal epithelium of rats (Liang et al., 2021) and established an *in vitro* multicellular gut epithelial system, examining the absorption and harmfulness of PS plastic of various dimensions (DeLoid et al., 2021). It was shown that smaller particles readily penetrate the cellular structure, leading to diminished cellular integrity and increased epithelial barrier penetrability. The GIT, the initial focal point of MNPs, and the research predominantly focus on marine species, but the toxicological data is largely absent. Furthermore, gastric fluid's impact on plastics' characteristics needs to be investigated.

By observation, prolonged contact with plastics among workers in polymer production facilities demonstrates an increased risk of various breathing disorders, with the possibility of developing pulmonary cancer. Excessive MNPs into the lower breathing passage of humans instigate the generation of ROS, cytokine release, and apoptosis, leading to detrimental effects on the pulmonary system and respiratory epithelial cell inflammation. This disturbs the lung epithelial barrier, adversely impacting the normal respiratory system functionality. A study shows that both healthy and asthma-prone laboratory rats, following MNP exposure through the pulmonary canal, displayed signs of swelling and asthma bouts in their pulmonary regions, invasion of immune cells, macrophage accumulation, and inflammatory mediator release (Lu et al., 2021).

Limited studies show the impact of MNP accumulation on the reproductive aspects. The studies show that MNPs accumulate in genitalia, thus posing potential risks. Laboratory rodents were orally fed MNPs for 28 days, and their tissue specimens indicated that they shed spermatocytes, and displayed reduced blood testosterone levels, increased testicular inflammatory cytokines, and deformed sperm cells (Jin et al., 2021). The same study observed that PS particles adversely affected the testicles of rodents, indicating deformed sperm cells, imbalanced hormonal concentrations, with reduced fertility. A study investigated the detrimental effects of PS particles on the genital system of female rodents, showing that the MNPs caused a decline in the follicular count and restrained ovum state and growth (Jin et al., 2021). Similar studies further demonstrate that there are notable changes in the white blood cell concentration in the fetal membrane, indicating that contact with MNPs may trigger immunodeficiency (Feng et al., 2023). Furthermore, studies on Fauna also show that extended contact with MNPs may cause harm to animal reproduction (Yang et al., 2023).

The central nervous system (CNS) analyzes and integrates various intra- and extra-personal information, as well as generates a coordinated response to these stimuli through complex mechanisms. Any detrimental effect on neural pathways will harm the CNS. Some studies have demonstrated that MNPs can penetrate the neurological system by penetrating the blood-brain barrier and aggregate in the brain of mice (Lee et al., 2022). The study further found that MNPs impacted learning and memory aspects in the brains of mice that were fed 2 μm of PS MPs (0.016 mg/g) for eight weeks, which can cause neuroinflammation in the hippocampus, which in turn would alter the genes and proteins contributing to synaptic plasticity (Lee et al., 2022). In a similar study, the research shows that PS particles prompt oxidative stress and reduce acetylcholine levels in mice, causing neurotoxicity (Wang et al., 2022), which also depends on dose, size, shape, and composition. Researchers have shown that zebrafish exhibited different neural patterns following ingestion of 70 nm PS particles (Jabri et al., 2024). These modifications are primarily observed as notable in their locomotive activity, aggression levels, and anti-predatory response. Contact with MNPs alone or conjoined with copper metal elevated the death ratio of zebrafish larvae, inducing neurological behavior disorders and diminished swimming tendency. Kwon and co-authors suggested that PS particles <2 μm may infiltrate the rodent nervous system, accumulate in nervous tissue cells, and trigger microglial inflammation, leading to cytokine secretion and programmed cell death (Kwon et al., 2022). Based on this limited research, it is instructive to conduct *in vivo* and epidemiological research to investigate the potential long-term effects of long-term exposure to MNPs on the human nervous system, including cognitive function and behavior, including on neural networks and observation of changes in neurotransmitter levels, gene expression, and synaptic plasticity following exposure to MNPs on specific populations, such as children, the elderly and individuals with pre-existing neurological disorders. In addition to the toxicological response of MNPs to certain aspects of biological systems, there are other aspects such as immune, endocrine, urinary, and motor systems, however, due to the limited scope of this writing and also the scarcity of research on this topic, research in this area must progress extensively, due to steep rise in micro- and nanoplastics pollution in our environment. This research will substantially impact the quality of life (QoL), especially for specific populations, such as children, the elderly, and individuals with pre-existing health conditions.

4.4. Security Risks of Micro Nanoplastics

The neustonic nature of MNPs causes them to float due to their hydrophobic nature. The spongy biofilm formed across MNPs tends to attract organisms and contaminants from the aquatic medium. It was reported that warfare agents, such as biomass of *Bacillus anthracis*, or similar biological agents of low cost and simple production technology can serve as a potential bioterrorism event (Stabnikov et al., 2020; Vaseashta, Stabnikov, et al., 2022). Potential biological agents can be spread in the atmosphere, freshwater, and seawater, and the dispersed spores of such agents, including *B. anthracis,* can spread in marine waters, on or around sea vessels, and in harbors, due to being heat resistant and other extreme environmental conditions. The seawater decontamination in the harbor in case of terrorism bio attack has not been studied yet and hence poses a serious security risk. The U.S. Environmental Protection Agency (EPA), in partnership with the Department of Homeland Security (DHS), evaluated the method of disinfection of solid surfaces and seawater with chlorine in case of an intentional release of such agents at port sites. However, due to the large volume, and the infeasibility of the method, it is logical to envision all possible options from a counter-terrorism standpoint. In practice, chlorination is not a realistic option due to the very large sea/harbor water volume, which is roughly 100–1000 million m^3 in a given harbor. Hence, until now, there is no affordable technology for decontamination of such large volumes of harbor after intentional infection with a biological agent. Figure 7 shows a mechanism of neustonic biofilm formation near water-air interphase.

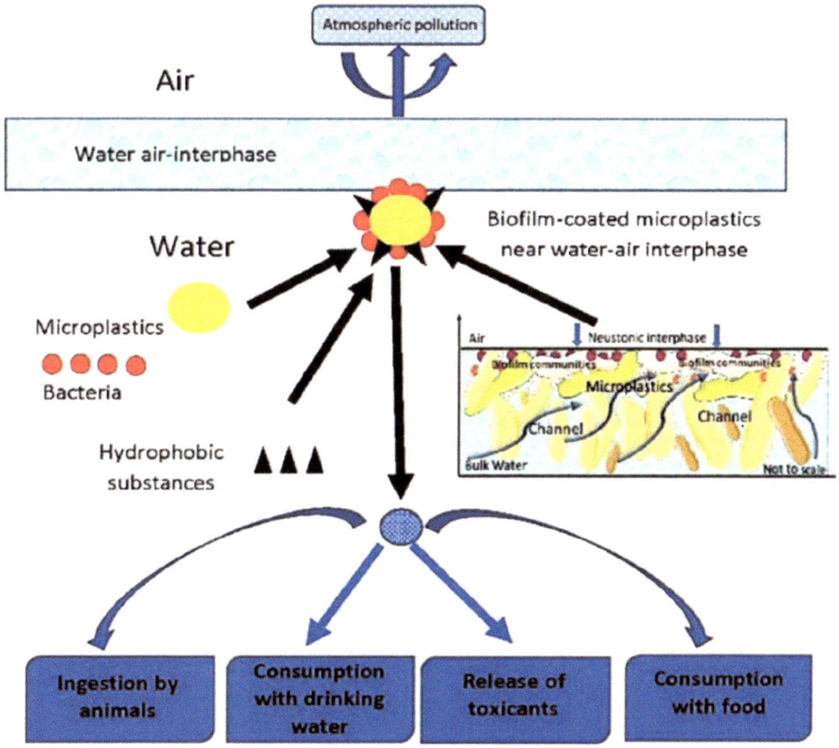

Figure 7: Mechanism of neustonic biofilm formation near water-air interphase (Vaseashta, Ivanov, et al., 2021).

5. Environmental Sustainability: Mitigation Strategies, and Use of Bioplastics

One of the major problems with MNPs is that they may take several years to deteriorate in aquatic environments due to being long-chain hydrocarbons. Their small size means they can travel easily, making them pervasive. Some researchers believe that MNPs might even be affecting the climate to an extent, although direct evidence is elusive, as of yet. However, studies have conclusively shown that MNPs alter agricultural land and aquatic environments. These minuscule plastic particles can contain a slew of chemicals, which, according to the PlastChem project report, has found more than 16,000 chemicals, approximately a quarter of which are classified as hazardous. In addition to containing these chemicals, plastic particles can also bind harmful pollutants in water and soil before being ingested, which makes corals and marine animals more susceptible to diseases. Scientists are still unraveling the fate of these tiny particles and the threat they pose to human health and the environment. Hence, it is necessary to design ways to minimize these particles, to the extent possible, through waste management, mitigation strategies, and filtration for environmental sustainability. In the section below, we present several strategies to minimize MNPs in the environment, not only for environmental sustainability but also to reduce adverse health impacts of these particles on marine animals and human health. Addressing the issue of mitigating microplastics using nanotechnologies offers a wide range of viable solutions. The methods include MNPs mitigation strategies that involve photocatalytic decomposition using nanomaterials, microbes-based solutions, nanofiber-based membranes with cryogels and other materials to capture hydrocarbons, and by using bio-based and biodegradable polymers.

5.1. Micro/Nano Plastics Mitigation

Recognizing the scope of pollution created by MNPs, there is an urgent need for comprehensive and practical strategies to mitigate these emerging pollutants. The effort is significantly prompted by leading researchers, policymakers, and environmental activists to act due to detrimental effects on organisms, such as disruptions in hormonal balance, and potential mortality. These particles have the potential to act as carriers for additional pollutants, thereby exacerbating the intricate ecological consequences. As discussed earlier, there is a growing concern regarding the potential transmission of detrimental chemicals and additives introduced during plastics production to humans, hence understanding their conjoined effects on ecosystems and human health is quintessential to develop efficient and focused mitigation strategies. Conventional waste management strategies frequently prove inadequate in effectively mitigating the MNPs pollution due to the challenges associated with its capture and identification, and the diverse sources from which they originate (Li, Zu, et al., 2023; Osman et al., 2023; Vanapalli et al., 2021). Various novel technologies and methodologies have been developed to capture and mitigate MNPs. While the research is ongoing to understand MPs' long-term effects on living organisms and nonliving components (Bhat, 2023; Chartres et al., 2024; Thacharodi et al., 2024), it is strategically important to mitigate these contaminants using multi-fold methods, viz., technology adaptation, policy implementation, and regulatory controls. In the following sections, we discuss some technology adaptations to dissociate/degrade MNPs in biodegradable or bio-friendly compounds. In addition, the footprint of MNPs can also be reduced by using bio-based (Haq et al., 2024) and biodegradable plastics (Moshood et al., 2022) and by strategic upcycling

(Sajwan et al., 2024; Zhao et al., 2022). Given the limited scope of this writing, a brief description of these processes is provided below.

5.1.1. Photodegradation by Nanophotonics

Nanomaterials have revolutionized photocatalysis by offering higher surface area-to-volume ratios and improved catalytic properties. Photocatalytic degradation is an environmentally friendly method to decrease MP pollution. The method uses UV photon energy, from a light source or solar radiation to decompose plastic to a large extent with minimal cost (Du et al., 2021). Titanium dioxide (TiO_2 and zinc oxide (ZnO) are the two most widely used photocatalysts due to their stability, low cost, and high activity. However, their efficiency under visible light is limited due to their wide band gaps. Recent advancements include doping with metals or non-metals (e.g., silver or nitrogen), which extend their absorption spectrum into the visible range (Thomas & Sandhyarani, 2013). Photocatalysis, under optimum conditions, usually shows significantly higher degradation efficiency compared to biodegradation and is carried out under milder conditions as compared to chemical processes (Li, Zhao, et al., 2023; Pan et al., 2023). With an appropriate light source and catalytic agent, the process of photocatalysis is suitable for the degradation of a wide range of plastics. Composite photocatalysts, such as TiO_2-graphene hybrids, have also shown promise, as graphene enhances electron mobility and reduces electron-hole recombination, thereby enhancing photocatalytic activity. These composites are particularly effective in degrading polyethylene terephthalate (PET), low-density polyethylene (LDPE), and other commonly used plastics found in MP pollution. Recently, nano-sized semiconductors of various compositions and nanocomposites based on them with high redox potentials, viz., WO_3, CdS, SnO_2 ZrO_2, ZnS, and others, have been used for photocatalytic degradation of plastics (Ariza-Tarazona et al., 2019). When they absorb photon energy ($E=h\upsilon$) exceeding the band gap energy of the semiconductor, participate in the formation of highly ROS, causing the polymer chain of the plastic to break, as shown in Figure 8. Factors influencing photocatalytic degradation include catalyst surface properties. Highly porous materials with large surface areas provide more active sites for ROS interaction with plastic surfaces. Additional parameters such as environmental conditions, viz., pH influence the generation of hydroxyl radicals and degradation, the temperature of photocatalytic reactions (normally between 20°C and 100°C), and light intensity. A higher light intensity enhances electron excitation and ROS production, thus improving degradation rates. For practical applications, photocatalysts must maintain their activity over multiple cycles without fouling and leaching. Since photocatalysis is a surface reaction, photodegradation with accompanying contaminants may be difficult if opaque contaminants prevent light irradiation covering MNPs (Hamd et al., 2022; Sundar & Kanmani, 2020).

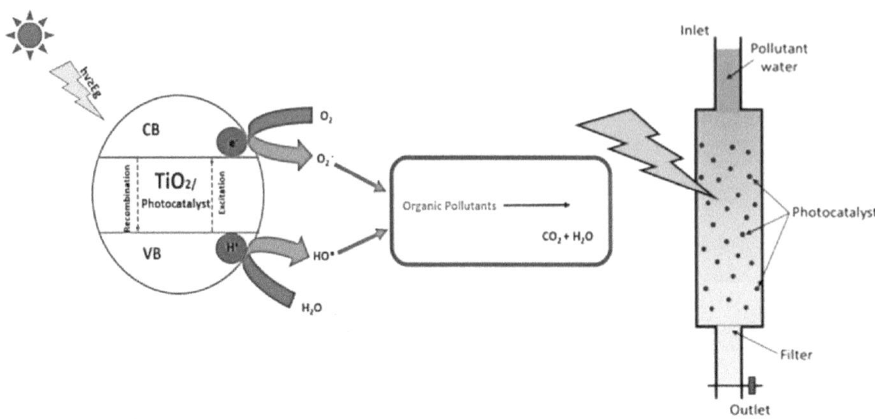

Figure 8: Photocatalytic dissociation of MNPS in biodegradable products.

5.1.2. Microbial Biodegradation

While waste management methods and substituting biodegradable, biobased plastics for fossil fuel-based plastics, as discussed later, could slow down the current trajectory and mitigate MP pollution, it will take several years before we see any substantial change. Biodegradation has emerged as one of the key solutions to break down plastics using enzymes (Zeenat et al., 2021). The method can be used as a standalone or in addition to nano-photocatalytic processes, as a sustainable and viable alternative for mitigating plastics. Recent evidence shows that some microbes are already learning how to work on petroleum-derived polymers (Zeenat et al., 2021). In 2016, Japanese scientists reported a bacterium that eats PET in samples from a recycling plant (Bethany Halford, 2016). It was also observed that enzymes that disintegrate PET into mono hydroxyethyl terephthalate attack the fraction of the amorphous region of the polymer more than the crystalline or rigid fraction, hence the polymers can be pretreated to minimize crystallinity. Hence, the researchers are now focused on artificially speeding up the evolutionary process with the microbes by isolating relevant enzymes from the microbes and engineering the enzymes to increase their activity. Two possible options are to use the enzymes to break down polymers into their original raw materials so they can be recycled; while the other approach is that the enzymes decompose into small polymer particles in compost heaps and wastewater treatment plants. When the polymer backbones align themselves and stack on top of each other, the enzymes find it difficult to attach to the polymer to break the bonds. Hence, pre-treating polymers enhances their surface area. Additionally, the ester linkages in the polymer provide a convenient point of attack for hydrolyzing enzymes. This is also true for the amide bonds in nylon. A company, Carbios, has developed an enzyme, derived from a natural digestive enzyme protease and is blended with PLA – a bio-based polymer that requires relatively warm (60°C) composting conditions. Several microprobes and enzymes that break down polymers have been proposed as a viable solution for plastic waste in water treatment and industrial composting. It was further demonstrated that a mutant version of the naturally occurring enzyme PETase, derived from *Ideonella sakaiensis* (Austin et al., 2018), is an enzyme that would break down PET fibers in simulated sewage sludge. An accidental discovery at Sandia National Research Labs revealed waxworms (caterpillars) to produce holes in plastic bags, only to determine that enzymes in the worm's saliva are responsible for the plastic degradation effect (Alderton, 2022; Sanluis-Verdes et al., 2022). Table 6 provides a list of organisms and enzymes, which are currently

Table 6: List of potential organisms and enzymes for mitigating microplastics

Type of Plastic	Organisms	Enzymes
Poly (3-hydroxybutyrate) (PHB)	*R. eutropha, R. rubrum, Pseudomonas, Alcaligenes*	Intracellular PHB depolymerase and extracellular PHA depolymerases
Poly (Ethylene Adipate)	*Penicillium* sp., *R. arrizus, R. delemar, Achromobacter* sp. and *Candida cylindracea*	Lipases
Poly (ε-Caprolactone)	*Penicillium* sp., *Aspergillus* sp., *R. arrhizus, Clostridium* sp.,	Lipases and esterases
Poly (hexamethylene carbonate)	*Roseateles depolymerans, Amycolatopsis* sp., *Candida cylindracea, Pseudomonas* sp., *Chromobacterium viscosus* and *R. arrhizus*	Cholesterol esterase, lipase, lipoprotein lipase
Polycaprolactone	*Pseudozyma jejuensis*	
Polyester polyurethane	*Pestalotiopsis microspora*	
Polyethylene (PE)	*Brevibacillus borstelensis, Acinetobacter* sp., *R. ruber*	Alkane hydroxylase, laccase
Polyethylene low density (LDPE)	*Aspergillus versicolor*	
Polyethylene terephthalate	*Ideonella sakaiensis*	
Polyethylene terephthalate	*Thermobifida fusca*	Lipase and esterase
Polystyrene	*Rhodococcus ruber*	
Polyurethanes (PU)	*R. delemar, Curvularia senegalensis, Comamonas acidovorans*	Lipase and esterase

in use. This discovery has inspired other researchers to scour the literature in search of plastic-degrading microbes, with more discoveries on their way, especially with artificial intelligence (AI) based platforms.

A microbial consortium or community consists of two or more bacterial or microbial groups working symbiotically. Such consortiums can be endosymbiotic or ectosymbiotic and may occasionally be applied collectively to explore the biodegradation potential of these microbial consortia in transforming plastics into biodegradable byproducts, as compared to individual strains. Specific focus is placed on the role of microbial consortia, as their synergistic interactions offer a scalable solution for addressing persistent plastic waste in diverse environmental conditions. While individual microorganisms such as *Pseudomonas fluorescens*, *Penicillium verrucosum*, and *Aspergillus flavus* have shown promising biodegradation capabilities, microbial consortia often exhibit superior efficiency (Cao et al., 2022; Park & Kim, 2019; Postolachi et al., 2022; Rastimesina, 2023). Despite promising results, several challenges remain in harnessing biodegradation for practical applications.

In addition to the above processes, several pretreatment steps are used either in parallel or as a standalone step to dissociate long hydrocarbon chains. One such process is photo-oxidation by Ultraviolet (UV) irradiation which induces oxidative changes on the polymer surface. This process breaks carbon chains in the polymer, forming carbonyl and hydroxyl groups that increase hydrophilicity and facilitate microbial attachment. Chemical treatment with acids, alkalis, or oxidizing agents (viz., H_2O_2) alters the chemical structure of the surface, creating functional groups, which renders microbial degradation easier. Yet another process is mechanical treatment, such as grinding or the introduction of microcracks increases surface area and creates defects, enabling easier penetration by microorganisms and their enzymes. Applying nanocomposite coatings as host catalytic materials, facilitates bond activation under light or chemical reactions, making the polymer more susceptible to enzymatic degradation. In addition, low-pressure plasma treatment alters the chemical and physical structure of the polymer surface, making it more hydrophilic, which promotes interactions with microorganisms. Pre-incubation in mineral solutions by immersing the film in nutrient solutions with microorganisms or enzymes can initiate the formation of bioactive layers on the polymer surface, enhancing its biodegradability. Each of these methods aims to facilitate microbial attachment, activate enzymes, and accelerate biodegradation processes.

5.1.3. Nanofiltration to Filter and Degrade Microplastics

Using recent advances in materials synthesis and nanotechnology, it is possible to make nanomaterials, nanofibers, and various other configurations of desired shapes and sizes (Nastase et al., 2006; Stamatin et al., 2006; Vaseashta, 2006; Vaseashta & Bölgen, 2022; Vaseashta & Dimova-Malinovska, 2005; Vaseashta & Khudaverdyan, 2013), including membranes. The filtration process consists of convection and diffusion transport, along with electromigration resulting from the charge associated with the membranes. Due to its size, nanofiltration (NF) is a pressure-driven process that removes contaminants based on size exclusion and molecular interactions. The pore size in NF membranes typically ranges from 0.1 – 10 nm, making them suitable for rejecting small particles, including MNPs, as well as dissolved salts, organic compounds, and other contaminants. Recently, with the addition of cryogels and other suitable additives in nanofibers, the authors were able to successfully remove long-chain hydrocarbons, including oils, pharmaceuticals, paints, and dyes (Bolgen et al., 2013; Demir, Bölgen, et al., 2021; Demir, Vaseashta, et al., 2021) and can be used for isolating MNPs.

Nanofiltration is widely used in the field of desalination and hence the removal of various inorganic ions from the solution to obtain clean and drinking water can also be accomplished. The NF processes can be used as a model for membrane characterization, process modeling, optimization, membrane design, and application (Hilal et al., 2007). Among several characteristics of membranes that determine their separation properties, include pore sizes, hydrophobicity, zeta potential, surface charge, and ion exchange capacity (Schäefer & Fane, 2021). The influence of specific solution conditions, such as pH, temperature and pressure, total component concentration, and ionic strength, is critical to predicting retention. These parameters, when interacting with dissolved substances, lead to membrane contamination. One of the major limitations of membrane is fouling which is an inevitable by-product of effective operation. In addition to MNPs, the typical contaminants are colloidal, biological, organic, and inorganic. Nanomembranes are typically used in water and wastewater industries (Criscuoli, 2022; Mijatović et al., 2004) and food industries such as dairy and sugar, whey concentration and demineralization, dextrose whey syrup purification and nutraceutical extraction, fruit juice concentration as well as wine industries (Cassano et al., 2019). Using recent advances in nanomembranes, the filtration can successfully integrate MNP filtration, due to the increased presence of MNPs in the aquatic environment. Using 3D printing technology, the authors have developed composite structures to capture polymers, and long-chain hydrocarbons, including pharmaceuticals. Although the investigation is in its preliminary stages, we anticipate using meta-materials as catalysts to detect and capture organic materials, volatile organic compounds (VOCs), as an extension of our work on capturing micro- nanoplastics in aquatic environments.

5.2. Environmental Impact Reduction

5.2.1. Use of Biopolymers

The etymologically correct way to define it is bio-based, bio-sourced, and plant-based. They should be referred to as 'bio-friendly' in the sense that they are consistent with the fauna and flora of the ecosystem. Biobased plastic or bioplastics are plastics derived from renewable biomass such as cellulose, hemicelluloses, lignin, vegetable fats and oils, corn starch, pea starch, potato starch, sugarcane, and sugar beets. Other recent sources include avocados, seaweeds, algae, shrimps, cactus, grapes, fish scales, insects, safflower, and mushrooms. As stated in the introduction, humans have been using natural polymers, such as rubber derived from the sap of rubber trees (*Hevea brasiliensis*), Shellac – a resin secreted by the lac bug, and Gutta-percha – a rubber-like material derived from trees native to Southeast Asia for a long time. Hence, bioplastics would mean polymers, since plastic only refers to an attribute. The global biopolymers market size was estimated at \$17.54B in 2023 and is projected to grow at a compound annual growth rate (CAGR) of 10.4% from 2024–2030. This is attributed to the expanding demand for PHA-based environmental plastics. Due to the heightened demand for eco-friendly packaging from a variety of end-users, major manufacturing companies are shifting to biodegradable products, as shown in Figure 9. Biopolymers are engaged in an array of end-use applications, including biomedical, pharmaceuticals, and food packaging (shown as inset). A second insert shows a Venn diagram of bio-based and bio-degradable plastics, highlighting overlap. In addition to the packaging, due to their excellent recovery, these plastics show a tremendous growth potential for biomedical applications.

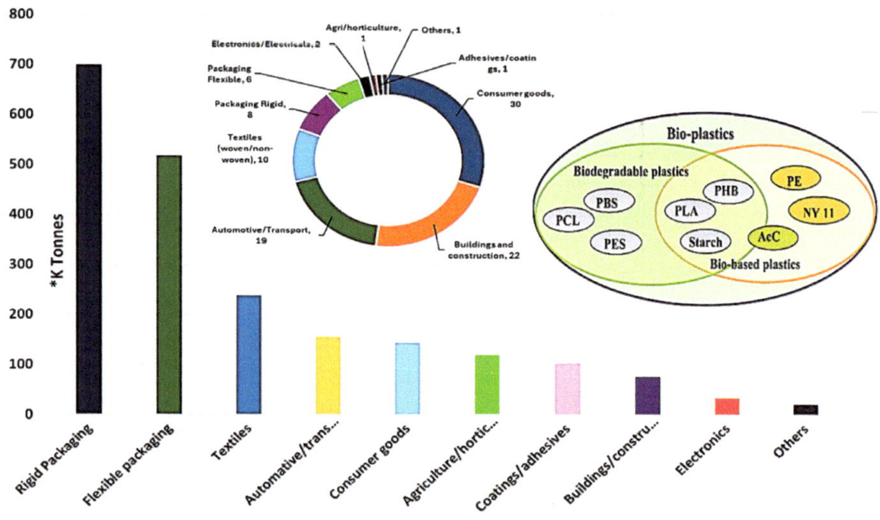

Figure 9: Bio-based and biodegradable plastics market-share with insets showing the use of plastic in various applications and a Venn diagram of bio-based and biodegradable plastics.

Transitioning from fossil-based plastics to bio-based plastics is a critical step forward in reducing environmental impact, conserving finite natural resources, and minimizing the effects of climate change, as part of Sustainable Development Goals (SDGs). As shown in Figure 10, bio-based plastics are derived from renewable biomass (e.g., corn, sugarcane, algae, lignin, etc.) and can be biodegradable or non-biodegradable, depending on the polymer structure. This shift involves significant technological, economic, and societal changes, however, there are

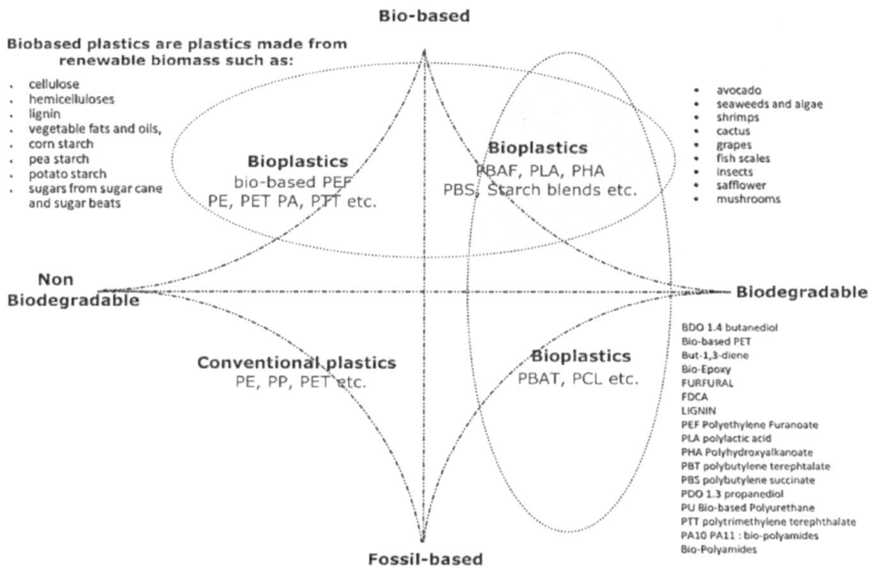

Figure 10: Transition matrix showing fossil-based to bio-based plastics with examples.

several challenges to realize this transition. As shown in Figure 9, considerable research and development activity is focused on reducing the cost of bio-based plastics resulting from raw material and processing differences. Furthermore, the durability and functionality need to be brought to par with fossil-based plastics. Other challenges include recycling compatibility for seamless integration into existing recycling schemes and the fact that using biomass for plastic production may compete with crop production and land use.

5.2.2. Recycling/Upcycling

As part of waste management, there are several well-documented and strategic mitigation schemes for plastic waste. Establishing a reliable recycling platform for packaging plastic waste is of significant importance considering its massive generation (Cheng et al., 2024; Singh & Walker, 2024). Implementing uniform and strategic means of mitigating plastic waste requires policy enactment and regulatory enforcement through legislative and legal processes. Some of these actions are discussed in the next section, as part of policy recommendations for plastic waste management. In addition to recycling, there is a significant effort in the upcycling of plastics, which is a process of converting plastic waste into new, higher-value products or materials. In a typical cycling process, plastic waste is treated by either pyrolysis or incineration, which is not only capital intensive but also creates a large carbon footprint. On the other hand, chemical upcycling of plastics is a catalytic strategy to convert waste plastics, such as PET, PE, PS, etc., to useful fuels, functionalized polymers, and other value-added chemicals having a direct impact on affordability, viability, and circular economy. In addition to chemical upcycling, mechanical, thermocatalytic (Martín et al., 2021; Pham & Cho, 2021), photocatalysis (Huang et al., 2022; Ohtani et al., 1992; Tofa et al., 2019; Zhou et al., 2022; Zhou et al., 2022), electrocatalysis (Li, Song, et al., 2023; Shi et al., 2021; Siddiqi & Sarlah, 2021; Wang et al., 2022; Zhang et al., 2023), enzyme catalysis (C.-C. Chen et al., 2020; Samak et al., 2020; Zhu et al., 2022), and photo electrocatalysis are also used as effective and efficient plastic upcycling technologies. Thermocatalysis of plastic wastes is generally carried out under harsh conditions with large thermal energy inputs. In electrocatalysis, the plastic wastes need to be converted into small molecules through pyrolysis or hydrolysis before the catalytic process and additional electricity input is required to trigger the catalytic reaction. Enzyme catalysis reaction takes place with a specific enzyme and at a specific site. Photocatalysis, as described earlier, can be conducted under ambient pressure and temperature with sunlight as the triggering energy form, and is considered a cheaper and greener route to transform plastic wastes into value-added products. Furthermore, the specific chemical bonds in the waste plastic can be precisely activated by controlling the electronic band structure of photocatalysts. There are other variations, such as repurposing plastic waste into art, jewelry, or home décor is referred to as artistic upcycling. Enzymatic and gasification of plastics to convert into useful raw materials are also employed for manufacturing new products. This is a new and upcoming area of research with a lot of potential and challenges. Due to the exponential growth of plastic products, there is a significant demand potential for upcycled valued added materials in projects such as construction, lumber, additive manufacturing, and even in the fashion industry. While additives in plastics complicate upcycling, an effort of this magnitude requires federal incentives and legislative actions to promote such initiatives.

6. Policy Recommendations, Waste Management, Regulatory Controls, and Ecological Restoration

MNP pollution is a complex problem to address, as the world is beyond the plastic overshoot threshold. It is critical to develop strategies to target plastic pollution at production, processing, use, and disposal sites. Reducing the impact of plastic pollution throughout the plastic lifecycle requires sustained and coordinated work across private sectors, local communities, government agencies, and stakeholders. It thus requires a strategic, multifaceted approach by way of policy recommendations, regulatory controls, technological innovations, and systematic and systemic ecological restoration. Within the scope of this publication, certain steps are highlighted to accomplish these tasks.

6.1. Policy Recommendations

Listed below are some actionable policy recommendations:

- *Improved waste disposal management, viz., to control pollution emanation from plastic production sites:* Extending EPA guidelines to reduce emissions of toxic air pollutants and harmful pollution that contributes to smog, expanding the Toxics Release Inventory Program (TRIP) to improve the public's understanding of releases associated with plastics production, and starting the process of prioritizing five chemicals used in plastic production, including vinyl chloride, for risk evaluations under the Toxic Substances Control Act (TSCA).
- *Recycling and upcycling of plastic waste products by eco-friendly design – so-called upcycling:* This is done through materials innovation to redesign plastics with circular economy as design criteria.
- *Use of bio-based polymers – over 90% of plastic is derived from fossil fuels:* To reduce pollution from the extraction of fossil fuels, the chemicals of concern and a range of hazardous air pollutants and volatile organic compounds, it is necessary to monitor air and water quality to understand the full extent of the environmental and human health risks of plastic production.
- *Eco-labeling programs:* Certify products based on their environmental impact and promote low-plastic alternatives. One of the key steps to decreasing the quantity of plastic waste generated is by limiting the initial use of materials that are unnecessary, difficult to manage, or likely to end up as pollutants in the environment.
- *Incentivize alternatives to plastic:* Material innovation and improved product design can help ensure that products are compatible with waste management systems and have minimal impacts on human health and the environment. Such actions include participating in the development of standards to promote recyclability and reuse, innovation in materials management, and additional research and development of materials that will create a more circular economy. One additional initiative is to phase out federal procurement of single-use plastics from food services by selecting reusable, compostable, and highly recyclable products.
- *Wastewater treatment efficiency:* Require wastewater plants to upgrade systems to filter out MNPs using membranes and advanced filtration mechanisms (Criscuoli, 2022; Vaseashta & Karagülle-Bölgen, 2018). Improve environmentally sound waste management by optimizing the collection of plastic, enhancing plastic recycling, and other measures to prevent plastic waste from entering the environment. To clean up existing plastic pollution and prevent additional plastic pollution from entering the environment, including the ocean, it is necessary to improve the capture of plastic, both before it enters the waste management system, and to prevent its escape during the waste management process.
- *Innovation grants:* Investing in infrastructure to improve reuse, recycling, and composting through competitive and transformative actions such as research grants for innovations by networking, technology accelerators, improvement of solid waste, and infrastructure, including pilot programs for removing large-scale marine debris using interception technologies to capture marine debris.

- *Promoting/incentivizing design with alternatives to plastics:* This can be done by providing subsidies for industries producing biodegradable or compostable materials.

6.2. Regulatory Controls

To the extent possible, the following are some of the regulatory controls to be implemented by local, state, and federal agencies.

- *Mandatory buyback of e-waste as part of producer responsibility:* That is, implement Extended Producer Responsibility (EPR), making producers accountable for plastic waste management.
- *Mandatory fines and penalties for discarding fishnets and fishing accessories in marine water:* This should be traceable by embedded barcodes on the nets, as part of a traceable inventory control mechanism.
- *Mandatory multiple reuses of a large variety of plastic containers:* This can be controlled by barcode.
- *Textile standards and innovations:* Introduce regulations for synthetic textiles to minimize fiber shedding. Incentivize the development of eco-friendly fabrics and filters in washing machines.
- *Tax incentive:* To use recycled plastics-based products.
- *Standardized measurement protocols:* Establish global or national protocols for MP assessment.
- *Microplastic monitoring programs:* Regularly measure plastic contamination in water, air, and soil.
- *Improvised tire design standards:* Encourage low-abrasion tires that generate fewer MP particles.
- *Stormwater management:* Instal filtering systems in urban drainage to capture tire-wear particles and roadside runoff.

6.3. Ecological Restoration

The escalating extent of human-induced damage to Earth's ecosystems necessitates a comprehensive strategy for ecosystem repair, positioning restoration ecology as a crucial component for our future survival. Restoration efforts need to focus on both removing existing pollution and preventing further contamination to restore ecosystems and biodiversity.

The approaches may consist of several steps, and some are listed below:

- *Pollution assessment and monitoring:* Baseline studies – map and quantify MNP contamination in ecosystems (e.g., coastal areas, rivers, and soils).
- *Long-Term monitoring:* Establish programs to monitor plastic levels over time and track the effectiveness of restoration efforts. The assessment must be plotted and updated in real-time. Such a real-time map would provide a visual indicator that would guide future actions.
- *Plastic cleanup and remediation:* It will consist of physical removal, such as

using filters, sieves, or booms to trap MNPs in rivers and coastal areas. Employ sediment vacuuming in water bodies to remove MPs from the benthic zone.

- *Soil remediation:* Use soil washing techniques to extract plastic particles. Test phytoremediation (plants that can bind or stabilize MPs) for terrestrial ecosystems.
- *Bioremediation:* Develop microbial communities and fungi capable of degrading certain plastics. Support research into enzymatic degradation technologies.
- *Ecosystem-based restoration:* Mangrove and wetland rehabilitation to restore these ecosystems, which act as natural filters, trapping plastic particles in sediments.
- *Reforestation and afforestation:* Increase vegetation cover along riverbanks to reduce soil erosion and trap microplastics.
- *Seagrass and coral restoration:* Reintroduce habitats that can help stabilize marine ecosystems affected by plastics.
- *Enhancing natural filtering systems:* Biofilters in wastewater treatment plants: Install advanced filtration systems (e.g., membrane bioreactors, nanofiltration) to reduce MP release into water bodies.
- *Stormwater management systems:* Use wetlands, retention ponds, and bioswales to capture plastics from urban runoff.
- *Promote circular economy and clean-up partnerships.*
- *Incentivize clean-ups:* Encourage partnerships between governments, NGOs, and communities to remove plastic waste from ecosystems.
- *Circular economy strategies:* Increase recycling and material recovery from waste streams to reduce the future load of MPs.
- *Restocking biodiversity and species recovery:* Reintroduce filter-feeding Species: Species like mussels and oysters, which naturally filter particles from water, can be reintroduced to support ecological recovery.
- *Improve fishery management:* Protect key marine species affected by plastic ingestion to restore ecological balance.
- *Regenerative agricultural practices:* Crop rotation, agroforestry, composting, etc.
- *Reduce microplastic use in agriculture:* Transition away from plastic mulch films and slow-release fertilizers encapsulated in plastic.
- *Compost management:* Monitor compost quality to ensure no MP contamination enters agricultural soil.
- *Awareness and policy support for restoration:* National parks, invasive species control, construction of bird nesting islands, etc.
- *Ecological literacy campaigns:* Raise awareness among citizens and industries about the importance of restoring ecosystems.
- *Legislative incentives:* Offer grants or subsidies to support restoration projects aimed at reducing MP contamination.
- *Research and innovation for future solutions:* For creating a smart future, commercializing new ideas, etc.

- *Innovate bio-degradable plastics:* Develop and promote materials that degrade safely in natural environments.
- *Fund research into ecological restoration from MNPs*: Support studies on the long-term ecological effects of microplastics and effective restoration interventions.

7. Conclusions and Path Forward

From the discussion above it is evident that plastic pollution is a critical global issue with widespread effects on humanity, ecosystems, and the planet as a whole. It very well may be classified as an existential risk and is rapidly reaching a crisis level. Plastic presents an interesting challenge, in that its versatility, durability, and affordability render it indispensable in modern life, but these same properties contribute to its environmental and health hazards. The production of plastic has grown much faster than the finding of solutions to mitigate its negative impact on the environment and human health. Chemically speaking, the plastics themselves are generally inert and do not have much effect on human health. One of the major problems is due to the sheer volume of plastic waste. Over 8 MT of plastic waste ends up in oceans annually, harming marine life, since marine animals mistakenly consume plastic for food, leading to ingestion, starvation, or entanglement. Another major issue is the degradation and fragmentation of plastics into micro- nanoplastics. It creates particles of smaller sizes, which are difficult to detect, and more reactive due to the large surface area. Additionally, many additives that are used to modify the properties of plastics are released into the environment. Currently, there is limited evidence that MNPs pose serious health issues in humans as it does in aquatic organisms, as the level of these plastics ingested is not at sufficient concentrations to cause the same effects as it does in aquatic animals. But, when we consider that many aquatic organisms are relevant to the fish industry and that they can ingest these tiny plastics, it is logical to hypothesize potential risks to human health and food security. In addition to the toxic chemicals absorbed, the MNPs in aquatic environments form biofilms around them and harmful pathogens get attracted to these particles from the contaminated ecosystem that have potentially negative effects on organisms and ecosystem health. It was shown that in addition to adverse health impacts to marine animals and humans through food chain supply, there are potential security risks. Since these studies are still in their preliminary stages, thorough studies on their impact on the terrestrial ecosystem, plant growth performance and yield, atmospheric interactions, etc. also should be conducted. Additional sources of MNPs are textiles, laundry, cosmetics, personal care products, abrasion of household plastics, etc.

To address this growing concern, a multi-faceted approach is suggested in this chapter, primarily since plastic is quintessential to our way of life. Thus, using technological platforms, it is necessary to reap the benefits of these materials while minimizing the negative health and environmental impacts.

These approaches include strategic plastic waste management through policy implementation, regulatory controls, and monitoring. In addition to recycling as part of waste management, it is necessary to upcycle plastic through a series of processes to convert it into useful new, higher-value products or materials. We also propose to gradually shift from fossil-based plastics to bio-based plastics, a trend consistent with the plastic industry, as there is now more focus on biomass-based plastics, especially for packaging. For fossil-based plastics, several processes are mentioned that can degrade plastic into environmentally friendly compounds. For bio-based plastics, it is conceivable that through design, the products will have an environmentally friendly end-of-life cycle.

One of the major problems with the degraded plastics is its detection. Several methods are mentioned in this chapter, and it is necessary to advance research in the detection and identification of MNPs in air and aquatic environments. This will also aid with toxicokinetic studies and mechanisms. There is insufficient and anecdotal evidence to suggest that MNPs cause oxidative stress through cell-barrier penetration. Also, there is limited toxicokinetic data using animal studies showing inflammation during liver histology and oxidative stress, energy, and lipid metabolism, however, more data is needed to conclusively link the MNPs with physiological response. The microplastics also travel by air, spreading their ill effects as airborne pollutants in the fiber form. The impact of MNPs on the lungs needs additional studies. There is a lack of studies on the effect of MPs and NPs on humans through the food chain which is a very serious threat to human food security. Therefore, in-depth studies should be carried out and immediate remedial measures should be taken before the MPs and NPs invade our daily diet.

For a comprehensive understanding of the current stage of plastic pollution in the aquatic ecosystem, there should be more synergistic expansion of databases on plastics of all size fractions, particularly in freshwater environments. Although there are a few datasets available in OSINT containing MNPs information collected from various oceans, regions, and sampling methods over several years; by leveraging powerful data analytic and visualization techniques, it is possible to uncover temporal patterns and insights that can help better understand the distribution and dynamics of marine MNPs. These data sets can analyze temporal trends, explore monthly and seasonal variations, compare sampling methods, visualize the geographical distribution, investigate density classes analysis, and assess the contributions of different organizations to the collection and analysis of MNPs data. Since plastics find their way to oceans mainly through rivers, the estimation of river plastic releases to the oceans is critical. Microplastics are also known as the carriers of microorganisms, hence more research is needed to identify the role of plastics as vectors for chemical and biological contaminants in the biota. Lastly, realizing that humanity, under no conceivable scenario, will abandon the production and use of plastics, the viable options are bio-based plastics, strategic waste management options, and pathways for large-scale mitigation of MNFPS in the atmosphere and hydrosphere.

References

Adegoke, K.A., Adu, F.A., Oyebamiji, A.K., Bamisaye, A., Adigun, R.A., Olasoji, S.O., & Ogunjinmi, O.E. (2023). Microplastics toxicity, detection, and removal from water/wastewater. *Marine Pollution Bulletin*, 187, 114546. https://doi.org/10.1016/j.marpolbul.2022.114546.

Adhikari, S., Kelkar, V., Kumar, R., & Halden, R.U. (2022). Methods and challenges in the detection of microplastics and nanoplastics: A mini-review. *Polymer International*, 71(5), 543–551. https://doi.org/10.1002/pi.6348.

Agboola, O.D., & Benson, N.U. (2021). Physisorption and chemisorption mechanisms influencing micro (nano) plastics-organic chemical contaminants interactions: A review. *Frontiers in Environmental Science*, 9, 678574. https://doi.org/10.3389/fenvs.2021.678574.

Alderton, M. (2022). *Scientists Discover Surprising Remedy to Plastic Pollution: Caterpillar Spit* [Treehugger]. https://www.treehugger.com/scientists-discover-surprising-remedy-to-plastic-pollution-caterpillar-spit-6748173.

Almeida, S., Ozkan, S., Gonçalves, D., Paulo, I., Queirós, C.S.G.P., Ferreira, O. et al. (2022). A brief evaluation of antioxidants, antistatics, and plasticizers additives from natural sources for polymers formulation. *Polymers*, 15(1), 6. https://doi.org/10.3390/polym15010006.

Amato-Lourenço, L.F., Carvalho-Oliveira, R., Júnior, G.R., Dos Santos Galvão, L., Ando, R.A., & Mauad, T. (2021). Presence of airborne microplastics in human lung tissue. *Journal of Hazardous Materials*, 416, 126124. https://doi.org/10.1016/j.jhazmat.2021.126124.

Andersen, I. (2024). *United Nations Environment Programme* (*UNEP*) (Annual Report 2023; p. 24). https://wedocs.unep.org/bitstream/handle/20.500.11822/44777/UNEP_Annual_Report_2023.pdf?sequence=19.

Ariza-Tarazona, M.C., Villarreal-Chiu, J.F., Barbieri, V., Siligardi, C., & Cedillo-González, E.I. (2019). New strategy for microplastic degradation: Green photocatalysis using a protein-based porous N-TiO_2 semiconductor. *Ceramics International*, 45(7), 9618–9624. https://doi.org/10.1016/j.ceramint.2018.10.208.

Austin, H.P., Allen, M.D., Donohoe, B.S., Rorrer, N.A., Kearns, F.L., Silveira, R.L. et al. (2018). Characterization and engineering of a plastic-degrading aromatic polyesterase. *Proceedings of the National Academy of Sciences*, 115(19). https://doi.org/10.1073/pnas.1718804115.

Baldwin, J., Noorali, S., & Vaseashta, A. (2023). Biology and behavior of severe acute respiratory syndrome coronavirus contagion with emphasis on treatment strategies, risk assessment, and resilience. *COVID*, 3(9), 1259–1303. https://doi.org/10.3390/covid3090089.

Berkner, S., Konradi, S., & Schönfeld, J. (2014). Antibiotic resistance and the environment—There and back again: Science & Society series on Science and Drugs. *EMBO Reports*, 15(7), 740–744. https://doi.org/10.15252/embr.201438978.

Besseling, E., Wegner, A., Foekema, E.M., Van Den Heuvel-Greve, M.J., & Koelmans, A.A. (2013). Effects of microplastic on fitness and PCB bioaccumulation by the lugworm *Arenicola marina* (L.). *Environmental Science & Technology*, 47(1), 593–600. https://doi.org/10.1021/es302763x.

Bethany Halford. (2016). World's first PET-munching microbe discovered. *C&EN Global Enterprise*, 94(49), 25–25. https://doi.org/10.1021/cen-09449-cover3.

Bhat, M.A. (2023). Indoor microplastics: A comprehensive review and bibliometric analysis. *Environmental Science and Pollution Research International*, 30(58), 121269–121291. https://doi.org/10.1007/s11356-023-30902-0.

Bhattacharjee, S., Lugger, J.A.M., & Sijbesma, R.P. (2017). Tailoring pore size and chemical interior of near 1 nm sized pores in a nanoporous polymer based on a discotic liquid crystal. *Macromolecules*, 50(7), 2777–2783. https://doi.org/10.1021/acs.macromol.7b00013.

Bhattacharya, P., Lin, S., Turner, J.P., & Ke, P.C. (2010). Physical adsorption of charged plastic nanoparticles affects algal photosynthesis. *The Journal of Physical Chemistry C*, 114(39), 16556–16561. https://doi.org/10.1021/jp1054759.

Bolea-Fernandez, E., Rua-Ibarz, A., Velimirovic, M., Tirez, K., & Vanhaecke, F. (2020). Detection of microplastics using inductively coupled plasma-mass spectrometry (ICP-MS) operated in single-event mode. *Journal of Analytical Atomic Spectrometry*, 35(3), 455–460. https://doi.org/10.1039/C9JA00379G.

Bolgen, N., Demir, D., & Vaseashta, A. (2013). Development of polymeric cryogels as potential matrices for removing antibiotics from wastewater. *In:* A. Vaseashta & S. Khudaverdyan (Eds.), *Advanced Sensors for Safety and Security* (pp. 243–249). Springer Netherlands. https://doi.org/10.1007/978-94-007-7003-4_21.

Brčeski, I., & Vaseashta, A. (2021). Environmental forensic tools for water resources. *In:* A. Vaseashta & C. Maftei (Eds.), *Water Safety, Security, and Sustainability* (pp. 333–370). Springer International Publishing. https://doi.org/10.1007/978-3-030-76008-3_15.

Butman, D., Stackpoole, S., Stets, E., McDonald, C.P., Clow, D.W., & Striegl, R.G. (2016). Aquatic carbon cycling in the conterminous United States and implications for terrestrial carbon accounting. *Proceedings of the National Academy of Sciences*, 113(1), 58–63. https://doi.org/10.1073/pnas.1512651112.

Campanale, C., Dierkes, G., Massarelli, C., Bagnuolo, G., & Uricchio, V.F. (2020). A relevant screening of organic contaminants present on freshwater and pre-production microplastics. *Toxics*, 8(4), 100. https://doi.org/10.3390/toxics8040100.

Cao, Z., Yan, W., Ding, M., & Yuan, Y. (2022). Construction of microbial consortia for microbial degradation of complex compounds. *Frontiers in Bioengineering and Biotechnology*, 10, 1051233. https://doi.org/10.3389/fbioe.2022.1051233.

Carbery, M., O'Connor, W., & Palanisami, T. (2018). Trophic transfer of microplastics and mixed contaminants in the marine food web and implications for human health. *Environment International*, 115, 400–409. https://doi.org/10.1016/j.envint.2018.03.007.

Cassano, A., Conidi, C., & Castro-Muñoz, R. (2019). Current and future applications of nanofiltration in food processing. *In: Separation of Functional Molecules in Food by Membrane Technology* (pp. 305–348). Elsevier. https://doi.org/10.1016/B978-0-12-815056-6.00009-7.

Chartres, N., Cooper, C.B., Bland, G., Pelch, K.E., Gandhi, S.A., BakenRa, A., & Woodruff, T. J. (2024). Effects of microplastic exposure on human digestive, reproductive, and respiratory health: A rapid systematic review. *Environmental Science & Technology*, acs.est.3c09524. https://doi.org/10.1021/acs.est.3c09524.

Chen, C.-C., Dai, L., Ma, L., & Guo, R.-T. (2020). Enzymatic degradation of plant biomass and synthetic polymers. *Nature Reviews Chemistry*, 4(3), 114–126. https://doi.org/10.1038/s41570-020-0163-6.

Chen, X., Mo, L., Zhang, L., Huang, L., Gao, Z., Peng, J. et al. (2024). Taxonomic diversity, predicted metabolic pathway, and interaction pattern of bacterial community in Sea Urchin *Anthocidaris crassispina*. *Microorganisms*, 12(10), 2094. https://doi.org/10.3390/microorganisms12102094.

Chen, Y., Wen, D., Pei, J., Fei, Y., Ouyang, D., Zhang, H., & Luo, Y. (2020). Identification and quantification of microplastics using fourier-transform infrared spectroscopy: Current status and future prospects. *Current Opinion in Environmental Science & Health*, 18, 14–19. https://doi.org/10.1016/j.coesh.2020.05.004.

Cheng, L., Chen, X., Gu, J., Kobayashi, N., Yuan, H., & Chen, Y. (2024). Chemical recycling of waste plastics: Current challenges and perspectives. *Fundamental Research*, S2667325824000578. https://doi.org/10.1016/j.fmre.2023.12.023.

Cole, M., Lindeque, P., Fileman, E., Halsband, C., Goodhead, R., Moger, J., & Galloway, T.S. (2013). Microplastic ingestion by zooplankton. *Environmental Science & Technology*, 47(12), 6646–6655. https://doi.org/10.1021/es400663f.

Correia, M., & Loeschner, K. (2018). Detection of nanoplastics in food by asymmetric flow field-flow fractionation coupled to multi-angle light scattering: Possibilities, challenges, and analytical limitations. *Analytical and Bioanalytical Chemistry*, 410(22), 5603–5615. https://doi.org/10.1007/s00216-018-0919-8.

Criscuoli, A. (2022). Membrane distillation for aqueous feeds treatment: *In:* A. Vaseashta, G. Duca, & S. Travin (Eds.), *Advances in Environmental Engineering and Green Technologies* (pp. 150–166). IGI Global. https://doi.org/10.4018/978-1-7998-7356-3.ch007.

Daniel, D.B., Ashraf, P.M., Thomas, S.N., & Thomson, K.T. (2021). Microplastics in the edible tissues of shellfishes sold for human consumption. *Chemosphere*, 264, 128554. https://doi.org/10.1016/j.chemosphere.2020.128554.

Das, A. (2023). The emerging role of microplastics in systemic toxicity: Involvement of reactive oxygen species (ROS). *Science of The Total Environment*, 895, 165076. https://doi.org/10.1016/j.scitotenv.2023.165076.

DeLoid, G.M., Cao, X., Bitounis, D., Singh, D., Llopis, P.M., Buckley, B., & Demokritou, P. (2021). Toxicity, uptake, and nuclear translocation of ingested micro-nanoplastics in an *in vitro* model of the small intestinal epithelium. *Food and Chemical Toxicology*, 158, 112609. https://doi.org/10.1016/j.fct.2021.112609.

Demir, D., Bölgen, N., & Vaseashta, A. (2021). Chitosan–gelatin cryogels as bio-sorbents for removal of dyes from aqueous solutions. *In:* A. Vaseashta & C. Maftei (Eds.), *Water Safety, Security, and Sustainability* (pp. 263–274). Springer International Publishing. https://doi.org/10.1007/978-3-030-76008-3_11.

Demir, D., Vaseashta, A., & Bölgen, N. (2021). Macroporous cryogels for water purification. *In:* A. Vaseashta & C. Maftei (Eds.), *Water Safety, Security, and Sustainability* (pp. 275–290). Springer International Publishing. https://doi.org/10.1007/978-3-030-76008-3_12.

Deng, Y., Zhang, Y., Lemos, B., & Ren, H. (2017). Tissue accumulation of microplastics in mice and biomarker responses suggest widespread health risks of exposure. *Scientific Reports*, 7, 46687. https://doi.org/10.1038/srep46687.

Desforges, J.-P.W., Galbraith, M., Dangerfield, N., & Ross, P.S. (2014). Widespread distribution of microplastics in subsurface seawater in the NE Pacific Ocean. *Marine Pollution Bulletin*, 79(1–2), 94–99. https://doi.org/10.1016/j.marpolbul.2013.12.035.

Di Pippo, F., Bocci, V., Amalfitano, S., Crognale, S., Levantesi, C., Pietrelli, L. et al. (2023). Microbial colonization patterns and biodegradation of petrochemical and biodegradable plastics in lake waters: Insights from a field experiment. *Frontiers in Microbiology*, 14, 1290441. https://doi.org/10.3389/fmicb.2023.1290441.

Dissanayake, P.D., Kim, S., Sarkar, B., Oleszczuk, P., Sang, M.K., Haque, M.N. et al. (2022). Effects of microplastics on the terrestrial environment: A critical review. *Environmental Research*, 209, 112734. https://doi.org/10.1016/j.envres.2022.112734.

Dong, C.-D., Chen, C.-W., Chen, Y.-C., Chen, H.-H., Lee, J.-S., & Lin, C.-H. (2020). Polystyrene microplastic particles: *In vitro* pulmonary toxicity assessment. *Journal of Hazardous Materials*, 385, 121575. https://doi.org/10.1016/j.jhazmat.2019.121575.

Du, H., Xie, Y., & Wang, J. (2021). Microplastic degradation methods and corresponding degradation mechanism: Research status and future perspectives. *Journal of Hazardous Materials*, 418, 126377. https://doi.org/10.1016/j.jhazmat.2021.126377.

Dubaish, F., & Liebezeit, G. (2013). Suspended microplastics and black carbon particles in the jade system, Southern North Sea. *Water, Air, & Soil Pollution*, 224(2), 1352. https://doi.org/10.1007/s11270-012-1352-9.

Enyoh, C.E., Wang, Q., Chowdhury, T., Wang, W., Lu, S., Xiao, K., & Chowdhury, Md. A.H. (2021). New analytical approaches for effective quantification and identification of nanoplastics in environmental samples. *Processes*, 9(11), 2086. https://doi.org/10.3390/pr9112086.

Eriksen, M., Lebreton, L.C.M., Carson, H.S., Thiel, M., Moore, C.J., Borrero, J.C. et al. (2014). Plastic pollution in the world's oceans: More than 5 trillion plastic pieces weighing over 250,000 tons afloat at sea. *PLoS ONE*, 9(12), e111913. https://doi.org/10.1371/journal.pone.0111913.

Farrell, P., & Nelson, K. (2013). Trophic level transfer of microplastic: *Mytilus edulis* (L.) to *Carcinus maenas* (L.). *Environmental Pollution*, 177, 1–3. https://doi.org/10.1016/j.envpol.2013.01.046.

Feng, Y., Tu, C., Li, R., Wu, D., Yang, J., Xia, Y. et al. (2023). A systematic review of the impacts of exposure to micro- and nano-plastics on human tissue accumulation and health. *Eco-Environment & Health*, 2(4), 195–207. https://doi.org/10.1016/j.eehl.2023.08.002.

Gagné, F., André, C., Turgeon, S., & Ménard, N. (2023). Evidence of polystyrene nanoplastic contamination and potential impacts in Mya arenaria clams in the Saint-Lawrence estuary (Canada). *Comparative Biochemistry and Physiology, Toxicology & Pharmacology: CBP*, 266, 109563. https://doi.org/10.1016/j.cbpc.2023.109563.

Gao, G., Zhao, X., Jin, P., Gao, K., & Beardall, J. (2021). Current understanding and challenges for aquatic primary producers in a world with rising micro- and nano-plastic levels. *Journal of Hazardous Materials*, 406, 124685. https://doi.org/10.1016/j.jhazmat.2020.124685.

Gao, N., Yang, L., Lu, X., Duan, Z., Zhu, L., & Feng, J. (2022). A review of interactions of microplastics and typical pollutants from toxicokinetics and toxicodynamics perspective. *Journal of Hazardous Materials*, 432, 128736. https://doi.org/10.1016/j.jhazmat.2022.128736.

Garcia, A.G., Suárez, D.C., Li, J., & Rotchell, J.M. (2021). A comparison of microplastic contamination in freshwater fish from natural and farmed sources. *Environmental Science and Pollution Research*, 28(12), 14488–14497. https://doi.org/10.1007/s11356-020-11605-2.

Gong, Y., Chen, X., & Wu, W. (2024). Application of fourier transform infrared (FTIR) spectroscopy in sample preparation: Material characterization and mechanism investigation. *Advances in Sample Preparation*, 11, 100122. https://doi.org/10.1016/j.sampre.2024.100122.

Gubała, D., Slastanova, A., Matthews, L., Islas, L., Wąsik, P., Cacho-Nerin, F. et al. (2024). Effects of erucamide on fiber "Softness": Linking single-fiber crystal structure and mechanical properties. *ACS Nano*, acsnano.4c00114. https://doi.org/10.1021/acsnano.4c00114.

Hahladakis, J.N., Velis, C.A., Weber, R., Iacovidou, E., & Purnell, P. (2018). An overview of chemical additives present in plastics: Migration, release, fate, and environmental impact during their use, disposal, and recycling. *Journal of Hazardous Materials*, 344, 179–199. https://doi.org/10.1016/j.jhazmat.2017.10.014.

Hamd, W., Daher, E.A., Tofa, T.S., & Dutta, J. (2022). Recent advances in photocatalytic removal of microplastics: Mechanisms, kinetic degradation, and reactor design. *Frontiers in Marine Science*, 9, 885614. https://doi.org/10.3389/fmars.2022.885614.

Haq, F., Kiran, M., Khan, I.A., Mehmood, S., Aziz, T., & Haroon, M. (2024). Exploring the pathways to sustainability: A comprehensive review of biodegradable plastics in the circular economy. *Materials Today Sustainability*, 101067. https://doi.org/10.1016/j.mtsust.2024.101067.

Hartmann, N.B., Rist, S., Bodin, J., Jensen, L.H., Schmidt, S.N., Mayer, P. et al. (2017). Microplastics as vectors for environmental contaminants: Exploring sorption, desorption, and transfer to biota. *Integrated Environmental Assessment and Management*, 13(3), 488–493. https://doi.org/10.1002/ieam.1904.

Hilal, N., Al-Zoubi, H., Darwish, N.A., & Mohammad, A.W. (2007). Performance of nanofiltration membranes in the treatment of synthetic and real seawater. *Separation Science and Technology*, 42(3), 493–515. https://doi.org/10.1080/01496390601120789.

Ho, D., Liu, S., Wei, H., & Karthikeyan, K.G. (2024). The glowing potential of Nile red for microplastics identification: Science and mechanism of fluorescence staining. *Microchemical Journal*, 197, 109708. https://doi.org/10.1016/j.microc.2023.109708.

Huang, Z., Shanmugam, M., Liu, Z., Brookfield, A., Bennett, E.L., Guan, R. et al. (2022). Chemical recycling of polystyrene to valuable chemicals via selective acid-catalyzed aerobic oxidation under visible light. *Journal of the American Chemical Society*, 144(14), 6532–6542. https://doi.org/10.1021/jacs.2c01410.

Ivar Do Sul, J.A., & Costa, M.F. (2014). The present and future of microplastic pollution in the marine environment. *Environmental Pollution*, 185, 352–364. https://doi.org/10.1016/j.envpol.2013.10.036.

Jabri, N.A., Abed, R.M.M., Habsi, A.A., Ansari, A., & Barry, M.J. (2024). The impacts of microplastics on zebrafish behavior depend on initial personality state. *Environmental Toxicology and Pharmacology*, 111, 104561. https://doi.org/10.1016/j.etap.2024.104561.

Jin, H., Ma, T., Sha, X., Liu, Z., Zhou, Y., Meng, X. et al. (2021). Polystyrene microplastics induced male reproductive toxicity in mice. *Journal of Hazardous Materials*, 401, 123430. https://doi.org/10.1016/j.jhazmat.2020.123430.

Kadac-Czapska, K., Ośko, J., Knez, E., & Grembecka, M. (2024). Microplastics and oxidative stress-current problems and prospects. *Antioxidants* (Basel, Switzerland), 13(5), 579. https://doi.org/10.3390/antiox13050579.

Kahn, L.G., Philippat, C., Nakayama, S.F., Slama, R., & Trasande, L. (2020). Endocrine-disrupting chemicals: Implications for human health. *The Lancet Diabetes & Endocrinology*, 8(8), 703–718. https://doi.org/10.1016/S2213-8587(20)30129-7.

Koelmans, A.A., Besseling, E., & Shim, W.J. (2015). Nanoplastics in the aquatic environment. Critical review. *In:* M. Bergmann, L. Gutow, & M. Klages (Eds.), *Marine Anthropogenic Litter* (pp. 325–340). Springer International Publishing. https://doi.org/10.1007/978-3-319-16510-3_12.

Kokilathasan, N., & Dittrich, M. (2022). Nanoplastics: Detection and impacts in aquatic environments: A review. *Science of The Total Environment*, 849, 157852. https://doi.org/10.1016/j.scitotenv.2022.157852.

Kooi, M., Nes, E.H.V., Scheffer, M., & Koelmans, A.A. (2017). Ups and downs in the ocean: Effects of biofouling on vertical transport of microplastics. *Environmental Science & Technology*, 51(14), 7963–7971. https://doi.org/10.1021/acs.est.6b04702.

Kwon, W., Kim, D., Kim, H.-Y., Jeong, S.W., Lee, S.-G., Kim, H.-C. et al. (2022). Microglial phagocytosis of polystyrene microplastics results in immune alteration and apoptosis *in vitro* and *in vivo*. *The Science of the Total Environment*, 807(Pt 2), 150817. https://doi.org/10.1016/j.scitotenv.2021.150817.

Lavoy, M., & Crossman, J. (2021). A novel method for organic matter removal from samples containing microplastics. *Environmental Pollution*, 286, 117357. https://doi.org/10.1016/j.envpol.2021.117357.

Lee, C.-W., Hsu, L.-F., Wu, I.-L., Wang, Y.-L., Chen, W.-C., Liu, Y.-J. et al. (2022). Exposure to polystyrene microplastics impairs hippocampus-dependent learning and memory in mice. *Journal of Hazardous Materials*, 430, 128431. https://doi.org/10.1016/j.jhazmat.2022.128431.

Leslie, H.A., Van Velzen, M.J.M., Brandsma, S.H., Vethaak, A.D., Garcia-Vallejo, J.J., & Lamoree, M.H. (2022). Discovery and quantification of plastic particle pollution in human blood. *Environment International*, 163, 107199. https://doi.org/10.1016/j.envint.2022.107199.

Li, M., Song, N., Luo, W., Chen, J., Jiang, W., & Yang, J. (2023). Engineering surface oxophilicity of copper for electrochemical CO_2 reduction to ethanol. *Advanced Science*, 10(2), 2204579. https://doi.org/10.1002/advs.202204579.

Li, W., Zhao, W., Zhu, H., Li, Z.-J., & Wang, W. (2023). State of the art in the photochemical degradation of (micro)plastics: From fundamental principles to catalysts and applications. *Journal of Materials Chemistry A*, 11(6), 2503–2527. https://doi.org/10.1039/D2TA09523H.

Li, W., Zu, B., Yang, Q., Guo, J., & Li, J. (2023). Sources, distribution, and environmental effects of microplastics: A systematic review. *RSC Advances*, 13(23), 15566–15574. https://doi.org/10.1039/D3RA02169F.

Liang, B., Zhong, Y., Huang, Y., Lin, X., Liu, J., Lin, L. et al. (2021). Underestimated health risks: Polystyrene micro- and nanoplastics jointly induce intestinal barrier dysfunction by ROS-mediated epithelial cell apoptosis. *Particle and Fibre Toxicology*, 18(1), 20. https://doi.org/10.1186/s12989-021-00414-1.

Lin, Y., Huang, X., Liu, Q., Lin, Z., & Jiang, G. (2020). Thermal fragmentation enhanced identification and quantification of polystyrene micro/nanoplastics in complex media. *Talanta*, 208, 120478. https://doi.org/10.1016/j.talanta.2019.120478.

Liu, G., Zhu, Z., Yang, Y., Sun, Y., Yu, F., & Ma, J. (2019). Sorption behavior and mechanism of hydrophilic organic chemicals to virgin and aged microplastics in freshwater and seawater. *Environmental Pollution*, 246, 26–33. https://doi.org/10.1016/j.envpol.2018.11.100

Lu, K., Lai, K.P., Stoeger, T., Ji, S., Lin, Z., Lin, X. et al. (2021). Detrimental effects of microplastic exposure on normal and asthmatic pulmonary physiology. *Journal of Hazardous Materials*, 416, 126069. https://doi.org/10.1016/j.jhazmat.2021.126069.

Lusher, A. (2015). Microplastics in the marine environment: Distribution, interactions and effects. *In:* M. Bergmann, L. Gutow, & M. Klages (Eds.), *Marine Anthropogenic Litter* (pp. 245–307). Springer International Publishing. https://doi.org/10.1007/978-3-319-16510-3_10.

Lv, L., He, L., Jiang, S., Chen, J., Zhou, C., Qu, J. et al. (2020). *In situ* surface-enhanced Raman spectroscopy for detecting microplastics and nanoplastics in aquatic environments. *Science of The Total Environment*, 728, 138449. https://doi.org/10.1016/j.scitotenv.2020.138449.

Malloy, A. (2011). Count, size, and visualize nanoparticles. *Materials Today*, 14(4), 170–173. https://doi.org/10.1016/S1369-7021(11)70089-X.

Martín, A.J., Mondelli, C., Jaydev, S.D., & Pérez-Ramírez, J. (2021). Catalytic processing of plastic waste on the rise. *Chem*, 7(6), 1487–1533. https://doi.org/10.1016/j.chempr.2020.12.006.

Masura, J., Baker, J., Foster, G., Arthur, C., & Herring, C. (2015). *Laboratory Methods for the Analysis of Microplastics in the Marine Environment* (NOAA Marine Debris Program Technical Memorandum NOS-OR&R-48). National Oceanic and Atmospheric Administration, Marine Debris Division, Silver Spring MD.

Mattsson, K., Johnson, E.V., Malmendal, A., Linse, S., Hansson, L.-A., & Cedervall, T. (2017). Brain damage and behavioural disorders in fish induced by plastic nanoparticles delivered through the food chain. *Scientific Reports*, 7(1), 11452. https://doi.org/10.1038/s41598-017-10813-0.

Mayer, P., Fernqvist, M.M., Christensen, P.S., Karlson, U., & Trapp, S. (2007). Enhanced diffusion of polycyclic aromatic hydrocarbons in artificial and natural aqueous solutions. *Environmental Science & Technology*, 41(17), 6148–6155. https://doi.org/10.1021/es070495t.

McNeil-Watson, F. (2013). Electrophoretic light scattering. *In:* G.C.K. Roberts (Ed.), *Encyclopedia of Biophysics* (pp. 648–654). Springer Berlin Heidelberg. https://doi.org/10.1007/978-3-642-16712-6_288.

Mijatović, I., Matošić, M., Hajduk Černeha, B., & Bratulić, D. (2004). Removal of natural organic matter by ultrafiltration and nanofiltration for drinking water production. *Desalination*, 169(3), 223–230. https://doi.org/10.1016/j.desal.2004.01.002.

Mizraji, R., Ahrendt, C., Perez-Venegas, D., Vargas, J., Pulgar, J., Aldana, M. et al. (2017). Is the feeding type related with the content of microplastics in intertidal fish gut? *Marine Pollution Bulletin*, 116(1–2), 498–500. https://doi.org/10.1016/j.marpolbul.2017.01.008.

Moshood, T.D., Nawanir, G., Mahmud, F., Mohamad, F., Ahmad, M.H., & AbdulGhani, A. (2022). Sustainability of biodegradable plastics: New problem or solution to solve the global plastic pollution. *Current Research in Green and Sustainable Chemistry*, 5, 100273. https://doi.org/10.1016/j.crgsc.2022.100273.

Munoz, M., Ortiz, D., Nieto-Sandoval, J., De Pedro, Z.M., & Casas, J.A. (2021). Adsorption of micropollutants onto realistic microplastics: Role of microplastic nature, size, age, and NOM fouling. *Chemosphere*, 283, 131085. https://doi.org/10.1016/j.chemosphere.2021.131085.

Murphy, J. (2018). *Introduction to Plastics Engineering*. Elsevier. https://doi.org/10.1016/C2014-0-03688-X.

Nastase, C., Nastase, F., Vaseashta, A., & Stamatin, I. (2006). Nanocomposites based on functionalized nanotubes in polyaniline matrix by plasma polymerization. *Progress in Solid State Chemistry*, 34(2–4), 181–189. https://doi.org/10.1016/j.progsolidstchem.2005.11.021.

Ohtani, B., Adzuma, S., Nishimoto, S., & Kagiya, T. (1992). Photocatalytic degradation of polyethylene film by incorporated extra-fine particles of titanium dioxide. *Polymer Degradation and Stability*, 35(1), 53–60. https://doi.org/10.1016/0141-3910(92)90135-R.

Osman, A.I., Hosny, M., Eltaweil, A.S., Omar, S., Elgarahy, A.M., Farghali, M. et al. (2023). Microplastic sources, formation, toxicity, and remediation: A review. *Environmental Chemistry Letters*, 21(4), 2129–2169. https://doi.org/10.1007/s10311-023-01593-3.

Pan, Y., Gao, S.-H., Ge, C., Gao, Q., Huang, S., Kang, Y. et al. (2023). Removing microplastics from aquatic environments: A critical review. *Environmental Science and Ecotechnology*, 13, 100222. https://doi.org/10.1016/j.ese.2022.100222.

Park, S.Y., & Kim, C.G. (2019). Biodegradation of micro-polyethylene particles by bacterial colonization of a mixed microbial consortium isolated from a landfill site. *Chemosphere*, 222, 527–533. https://doi.org/10.1016/j.chemosphere.2019.01.159.

Peñalver, R., Arroyo-Manzanares, N., López-García, I., & Hernández-Córdoba, M. (2020). An overview of microplastics characterization by thermal analysis. *Chemosphere*, 242, 125170. https://doi.org/10.1016/j.chemosphere.2019.125170.

Perreard, S., Boucher, J., Gallato, M., Gomis, M., Mazzatorta, P., Gaboury, A., & Vorin, N. (2024). Plastic Overshoot Day (p. 264). *Earth Action*. https://plasticovershoot.earth/wp-content/uploads/2024/04/EA_POD_report_2024.pdf.

Pham, D.D., & Cho, J. (2021). Low-energy catalytic methanolysis of poly(ethyleneterephthalate). *Green Chemistry*, 23(1), 511–525. https://doi.org/10.1039/D0GC03536J.

Phuong, N.N., Duong, T.T., Le, T.P.Q., Hoang, T.K., Ngo, H.M., Phuong, N.A. et al. (2022). Microplastics in Asian freshwater ecosystems: Current knowledge and perspectives. *Science of The Total Environment*, 808, 151989. https://doi.org/10.1016/j.scitotenv.2021.151989.

Picó, Y., & Barceló, D. (2020). Pyrolysis gas chromatography-mass spectrometry in environmental analysis: Focus on organic matter and microplastics. *TrAC Trends in Analytical Chemistry*, 130, 115964. https://doi.org/10.1016/j.trac.2020.115964.

Pikuda, O., Roubeau Dumont, E., Chen, Q., Macairan, J.-R., Robinson, S.A., Berk, D., & Tufenkji, N. (2023). Toxicity of microplastics and nanoplastics to Daphnia magna: Current status, knowledge gaps, and future directions. *TrAC Trends in Analytical Chemistry*, 167, 117208. https://doi.org/10.1016/j.trac.2023.117208.

Postolachi, O., Rastimesina, I., Vorona, V., & Mamaliga, V. (2022). Screening of cultivation media for LDPE biodegradation by *Pseudomonas fluorescens*. *Scientific Bulletin. Series F. Biotechnologies*, XXVI(2), 151–157.

Prakash, R., Majumder, S.K., & Singh, A. (2018). Flotation technique: Its mechanisms and design parameters. *Chemical Engineering and Processing – Process Intensification*, 127, 249–270. https://doi.org/10.1016/j.cep.2018.03.029.

Prattichizzo, F., Ceriello, A., Pellegrini, V., La Grotta, R., Graciotti, L., Olivieri, F. et al. (2024). Micro-nanoplastics and cardiovascular diseases: Evidence and perspectives. *European Heart Journal*, 45(38), 4099–4110. https://doi.org/10.1093/eurheartj/ehae552.

Rani, M., Ducoli, S., Depero, L.E., Prica, M., Tubić, A., Ademovic, Z. et al. (2023). A complete guide to extraction methods of microplastics from complex environmental matrices. *Molecules*, 28(15), 5710. https://doi.org/10.3390/molecules28155710.

Rastimesina, I. (2023). Isolation of microbial consortia from polyethylene films surface. *Book of Abstracts SIMI 2023*, 84–85. https://doi.org/10.21698/simi.2023.ab31.

Ritchie, H., Samborska, V., & Roser, M. (2023). *Plastic Pollution* [Our World Data]. https://ourworldindata.org/plastic-pollution.

Saikumar, S., Mani, R., Ganesan, M., Dhinakarasamy, I., Palanisami, T., & Gopal, D. (2024). Trophic transfer and their impact of microplastics on estuarine food chain model. *Journal of Hazardous Materials*, 464, 132927. https://doi.org/10.1016/j.jhazmat.2023.132927.

Sajwan, D., Sharma, A., Sharma, M., & Krishnan, V. (2024). Upcycling of plastic waste using photo-, electro-, and photoelectrocatalytic approaches: A way toward circular economy. *ACS Catalysis*, 14(7), 4865–4926. https://doi.org/10.1021/acscatal.4c00290.

Salvia, R., Rico, L.G., Bradford, J.A., Ward, M.D., Olszowy, M.W., Martínez, C. et al. (2023). Fast-screening flow cytometry method for detecting nanoplastics in human peripheral blood. *MethodsX*, 10, 102057. https://doi.org/10.1016/j.mex.2023.102057.

Samak, N.A., Jia, Y., Sharshar, M.M., Mu, T., Yang, M., Peh, S., & Xing, J. (2020). Recent advances in biocatalysts engineering for polyethylene terephthalate plastic waste green recycling. *Environment International*, 145, 106144. https://doi.org/10.1016/j.envint.2020.106144.

Sanluis-Verdes, A., Colomer-Vidal, P., Rodriguez-Ventura, F., Bello-Villarino, M., Spinola-Amilibia, M., Ruiz-Lopez, E. et al. (2022). Wax worm saliva and the enzymes therein are the key to polyethylene degradation by Galleria mellonella. *Nature Communications*, 13(1), 5568. https://doi.org/10.1038/s41467-022-33127-w.

Schäefer, A.I., & Fane, A.G. (Eds.). (2021). *Nanofiltration: Principles, Applications, and New Materials* (1st Edn.). Wiley. https://doi.org/10.1002/9783527824984.

Schuyler, Q., Wilcox, C., Townsend, K., Hardesty, B., & Marshall, N. (2014). Mistaken identity? Visual similarities of marine debris to natural prey items of sea turtles. *BMC Ecology*, 14(1), 14. https://doi.org/10.1186/1472-6785-14-14.

Schwabl, P., Köppel, S., Königshofer, P., Bucsics, T., Trauner, M., Reiberger, T., & Liebmann, B. (2019). Detection of various microplastics in human stool: A prospective case series. *Annals of Internal Medicine*, 171(7), 453–457. https://doi.org/10.7326/M19-0618.

Shi, R., Liu, K.-S., Liu, F., Yang, X., Hou, C.-C., & Chen, Y. (2021). Electrocatalytic reforming of waste plastics into high value-added chemicals and hydrogen fuel. *Chemical Communications*, 57(94), 12595–12598. https://doi.org/10.1039/D1CC05032J.

Shi, X., Chen, Z., Wei, W., Chen, J., & Ni, B.-J. (2023). Toxicity of micro/nanoplastics in the environment: Roles of plastisphere and eco-corona. *Soil & Environmental Health*, 1(1), 100002. https://doi.org/10.1016/j.seh.2023.100002.

Shi, Y., Yan, C., Zhou, Y., Wu, J., Wang, Y., Yu, S., & Chen, Y. (2021). Polymer materials for additive manufacturing—Powder materials. *In: Materials for Additive Manufacturing* (pp. 9–189). Elsevier. https://doi.org/10.1016/B978-0-12-819302-0.00002-X.

Shrivastava, A. (2018). *Introduction to Plastics Engineering*. Elsevier. https://doi.org/10.1016/C2014-0-03688-X.

Siddiqi, Z., & Sarlah, D. (2021). Electrochemical dearomatization of commodity polymers. *Journal of the American Chemical Society*, 143(50), 21264–21269. https://doi.org/10.1021/jacs.1c11546.

Singh, N., & Walker, T.R. (2024). Plastic recycling: A panacea or environmental pollution problem. *Npj Materials Sustainability*, 2(1), 17. https://doi.org/10.1038/s44296-024-00024-w.

Singh, S., Kumar Naik, T.S.S., Anil, A.G., Dhiman, J., Kumar, V., Dhanjal, D.S. et al. (2022). Micro (nano) plastics in wastewater: A critical review on toxicity risk assessment, behaviour, environmental impact, and challenges. *Chemosphere*, 290, 133169. https://doi.org/10.1016/j.chemosphere.2021.133169.

Singh, V.K., Sharma, J., Pathak, A.K., Ghany, C.T., & Gondal, M.A. (2018). Laser-induced breakdown spectroscopy (LIBS): A novel technology for identifying microbes causing infectious diseases. *Biophysical Reviews*, 10(5), 1221–1239. https://doi.org/10.1007/s12551-018-0465-9.

Sjollema, S.B., Redondo-Hasselerharm, P., Leslie, H.A., Kraak, M.H.S., & Vethaak, A.D. (2016). Do plastic particles affect microalgal photosynthesis and growth? *Aquatic Toxicology*, 170, 259–261. https://doi.org/10.1016/j.aquatox.2015.12.002.

Song, J., Wang, C., & Li, G. (2024). Defining primary and secondary microplastics: A connotation analysis. *ACS ES&T Water*, 4(6), 2330–2332. https://doi.org/10.1021/acsestwater.4c00316.

Sooriyakumar, P., Bolan, N., Kumar, M., Singh, L., Yu, Y., Li, Y. et al. (2022). Biofilm formation and its implications on the properties and fate of microplastics in aquatic environments: A review. *Journal of Hazardous Materials Advances*, 6, 100077. https://doi.org/10.1016/j.hazadv.2022.100077.

Stabnikov, V., Ivanov, V., & Vaseashta, A. (2020). Biotechnological immobilization of chemical, biological, and radioactive pollutants on land and infrastructure demolition waste after industrial accident, military action, or terrorist attack. *In:* Fernando Pacheco-Torgal Volodymyr Ivanov, Daniel C.W. Tsang (Eds.), *Bio-Based Materials and Biotechnologies for Eco-Efficient Construction* (pp. 377–393). Elsevier. https://doi.org/10.1016/B978-0-12-819481-2.00018-0.

Stabnikova, O., Stabnikov, V., Marinin, A., Klavins, M., Klavins, L., & Vaseashta, A. (2021). Microbial life on the surface of microplastics in natural waters. *Applied Sciences*, 11(24), 11692. https://doi.org/10.3390/app112411692.

Stamatin, I., Berlic, C., & Vaseashta, A. (2006). On the computer-aided modelling of analyte–receptor interactions for an efficient sensor design. *Thin Solid Films*, 495(1–2), 312–315. https://doi.org/10.1016/j.tsf.2005.08.299.

Sundar, K.P., & Kanmani, S. (2020). Progression of photocatalytic reactors and its comparison: A review. *Chemical Engineering Research and Design*, 154, 135–150. https://doi.org/10.1016/j.cherd.2019.11.035

Tawfik, S.Y. (2017). Flame retardants: Additives in plastic technology. *In:* L.S. Palsule (Ed.), *Polymers and Polymeric Composites: A Reference Series* (pp. 1–27). Springer Berlin Heidelberg. https://doi.org/10.1007/978-3-642-37179-0_76-2.

Thacharodi, A., Hassan, S., Meenatchi, R., Bhat, M.A., Hussain, N., Arockiaraj, J. et al. (2024). Mitigating microplastic pollution: A critical review on the effects, remediation, and utilization strategies of microplastics. *Journal of Environmental Management*, 351, 119988. https://doi.org/10.1016/j.jenvman.2023.119988.

Thomas, R.T., & Sandhyarani, N. (2013). Enhancement in the photocatalytic degradation of low density polyethylene–TiO_2 nanocomposite films under solar irradiation. *RSC Advances*, 3(33), 14080. https://doi.org/10.1039/c3ra42226g.

Tofa, T.S., Kunjali, K.L., Paul, S., & Dutta, J. (2019). Visible light photocatalytic degradation of microplastic residues with zinc oxide nanorods. *Environmental Chemistry Letters*, 17(3), 1341–1346. https://doi.org/10.1007/s10311-019-00859-z.

Torres, F.G., Dioses-Salinas, D.C., Pizarro-Ortega, C.I., & De-la-Torre, G.E. (2021). Sorption of chemical contaminants on degradable and non-degradable microplastics: Recent progress and research trends. *Science of The Total Environment*, 757, 143875. https://doi.org/10.1016/j.scitotenv.2020.143875.

Usman, S., Abdull Razis, A.F., Shaari, K., Amal, M.N.A., Saad, M.Z., Mat Isa, N. et al. (2020). Microplastics pollution as an invisible potential threat to food safety and security, policy challenges, and the way forward. *International Journal of Environmental Research and Public Health*, 17(24), 9591. https://doi.org/10.3390/ijerph17249591.

Van Cauwenberghe, L., Devriese, L., Galgani, F., Robbens, J., & Janssen, C.R. (2015). Microplastics in sediments: A review of techniques, occurrence and effects. *Marine Environmental Research*, 111, 5–17. https://doi.org/10.1016/j.marenvres.2015.06.007.

Vanapalli, K.R., Dubey, B.K., Sarmah, A.K., & Bhattacharya, J. (2021). Assessment of microplastic pollution in the aquatic ecosystems: An Indian perspective. *Case Studies in Chemical and Environmental Engineering*, 3, 100071. https://doi.org/10.1016/j.cscee.2020.100071.

Vaseashta, A. (2006). Carbon nanotubes based devices and sensors. *Materials and Manufacturing Processes*, 21(7), 710–716. https://doi.org/10.1080/10426910600613595.

Vaseashta, A. (Ed.). (2015). *Life Cycle Analysis of Nanoparticles: Risk, Assessment, and Sustainability*. DEStech Publications, Inc.

Vaseashta, A. (2018). Roadmapping the future in defense and security: Innovations in technology using multidisciplinary convergence. *In:* P. Petkov, D. Tsiulyanu, C. Popov, & W. Kulisch (Eds.), *Advanced Nanotechnologies for Detection and Defence against CBRN Agents* (pp. 3–14). Springer Netherlands. https://doi.org/10.1007/978-94-024-1298-7_1.

Vaseashta, A., & Dimova-Malinovska, D. (2005). Nanostructured and nanoscale devices, sensors and detectors. *Science and Technology of Advanced Materials*, 6(3–4), 312–318. https://doi.org/10.1016/j.stam.2005.02.018.

Vaseashta, A., & Irudayaraj, J. (2005). Nanostructured and nanoscale devices and sensors. *Journal of Optoelectronics and Advanced Materials*, 7(1), 35–42.

Vaseashta, A., & Khudaverdyan, S. (Eds.). (2013). *Advanced Sensors for Safety and Security*. Springer Netherlands. https://doi.org/10.1007/978-94-007-7003-4.

Vaseashta, A., & Karagülle-Bölgen, N. (2018). Loaded nanofibers: Force protection, filtration, decontamination. *In:* P. Petkov, D. Tsiulyanu, C. Popov, & W. Kulisch (Eds.), *Advanced Nanotechnologies for Detection and Defence against CBRN Agents* (pp. 241–252). Springer Netherlands. https://doi.org/10.1007/978-94-024-1298-7_23.

Vaseashta, A., Ivanov, V., Stabnikov, V., & Marinin, A. (2021). Environmental safety and security investigations of neustonic microplastic aggregates nearwater-air interphase. *Polish Journal of Environmental Studies*, 30(4), 3457–3469. https://doi.org/10.15244/pjoes/131947.

Vaseashta, A., Gevorgyan, G., Kavaz, D., Ivanov, O., Jawaid, M., & Vasović, D. (2021). Exposome, biomonitoring, assessment, and data analytics to quantify universal water quality. *In:* A. Vaseashta & C. Maftei (Eds.), *Water Safety, Security, and Sustainability* (pp. 67–114). Springer International Publishing. https://doi.org/10.1007/978-3-030-76008-3_4.

Vaseashta, A., & Bölgen, N. (Eds.). (2022). *Electrospun Nanofibers: Principles, Technology, and Novel Applications*. Springer International Publishing. https://doi.org/10.1007/978-3-030-99958-2.

Vaseashta, A., Duca, G., & Travin, S.O. (Eds.). (2022). *Handbook of Research on Water Sciences and Society*. Engineering Science Reference. DOI: 10.4018/978-1-7998-7356-3.

Vaseashta, A., Stabnikov, V., Klavins, M., & Ivanov, V. (2022). Decontamination of seawater in a harbor: Case study of potential bioterrorism attack. *In:* Á. Rocha, C.H. Fajardo-Toro, & J.M.R. Rodríguez (Eds.), *Developments and Advances in Defense and Security* (Vol. 255, pp. 217–226). Springer Singapore. https://doi.org/10.1007/978-981-16-4884-7_17.

Velzeboer, I., Kwadijk, C.J.A.F., & Koelmans, A.A. (2014). Strong sorption of PCBs to nanoplastics, microplastics, carbon nanotubes, and fullerenes. *Environmental Science & Technology*, 48(9), 4869–4876. https://doi.org/10.1021/es405721v.

Vethaak, A.D., & Legler, J. (2021). Microplastics and human health. *Science*, 371(6530), 672–674. https://doi.org/10.1126/science.abe5041.

Virtanen, T., Rudolph, G., Lopatina, A., Al-Rudainy, B., Schagerlöf, H., Puro, L. et al. (2020). Analysis of membrane fouling by Brunauer-Emmett-Teller nitrogen adsorption/desorption technique. *Scientific Reports*, 10(1), 3427. https://doi.org/10.1038/s41598-020-59994-1.

Wagner, M., & Lambert, S. (Eds.). (2018). *Freshwater Microplastics: Emerging Environmental Contaminants?* (Vol. 58). Springer International Publishing. https://doi.org/10.1007/978-3-319-61615-5.

Wang, J., Li, X., Zhang, T., Chen, Y., Wang, T., & Zhao, Y. (2022). Electro-reforming polyethylene terephthalate plastic to co-produce valued chemicals and green hydrogen. *The Journal of Physical Chemistry Letters*, 13(2), 622–627. https://doi.org/10.1021/acs.jpclett.1c03658.

Wang, S., Han, Q., Wei, Z., Wang, Y., Xie, J., & Chen, M. (2022). Polystyrene microplastics affect learning and memory in mice by inducing oxidative stress and decreasing the level of acetylcholine. *Food and Chemical Toxicology*, 162, 112904. https://doi.org/10.1016/j.fct.2022.112904.

Wang, W., Gao, H., Jin, S., Li, R., & Na, G. (2019). The ecotoxicological effects of microplastics on aquatic food web, from primary producer to human: A review. *Ecotoxicology and Environmental Safety*, 173, 110–117. https://doi.org/10.1016/j.ecoenv.2019.01.113.

Wotton, R.S., & Malmqvist, B. (2001). Feces in aquatic ecosystems. *BioScience*, 51(7), 537. https://doi.org/10.1641/0006-3568(2001)051[0537:FIAE]2.0.CO;2.

Yang, C.Z., Yaniger, S.I., Jordan, V.C., Klein, D.J., & Bittner, G.D. (2011). Most plastic products release estrogenic chemicals: A potential health problem that can be solved. *Environmental Health Perspectives*, 119(7), 989–996. https://doi.org/10.1289/ehp.1003220.

Yang, J., Kamstra, J., Legler, J., & Aardema, H. (2023). The impact of microplastics on female reproduction and early life. *Animal Reproduction*, 20(2), e20230037. https://doi.org/10.1590/1984-3143-ar2023-0037.

You, X., You, M., Lyu, Y., Peng, G., & Sun, W. (2022). Single and combined exposure to micro(nano)plastics and azithromycin disturbing the photosynthetic carbon fixation of *Synechocystis* sp. *Environmental Science: Nano*, 9(12), 4354–4366. https://doi.org/10.1039/D2EN00204C.

Zeenat, Elahi, A., Bukhari, D.A., Shamim, S., & Rehman, A. (2021). Plastics degradation by microbes: A sustainable approach. *Journal of King Saud University – Science*, 33(6), 101538. https://doi.org/10.1016/j.jksus.2021.101538.

Zhang, H., Wang, C., Luo, H., Chen, J., Kuang, M., & Yang, J. (2023). Iron nanoparticles protected by chainmail-structured graphene for durable electrocatalytic nitrate reduction to nitrogen. *Angewandte Chemie International Edition*, 62(5), e202217071. https://doi.org/10.1002/anie.202217071.

Zhao, X., Korey, M., Li, K., Copenhaver, K., Tekinalp, H., Celik, S. et al. (2022). Plastic waste upcycling toward a circular economy. *Chemical Engineering Journal*, 428, 131928. https://doi.org/10.1016/j.cej.2021.131928.

Zheng, H., Vidili, G., Casu, G., Navarese, E.P., Sechi, L.A., & Chen, Y. (2024). Microplastics and nanoplastics in cardiovascular disease: A narrative review with worrying links. *Frontiers in Toxicology*, 6, 1479292. https://doi.org/10.3389/ftox.2024.1479292

Zhou, D., Wang, L., Zhang, F., Wu, J., Wang, H., & Yang, J. (2022). Feasible degradation of polyethylene terephthalate fiber-based microplastics in alkaline media with Bi_2O_3 @N-TiO_2 Z-scheme photocatalytic system. *Advanced Sustainable Systems*, 6(5), 2100516. https://doi.org/10.1002/adsu.202100516.

Zhou, H., Wang, Y., Ren, Y., Li, Z., Kong, X., Shao, M., & Duan, H. (2022). Plastic waste valorization by leveraging multidisciplinary catalytic technologies. *ACS Catalysis*, 12(15), 9307–9324. https://doi.org/10.1021/acscatal.2c02775.

Zhou, Z., Maxeiner, K., Ng, D.Y.W., & Weil, T. (2022). Polymer chemistry in living cells. *Accounts of Chemical Research*, 55(20), 2998–3009. https://doi.org/10.1021/acs.accounts.2c00420.

Zhu, B., Wang, D., & Wei, N. (2022). Enzyme discovery and engineering for sustainable plastic recycling. *Trends in Biotechnology*, 40(1), 22–37. https://doi.org/10.1016/j.tibtech.2021.02.008.

2 | Microplastic Pollution Mitigation from Fresh and Marine Water Bodies

Olena Stabnikova[1*(0000-0003-4156-3021)], Viktor Stabnikov[2], Andrii Marynin[1], Maris Klavins[3] and Ashok Vaseashta[4,5,6(0000-0002-5649-0067)]

[1] Advanced Research Laboratory, National University of Food Technologies, Kyiv, Ukraine

[2] Department of Biotechnology and Microbiology, National University of Food Technologies, Kyiv, Ukraine

[3] Department of Environmental Science, University of Latvia, House of Nature, Riga, Latvia

[4] International Clean Water Institute, Office of Strategic Research, Manassas, VA, 20112 USA

[5] Institutul de Cercetare al Universităţii din Bucureşti, Strada Atomiştilor, Magurele, Romania

[6] Ghitu Institute of Electronic Engineering and Nanotechnology, Technical University of Moldova, and Academy of Sciences of Moldova, Chisinau, Moldova

Abstract: Microplastics constitute a widespread and ever-increasing pollutant in global water systems. Notably, the estimated amount of microplastics in the ocean is much less than the estimated input. The natural processes that make microplastics disappear from the water environment and ways to mitigate the extent of microplastics entering into water systems with treated wastewater are discussed. This chapter describes in detail microplastic contamination mitigation methods from fresh and marine water bodies.

1. Introduction

Microplastics (MPs) constitute one of the most harmful pollutants in aquatic environments and are observed almost everywhere in aquatic ecosystems. According to Eriksen et al. (2014), more than 5 trillion plastic pieces were

*Corresponding author: stabstab6@gmail.com

floating in the world's oceans in 2014, with a total weight of over 250,000 tons. It was reported that the quantity of microplastic particles in marine aquatic systems can reach 140 particles/m^3 and 8,766 particles/m^3 in water and sediment, respectively (Thushari & Senevirathna, 2020). In an aquatic environment, MPs serve as a habitat for microorganisms, including pathogens, a site for the formation of biofilms, and a means of transporting microbes to new territories (Dudek et al., 2020; Oberbeckmann & Labrenz, 2020; Stabnikova et al., 2021). At the same time, MPs can act as sorbents for other environmental pollutants such as persistent organic pollutants, hydrocarbons, pharmaceuticals, heavy metals, and others (Fan et al., 2021; Klavins et al., 2022; Vaseashta et al., 2022). One of the mechanisms of microplastic toxicity is the desorption of pollutants when particles enter a living organism.

In 2019, annual plastic production was 368 million tons, and it is expected to grow up to 33 billion tons by 2050 (Bellasi et al., 2020). Meanwhile, it is estimated that 76% of the total plastics produced are landfilled or spread in the natural environment (Geyer, 2020). However, it was found that a large part of plastics getting into the ocean was missing. An assessment of the global floating plastic on the surface of the world's oceans showed an amount much less than that which should have gotten there (Cózar et al., 2014). According to Kane et al. (2020), 99% of plastics entering into the ocean are accumulated on the bottom. The reasons for this may be the fragmentation of plastic into small micro- and nanoparticles (NPs), utilization of the latter as food by marine life, deposition of MPs from the water column to the ocean floor, and possibly other yet unknown processes. MP particles of 0.3–5 mm present from 13–32% of the total weight of plastics floating in the ocean (Gerritse et al., 2021). There are some processes that can occur spontaneously in natural water bodies and ensure the removal of MPs from the aquatic systems. Thus, analysis of sediment cores collected in Japan, Thailand, Malaysia, and South Africa showed the presence of MPs sized between 315 μm–5 mm (most around 315 μm) from materials such as polyethylene (PE), polypropylene (PP), polystyrene (PS), polyethylene terephthalate (PET), polyethylene-polypropylene copolymer (PEP), and polyacrylates (PAK) in amounts from 100–1,900 pieces/kg of dry sediment that could be compared with surface water samples (Matsuguma et al., 2017) or, in some cases, even several orders of magnitude higher (Thushari & Senevirathna, 2020). MP particles sized from 75–161 μm were found in depths from 1,176–4,844 m in 11 samples of the top sediment layer with an average abundance of up to 1 MP piece per 25 cm^3 in several locations in the Atlantic Ocean and the Mediterranean Sea (Van Cauwenberghe et al., 2013). In a long-term microcosmic experiment with constantly circulating seawater, MPs have been observed in surface water and on the bottom, proving that they do not normally occur in the water column (Gerritse et al., 2021). So, generally, the MPs are removed from upper water layers due to their aggregation followed by sedimentation (Woodall et al., 2014). However, a comparative study using Fourier Transform Infrared (FTIR) spectroscopy on the effect of three years of exposure of PP, LDPE, and HDPE plastic pellets and

MP particles, which have been collected from North Pacific surface waters, to seawater under sunlight or in the dark showed that most of the polyethylene microplastics from the ocean were exposed to weathering within 18–30 months, which suggests their stay in the upper layers of water for a long time (Brandon et al., 2016). MPs could be removed from the ocean in different ways.

2. Natural Ways of Removal of Microplastics from Water Bodies

2.1. Removal of Microplastics with Marine Snow

The role of marine snow in the vertical movement of organic debris in a water column in the marine environment is well known (Engel, 2000). It has been hypothesized that MPs could be removed from the ocean's surface with marine snow, a continuous shower of organic aggregates (particles sized >200 μm) consisting mainly of biological debris falling from the top layers of water to the seafloor (Kvale et al., 2020; Zhao et al., 2017). It was shown that MP particles of 6 plastic types with different densities and with sizes ranging from 7–3,000 μm, which were incorporated into laboratory-made marine snows, that they settled down in the seawater with a higher sinking rate than buoyant ones (Porter et al., 2018). So, movement with marine snow helped to cleanse the ocean's water from MPs. However, the MPs included in the marine snow aggregates became more bioavailable to benthic organisms. For example, the indigestion of free MP particles by mussels ranged from 0–340 MPs per individual, while when the MP particles were included in the sea snow, this number increased to 1.6×10^5 per individual (Porter et al., 2018). So, on the one hand, the incorporation of MPs in marine snow is an important way to diminish the amount of buoyant MP particles in the global ocean; on the other hand, this can have unpredictable adverse effects on marine inhabitants. MPs could also be removed from the bulk of water by incorporating them into bio-deposits like zooplankton fecal pellet-bound plastic. The incorporation of MP into fecal pellets increased their sinking velocity by $1\cdot10^3$–$1\cdot10^4$ times (Piarulli & Airoldi, 2020). However, the aggregation of MPs into marine snow as well as their removal with fecal pellets are only temporary fixes, because with demineralization free MPs are released and possibly return to the euphotic zone (Kvale et al., 2020).

2.2. Removal of Microplastics due to Incorporation into Iron-containing Flocs

The process of formation of amorphous particles containing ferric oxide due to oxidation of Fe(II) in lakes is well known (Tipping et al., 1981). These particles are usually formed during lake stratification due to the contact of anoxic water containing ferrous iron with oxic water followed by ferrous oxidation and the production of ferric ions. Produced particles consist not only of iron (30–40% by weight) but also of carbon compounds, humic substances, and mineral elements, such as P, N, Mn, Si, S, Ca, and Mg, and they also have the ability to flocculate and precipitate. It was shown that the removal of burial MPs could occur in freshwater reservoirs with low velocity due to their incorporation into sinking iron-organic aggregates followed by precipitation (Leiser et al., 2021a). The authors experimentally showed that the addition of Fe(II) to natural freshwater resulted in the formation of sinking flocs containing iron, microbial cells, mostly bacteria, and organic compounds presented mainly by extracellular polymeric substances. The reddish color of flocs indicated the presence of Fe (oxy) hydroxides due to the convection of soluble Fe(II) into insoluble Fe(III). These sinking flocs included polyethylene microplastics, which were removed with flocs from the water column and transported to the sediment. Despite the fact that iron reduction occurred, a small amount of MPs was released from the sediment during 63 experiment days.

2.3. Removal of Microplastics from Ocean Waters by Coral Activity

Reef-forming corals can play a certain role in the removal of MPs from ocean waters. In the last few years, MP pollution has been detected in global coral reef regions (Huang et al., 2021). Corals can ingest MP particles or be a temporary sink for them. For example, over 100 MP pieces per polyp were detected in the temperate coral *Astrangia poculata* in Rhodes, Greece (Rotjan et al., 2019), while removal of MPs by Red Sea coral species through adhesion to their surface was 40 times more efficient than through feeding (Martin et al., 2019). However, the participation of corals in decreasing the quantity of MP particles in the ocean is more significant because their skeleton can serve as a permanent deposition site for MPs. Because corals produce a skeleton made up of aragonite, a form of calcium carbonate, it is possible to predict the possibility of capture of MPs present in water during coral skeleton construction, whereby they become part of the newly formed skeleton. The rate of coral growth determines the quantity of MPs accumulating in the coral skeleton. According to Reichert et al. (2021), the reef-building corals may remove 0.09–2.82% of the bioavailable MPs from tropical shallow-reef waters per year. As global coral reefs occupy an area of approximately 250,000 km^2 (Hochberg & Gierach, 2021), their role in removing part of the MPs will be tangible.

2.4. Removal of Microplastics from Bulk Water by Phytoplankton Aggregates

MPs can form aggregates with microalgae and be removed from the bulk of water to the bottom of the ocean (Cunha et al., 2019; Long et al., 2015). These so-called hetero aggregates consist of MPs, microalgae, and exopolymeric substances (EPSs) excreted by microalgae. The density of MPs can change due to processes occurring in the surrounding environment leading to their precipitation. This possibility was demonstrated with polypropylene (Lagarde et al., 2016). One of the most widely used plastics in industry, polypropylene, has a low density of 0.91 g/cm^3 (Stabnikova et al., 2021). It represented a large part (16.5%) of the total quantity of MP debris floating in the ocean (Isaac & Kandasubramanian, 2021). However, its density has changed after the interaction with model freshwater microalgae *Chlamydomas reinhardtii* (Lagarde et al., 2016). MP particles sized from 0.4–1 mm were added, in an amount of 100 mg/100 ml of microalgae suspension with a cell concentration of 4×10^6 per ml and stayed in contact for 20 days. The hetero aggregates produced consisted of fragments of polypropylene and microalgae with exopolysaccharides in a ratio of approximately 1:1 with a density of 1.2 g/cm^3. The size of the aggregates increased with time due to increased adhesion of PP fragments between one another. These hetero aggregates play the role of a pathway for the transport of MP particles from the water surface to the sediment. However, the high-density polyethylene (HDPE) MP did not form aggregates, which may be due to the lower cohesive ability of EPS produced by the microalgae attached to MPs. The authors concluded that the chemical nature of MPs determines its fate in the environment.

The properties of the aggregates of MPs with microalgae also depend on the characteristics of the microalgae. Thus, the MPs aggregated with diatom *Chaetoceros neogracile* and cryptophyte *Rhodomonas salina* differed by their ability to sink in seawater: the sinking rate of the diatom aggregates was higher due to the presence of biogenic silica, which has a higher density than the organic matter. Generally, the diatom aggregates had a bigger size than the aggregates with *R. salina* (Long et al., 2015). The same fate was shown for the polypropylene MP covered with an algae biofilm in freshwater lakes in China (Chen et al., 2019). The biofilm growth depended on the season and required different times to change the density of plastics required for their sinking. In general, during the warm season, the plastics lost buoyancy faster than during the cold season, and their sedimentation ability was negatively correlated with the size, i.e., smaller particles were removed from the water surface faster than larger ones.

2.5. Removal of Microplastics from Bulk Water by Aggregation of Microplastics due to Covering with Biofilms and Biofouling

One of the reasons for the partial disappearance of MPs from the ocean's surface is change of MP density due to the formation of biofilms and biofouling (Amaral-Zettler et al., 2021; Harrison et al., 2011; Kaiser et al., 2017; Leiser et al., 2021b; Michels et al., 2018; Rummel et al., 2017; Stabnikova et al., 2021). It was experimentally proved that polystyrene beads with a 700–900 mm diameter rapidly coagulated with biogenic materials contained in seawater and formed aggregates within a few days (Michels et al., 2018). The process of aggregate formation was more rapid when the plastic surface was covered with biofilms. In contrast to that, low aggregation of MPs was observed when the biogenic particles were absent. Among the factors that determined the formation of aggregates, the authors included: (a) the number of particles present in the water and their size at the initial step of the process, and (b) their stickiness, which increased greatly when particles were covered with biofilm due to the presence of polysaccharides and extracellular DNA (Michels et al., 2018). The produced experimental aggregates resulting from composting of MPs and biogenic particles were stable for several days; so, it is likely that such aggregates can remain stable during sinking in the water column and reach the ocean floor without disintegration.

It was shown that the presence of EPSs in seawater increased the coagulation of biogenic particles and the formation of larger aggregates, enhancing the vertical movement and trapping in sediments (Chow et al., 2015; Engel, 2000). This was demonstrated for the EPSs synthesized by coccolithophore algae *Emiliania huxleyi* (Chow et al., 2015) and diatom *Skeletonema costatum* (Engel, 2000). However, it was also shown that the EPSs synthesized by microorganisms in the biofilm were able to participate in the aggregation of microbial cells and the development of high cell density (Flemming & Wingender, 2010). According to Ian Sutherland (2001), microbial EPSs make a strong and sticky framework. MPs floating in natural waters are rapidly covered with biofilms, the microbial habitats of which produce EPSs that can participate in aggregate formation (Stabnikova et al., 2021; 2022).

Diatom algae and cyanobacteria are commonly found in biofilms covering microplastics in marine environments (Kaiser et al., 2017; Zettler et al., 2013). At the same time, heterotrophic bacteria can consume the organic compounds excreted by algae and produce nutrients needed for algae growth. So, the growth of diatoms is closely associated with the presence of heterotrophic bacteria in biofilms (Du et al., 2022; Kouzuma & Watanabe, 2015; Mayali, 2018). It was demonstrated that polyethylene MP placed in a eutrophic reservoir in summertime accrued a biofilm cover consisting of common phototrophic sessile cyanobacteria (also called blue-green algae) *Chamaesiphon* spp. and *Leptolyngbya* spp. While forming biofilms on MPs, these cyanobacteria precipitated calcite that made the formed aggregates sink in a freshwater reservoir (Leiser et al., 2021b). An interesting finding has been reported for the polyethylene MPs covered with biofilms: the ratio of surface area to volume (SA:V) of MPs plays an important role in the sinking processes (Amaral-Zettler et al., 2021). It was suggested that the covering of MPs sized 1 mm and above with biofilm consisting of microorganisms and EPSs is not enough for precipitation. However, smaller plastic particles (particularly 5 μm) in the bacterial biofilm covering could have their density changed. This was especially the case for the particles below the euphotic zone, where heterotrophs dominate in biofilm composition. The authors believed that the ratio of SA:V determines the behavior of MPs covered with biofilm: small pieces of plastic particles or shapes with higher surface area to volume ratios (SA/V above 100), such as films, maybe sinking due to the formation of microbial biofilms or incorporation into aggregates, whereas larger pieces with SA/V below

100 will be sinking only when invertebrates dominate in the biofilm. Extensive research has been done recently to clearly understand the process of sinking of MPs covered with biofilms and its dependence on the type, size, and shape of MP particles (Fazey & Ryan, 2016; Jalón-Rojas et al., 2022; Liu et al, 2020; Zhao et al., 2022). It was shown, for example, that a small LDPE film (5 × 5 × 0.1 mm) sank in 17 days in Cape coastal waters, while buoyant larger fragments of HDPE sank after 60 days (Fazey & Ryan, 2016). Particle shapes determine the surface area to volume ratio, which characterizes the ability of particles to sink, and flat, thin fragments have higher ratios than spherical ones (Fazey & Ryan, 2016).

MP biofouling is also attracting attention as an effective means of removing MPs from surface waters. It is supposed that the formation of biofilms can increase the density of plastic particles leading to their sinking. The ratio of surface to volume is higher for MPs than for macroplastics; therefore, biofouling could help remove millimeter-sized fragments of floating plastics from the water surface (Cózar et al., 2014; Eriksen et al., 2014; Gerritse et al., 2021; Kaiser et al., 2017; Kooi et al., 2017). The model proposed by Kooi et al. (2017) predicts that all particles floating in the ocean can settle due to biofouling, and the smaller particles start settling first, and their maximum concentration will be observed at intermediate depths and become a risk for species living there. However, the fate of these MP particles settled due to fouling is still unclear: do they remain on the seabed, or, in the event of biofilm destruction, rise to the surface again? (Andrady, 2011; Wayman & Niemann, 2021).

Consequently, the progressive fragmentation of plastic objects into more and smaller pieces will lead to a gradual quantitative increase of fragments of smaller sizes. According to Fischer et al. (2022), biofouling includes the attachment, growth, and loss of algae on MP particles. An increase in biofilm mass is observed in the euphotic zone, while its decrease is going on below the euphotic zone. Biofouling strongly depends on the season and water temperature. Latva et al. (2022) showed that plastics in the earliest minutes or hours after their entering the coastal water environment of the Mediterranean Sea in the springtime were colonized by representatives of the *Bacteroidetes* group, later by the *Gammaproteobacteria*, and then by the *Alphaproteobacteria* groups. In case when plastics were exposed to sunlight, phototrophic organisms from the phylum *Ochrophyta*, first diatoms and then brown algae, dominated among the eukaryotic community attached to the plastic particles during their three-day incubation in seawater. Representatives of the phylum *Bacteroidetes* and diatoms were also most abundant on the polyethylene terephthalate samples incubated for 5–6 weeks in the North Sea in different seasons (Oberbeckmann et al., 2016). However, the bacterial as well as eukaryotic species in the composition of the biofilm covering plastic particles may vary depending on the type of plastic, season, geographical location, and conditions in aquatic environments (Gong et al., 2019; Kaiser et al., 2017, 2019; Kirstein et al., 2018; Miao et al., 2021; Oberbeckmann et al., 2014; Stabnikova et al., 2021; Sudhakar et al., 2007). It was shown that a biofilm with 28–34 mg dry weight/cm^2 of surface developed on such synthetic polymers as polycarbonate, low-density polyethylene, high-density polyethylene, and polypropylene, which have been immersed in the Bay of Bengal, India, for one year (Artham et al., 2009). It was also shown that the bacteria from biofilm covering plastics differed taxonomically from the planktonic microbes present in the surrounding environment (Bryant et al., 2016).

All of the processes described above led to the sinking of MPs and ensured their accumulation in deep ocean sediments. It should be stressed, however, that this is not a solution to MP pollution. The MPs transferred from the upper layers and incorporated in aggregates including biogenic matter serve as attractive food to benthic organisms and may accumulate in them with unpredictable consequences.

2.6. Adsorption of Microplastic on Algae and Macrophytes

One of the ways for MP particles to disappear from seawater is their adsorption on the micro- and macroalgae. It was shown that green microalgae *Scenedesmus* and *Chlorella* adsorbed positively charged nanosized polystyrene beads, and this process impeded algal photosynthesis possible due to the prevention of light from reaching the photosynthesis centers (Bhattacharya et al., 2010). Similar results were reported by Notle et al. (2017), who showed also that the binding of polystyrene NPs onto unicellular green alga *Pseudokirchneriella subcapitata* was stronger for neutral and positively charged particles than that of negatively charged ones. It is in accordance with the fact that most algae bear negative surface charges due to the presence of carboxylic (–COOH) and amino ($–NH_2$) functional groups on the cytomembrane (Li et al., 2022). Gutow et al. (2015) showed that macroalgae *Fucus vesiculosus* retained suspended polystyrene MPS and the number of MPs attached to the seaweed correlated with the concentrations of suspended particles in the water. In a study by Sundæk et al. (2018), edible seaweed *Fucus vesiculosus*, placed for 2 hours in filtered seawater containing fluorescent polystyrene MP particles, 597 MPs/mL, adsorb in an average of 74% of the total number of particles resulted in their content 3.99 MPs/mm^2 of seaweed surface. The content of MPs was reduced on average to 0.22 MPs/mm^2 by washing, which corresponds to an average reduction of 94.5%, so on average, 4% of the particles present in seawater were left on the seaweed. Finally, an average of 515 MPs/g wet weight or 2787 MPs/g dry weight remained on the seaweed after washing (Sundæk et al., 2018). According to the Food and Agriculture Organization (FAO), 36 million tons (wet weight) of seaweed were produced in 2020 (FAO, 2022). Thus, it is evident that adsorption on the macroalgae surface participates in the MP removal from seawater, and consumption of seaweeds can be a route of human exposure to MPs.

It was shown that MPs can interact via adsorption with macrophytes seagrass, and common duckweed (Kalčíková, 2023). Polystyrene MPs adhered to the surface of leaves, rhizomes, and roots of the submerged marine macrophytes seagrass *Cymodocea nodosa* (Ucria) Ascherson (Menicagli et al., 2022) and polyethylene MPs adhere to the aquatic free-floating plant common duckweed *Lemna minor*

(L.) (Mateos-Cárdenas et al., 2019; Rozman et al., 2023). Phytoremediation is proposed as one of the possible technologies for MPs removal from aquatic systems: polyethylene MPs, which due to their low density concentrated at the air-water interface in aquatic systems, could adhere to the floating macrophyte *Lemna minor*, and the plant biomass with attached MPs must be further processed using direct combustion, pyrolysis, or anaerobic digestion (Rozman et al., 2023).

2.7. Accumulation of Microplastics in Marine Inhabitants

MPs present in water environments may be consumed and accumulated in aquatic organisms from zooplankton to higher trophic levels (fish and mammals) (Costa et al., 2020). Contamination of the food web begins with the consumption of MPs by zooplankton (primary consumers in aquatic food chains) feeding phytoplankton representatives containing MPs (Figure 1) (Cole et al., 2013).

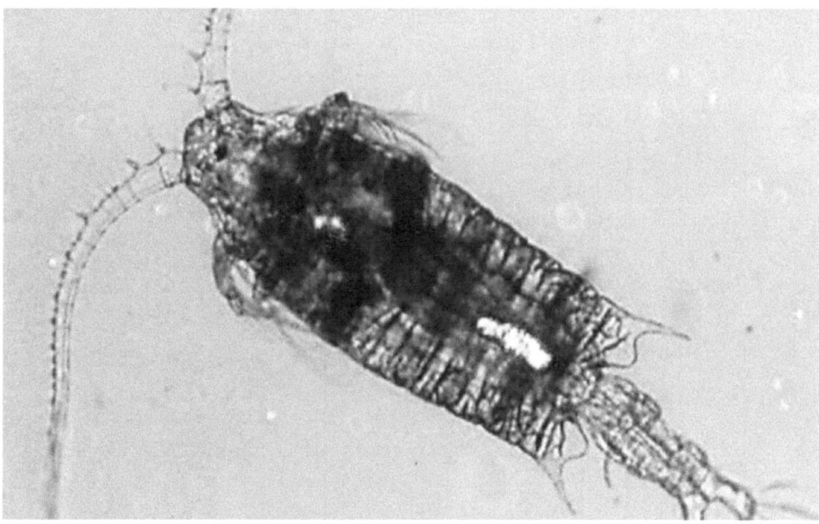

Figure 1: The copepod *Centropages typicus* containing 7.3 μm polystyrene beads (adopted from Cole et al., 2013).

MPs can also accumulate in other marine inhabitants that feed on phytoplankton or algae contaminated with MPs. Thus, it was shown that the periwinkle *Littorina littorea* did not recognize polystyrene particles attached to the algae *Fucus vesiculosus*, and fed without disassembling clean or MP-contaminated algae (Gutow et al., 2015). Plastic debris has been found in wild-caught and commercial fish, mussels, and shellfish destined for human consumption (Rochman et al., 2020). In fact, MPs have been shown to be present in around 80% of major fish species caught (Walkinshaw et al., 2020).

Thus, due to the mistaken consumption of MPs as food, its amount present in the water environment decreases. Beyond the ingestion of MPs, it was recently shown for mussels that adherence of MPs to their surface also plays a significant role in MP accumulation by aquatic organisms (Kolandhasamy et al., 2018). MPs are found in different marine species (Table 1).

Table 1: Accumulation of microplastics in marine species

Marine Inhabitants	Amounts of Microplastics	Location	Reference
Periwinkle	20 MPs/g dw	The North Sea, the Netherlands	Leslie et al., 2017
Sand hopper	11 MPs/g dw		
Pacific oyster	87 MPs/g dw		
Blue mussel	105 MPs/g dw		
Pacific oyster	30 MPs/g dw		
Blue mussel	19 MPs/g dw		
Yellowfin bream	0.6 MPs/fish	Sydney Harbour, Australia	Halstead et al., 2018
Sea mullet	2.5 MPs/fish		
Silver biddy	0.1 MPs/fish		
Mussels	3.0±0.9 MPs/g ww	Scotland, UK	Catarino et al., 2018
Mussels	0.11-4.58 MPs/g fw	The Marmara Sea, Turkey	Gedik et al., 2022
Fish	1.90±0.10 MPs/fish	The English Channel, UK	Lusher et al., 2013
Fish	1.58±0.23 MPs/fish	Brisbane, Sydney, Australia, fish markets	Wootton et al., 2021
Fish	0.86±0.14 MPs/fish	Suva, Fiji, fish markets	
Prawns	0.8±0.1 MPs/prawn	Australian, seafood markets	Ogunola et al., 2022
Crabs	1.6±0.1 MPs/crab		
Crabs	4.2 MPs/crab	Mosquito Lagoon, Florida, USA	Waite et al., 2018
Eastern oysters	16.5 MPs/oyster		
Bivalves	2.1-10.5 MPs/g fw	Shanghai, China, fish market	Li et al., 2015
Anchovy	0.15±0.04 MPs/fish	The Black Sea, Turkey	Eryaşar et al., 2022
Whiting mullet	0.28±0.06 MPs/fish		
Red mullet	0.40±0.07 MPs/fish		

Fish	2.2±0.89 MPs/fish	The Bay of Bengal, Bangladesh	Ghosh et al., 2021
Fish	2.83±1.84 MPs/fish	Mangrove Wetland, China	Huang et al., 2020
Fish	1.25±0.13 MPs/fish	The Pacific Ocean, French Polynesia	Garnier et al., 2019
Juvenile Chinook salmon	1.2±1.4 MPs/fish	Vancouver Island, British Columbia	Collicutt et al., 2019
Flounder	0.18±0.55 MPs/fish	The Mondego Estuary, Portugal	Bessa et al., 2018
Sea bream	3.14±3.25 MPs/fish		
Sea bass	0.30±0.61 MPs/fish		
Sea salt	230-575 MPs/kg	Gujarat, India	Vidyasakar et al., 2021
Sea salt	115-505 MPs/kg	Tamil Nadu, India	
Sea salt	50-280 MPs/kg	Spain	Iñiguez et al., 2017

Abbreviations: dw – Dry weight; ww – Wet weight; fw – Fresh weight.

Trophic transfer, bioaccumulation, and biomagnification of MPs in seafood provide a viable pathway for their entry into the human body (Vázquez-Rowe et al., 2021). MPs have negative effects on aquatic living organisms due to reproductive disorders, malnutrition, external abrasions, and blockages (Kumar et al., 2020; Sun et al., 2021), and also revealed adverse effects on human health (Ghosh et al., 2023; Kirshtein et al., 2016; Smith et al., 2018). For example, cytotoxic effects of MPs (mainly 10 μm thick polystyrene) on human cerebral and epithelial cells have been shown (Schirinzi et al., 2017), and it has been reported that polystyrene particles (0.25±0.06 μm) have been taken up by human keratinocytes (Triebskorn et al., 2019). It is believed that MPs can cause inflammation of the gastrointestinal system and, when MP particles act as carriers of environmental chemicals, they can cause toxic effects and organ damage in humans and increase the risk of cancer (Li et al., 2023). Meanwhile, drinking water from freshwater natural resources contains MP particles and is a source of these contaminants in the human diet, since water treatment using coagulation-flocculation removes only up to 88% of MPs (Pivokonsky et al., 2018). Tap water in China contains MP particles, mainly polyethylene and polypropylene, ranging from 440±75 particles/L (Tong et al., 2020). Determination of MPs in water taken from various natural sources showed that their content ranged from 1473–3605 particles per 1 L, and the content of MPs in purified drinking water varied from 338–628 particles per 1 L (Pivokonsky et al., 2018). However, only MP particles smaller than 20 μm can penetrate organs, and particles smaller than 10 μm can penetrate all organs and penetrate cell membranes (Vaseashta et al., 2021). Although the impact of MP consumption on human health is still not well understood and there is currently no clear scientific evidence that MPs negatively affect human health, their removal from organisms is difficult, and MP particles can accumulate in the body.

2.8. Degradation of Microplastics in Aquatic Environments

It is considered that MPs are most likely formed on the beaches of natural water bodies. MPs can enter natural water systems directly as tiny particles generated by commercial commodities (primary MPs) or as a result of larger plastic fragments breaking down by the aging process, action of wind and waves, ultraviolet radiation, and such environmental conditions as humidity and temperature (secondary MPs) (Andrady, 2011; Waymana & Niemann, 2021). The synergetic effects of photodegradation, mechanical degradation, and hydrolysis on the disintegration of LDPE and PP plastics have been demonstrated in laboratory experiments (Lestari et al., 2022). However, weight loss was not high in both cases and was 3.07% for LDPE after the combined exposure of UV and high-water velocity for 480 h and 2.37% for PP under high water velocity only for the same period. A study of the aging of PE, PP, and PS placed in a microcosm with seawater circulation for 1 year showed a negligible gravimetric weight loss of ≤1% (Gerritse et al., 2020). A long-term monitoring (6 months from March–September 2022) of the characteristics of industrial filament plastic granules in special cages plunged in the Mediterranean Sea at a depth of 10 m near the sea floor showed the formation of a porous surface, and cavities were noted in HDPE and PP granules. However, no significant reduction in the weight of the plastic granules was observed (De Monte et al., 2022).

In general, it is believed that plastics, such as polystyrene, polyamide, polyvinyl chloride, and polypropylene, which make up more than 80% of the annual production of plastics, are resistant to the action of microorganisms due to: (a) their high molecular weight, which makes transportation of large molecules through the cell membrane of microorganisms impossible; (b) strong C–C bonds; (c) high C/N ratio, (d) hydrophobicity of surface, which prevents attachment of microbial cells to hydrophilic surfaces, and (e) lack of proper functional groups (Cai et al., 2023; Danso et al., 2019; Stabnikova et al., 2021; Urbanek et al., 2018). All this causes a constant increase in the amount of MPs in the environment. However, a minor weight loss of plastics placed in natural water bodies has been observed (Table 2).

LDPE, HDPE, and PP sheets of 1.5 mm thickness, rectangular shapes, and 10–15 cm sizes were immersed in a marine environment at a depth of 3 m for 6 months from April–September (Sudhakar et al., 2007). The inhabitants of plastic (LDPE, HDPE, and PP) biofilms that have developed within this period in the marine environment included photosynthetic organisms containing chlorophyll and prokaryotic bacteria, such as *Pseudomonas* sp., anaerobic, heterotrophic, and iron-reducing bacteria (Sudhakar et al., 2007). The average increase of the amount of biofilm per cm^2 of plastic surface was 0.19 $mg\cdot10^{-2}$ for low-density polyethylene, 0.22 $mg\cdot10^{-2}$ for high-density polyethylene, and 0.29 $mg\cdot10^{-2}$ for polypropylene. The biofilms were heterogeneous and consisted of cell

Table 2: Degradation of plastics in natural water bodies

Plastics	Weight Loss, %	Location, Degradation Time	Reference
HDPE	0.5–0.8	The Bay of Bengal near Chennai Port, India, at a depth of 3 m, 6 months	Sudhakar et al., 2007
LDPE	1.5-2.5		
PP	0.5–0.6		
PE	0.6	The Baltic Sea, Poland, at a depth of 2 m, 12 months	Rutkowska et al., 2002
LDPE	1.9	The Bay of Bengal, East coast of India, a depth of 3 m, 12 months	Artham et al., 2009
HDPE	1.6		
PC	0.69		
PP	0.65		
PET	7.49	Ennore Port, India, at a depth of 1 m, 12 months	Muthukumar et al., 2011
PU	4.25		
PU-A	19.0	Baltic Sea, Gdynia Harbour, at a depth of 2 m, 10°C, pH≈8.2, 12 months	Krasowska et al., 2015
PU-C	4.0		

Abbreviations: HDPE – High density polyethylene; LDPE – Low density polyethylene; PP – Polypropylene; PE – Polyethylene; PC – Polycarbonate; PET – Polyester; PU – Polyurethane; PU-A – poly(ethylene-butylene-adipate); PU-C – poly(caprolactone).

clusters with exopolysaccharides. The microbial population of biofilms included representatives of the *Pseudomonas* and *Clostridium* genera, which are known to be able to degrade polyethylene (Ghatge et al., 2020). The weight loss of polyethylene films was lower than 0.6% after 12 months of incubation in the Baltic Sea (Rutkowska et al., 2002). Two types of polyurethanes (pieces of sheets with a thickness of 2 mm) were incubated in the Baltic Sea at 2 m depth at an average annual temperature of 10°C (changes were from 1–20°C) and average pH of about 8.2 (Krasowska et al., 2015). After 12 months, degradation by the natural microbial community of poly(ethylene-butylene-adipate) (PU-A) was 19%, and poly(caprolactone) (PU-C) was only 4%. There is some research devoted to the potential ability of bacteria isolated from marine environments to degrade plastic materials (Table 3).

The biomass of strain AIIW2 isolated from the water of the plastic-polluted coastal area of Arambhada, Gujarat, India, which showed a 97.39% similarity with the *Bacillus* species, was used to treat PVC, LDPE, and HDPE films placed in sterile Bushnell-Haas minimal medium for 90 days (Kumari et al., 2019). Bacteria colonized the plastic films causing local deterioration of their surface and some structural changes, while weight loss was insignificant: 0.26% for PVC, 0.96% for LDPE, and 1% for HDPE.

Table 3: Degradation of plastic films by marine bacteria

Plastics	Microorganism	Conditions, Time	Weight Loss, %	Reference
PVC	*Bacillus* sp. strain AIIW2	Bushnell-Haas medium, AIIW2, 30°C, at 180 rpm, for 90 days	0.26	Kumari et al., 2019
LDPE			0.96	
HDPE			1.00	
LDPE	*Bacillus sphaericus*	Sea water, pH 7.5, 30°C, at 180 rpm, for 1 year	10.0	Sudhakar et al., 2008
HDPE			3.5	
LDPE	*Bacillus cereus*	Sea water, pH 7.5, 30°C, at 180 rpm, for 1 year	5.0	Sudhakar et al., 2008
HDPE			2.0	
PE	*Bacillus cereus*	Synthetic medium, 300 mL with 0.5 g of UV-treated MPs, 29°C at 150 rpm, 40 days	1.6	Auta et al., 2017
PET			6.6	
PS			7.4	
PE	*Bacillus gottheilii*		6.2	
PET			3.0	
PS			5.8	
HDPE	*Brevibacillus borstelensis*	Bushnell-Haas medium, 37°C, 30 days	11.4	Mohanrasu et al., 2018
PS	Marine consortia	Enriched filtered saline water, 5 months	11.0	Syranidou et al., 2019
PE			7.0	

Abbreviations: PVC – Polyvinyl chloride; LDPE – Low-density polyethylene; HDPE – High-density polyethylene; PE – Polyethylene; PET – Polyethylene terephthalate; PS – Polystyrene.

LDPE and HDPE films (80 × 25 mm) were treated with marine bacteria *Bacillus sphaericus*, isolated from shallow waters of the Indian Ocean, at pH 7.5 and temperature 30°C, with the polymer as the sole carbon source in the seawater for 1 year. The weight losses of LDPE and HDPE samples were 10% and 3.5%, respectively (Sudhakar et al., 2008). The minor changes in the physical and mechanical properties of LDPE and HDPE initially/after 1 year were the following: contact angle – 90/86 and 73.0/71.9; crystallinity, % – 75/73 and 85/86; tensile strength, N/mm^2, MPa – 13.71/13.18 and 18.90/17.06; elongation, % – 163.64/168.13 and 201.12/219.10; breaking load, N – 48/44 and 56/53, respectively; surface morphological changes were observed as well (Sudhakar et al., 2008).

The bacterial strains *Bacillus cereus* and *B. gottheilii* isolated from mangrove sediment in Peninsular Malaysia showed the ability to grow in a medium containing different UV-treated MPs as the sole carbon source; the calculated weight loss, %, of the MP particles after 40 days were: for *B. cereus*, 1.6, 6.6, and 7.4% for PE, PET, and PS, respectively; for *B. gottheilii*, 6.2, 3.0, and 5.8% for PE, PET, and PS, respectively (Auta et al., 2017).

There was a study aimed at creating conditions for increasing the content of microorganisms capable of degradation of naturally weathered PE and PS plastic films in enriched filtered saline water (C:N:P ratio of 100:10:1) (Syranidou et al., 2019). Microbial biofilms developed on these plastic particles in the same medium were used as inoculum for an experiment lasting for 5 months. Despite the fact that, among the microbial community of newly formed bioaugmented biofilms, the bacteria belonging to the orders *Rhodobacterales*, *Oceanospirillales*, and *Burkholderiales*, representatives of which are able to degrade plastic or hydrocarbons (Danso et al., 2019; Stabnikova et al., 2021; Vaksmaa et al., 2021), were abundant, the weight loss was not too significant, being 11% for PS and 7% for PE pieces (Syranidou et al., 2019). According to the functional analysis, the authors make a prediction that the cells carrying xenobiotic and hydrocarbon degradation genes will be overrepresented in the biofilms that cover plastics.

The degradation of MPs in the aquatic environment occurs in the same way as in the case of larger pieces of plastic, that is, abiotic (physical and chemical) treatment or biotic treatment through microbial activity, or a combination of these processes. Biodegradation of MPs to diminish their amount entering the global ocean has not been studied well, but it is supposed that its effectiveness is very low.

However, recent studies found that some marine fungi such as *Zalerion maritimum* and *Alternaria alternata* are promising degraders of MPs. Thus, the fungus *Zalerion maritimum* growing for 14 days in a minimum medium added with polyethylene microplastic particles, degraded them with an efficiency of 43% (Paço et al., 2017). Fungi *Alternaria alternata,* isolated from Huiquan bay of Qingdao, China, colonized and degraded polyethylene film, forming numerous holes across the film and decreasing its molecular weight by 95% after 120 days of treatment (Gao et al., 2022). With the aid of stable isotope assays, it was shown that marine yeasts *Rhodotorula mucilaginosa,* isolated from plastic debris from North Sea water, were capable of degrading UV-treated ^{13}C-polyethylene and polyethylene-derived carbon was incorporated in biomass (Vaksmaa et al., 2023). *R. mucilaginosa* can also degrade initial polyethylene without UV treatment, but in this case, biomass contains less ^{13}C. The results of this study demonstrated that plastic photooxidation is an important step making it available for microbial degradation.

3. Ways to Diminish Microplastic Pollution of Water Basins

3.1. Removal of Microplastics from Treated Wastewaters

Although the amount of MPs in the ocean is much less than its input into rivers due to its partial disappearance during the passage of estuaries (mainly as a result of beaching) (Lopez et al., 2021), rivers, together with coastal beaches, are considered the main sources of marine plastic pollution (Chenillat et al., 2021). It is estimated that about 2 million tons of plastic waste enter the ocean annually from 40,760 rivers (Lebreton et al., 2017).

It was proposed to classify the MPs present in wastewaters in two groups: (1) primary MPs coming from personal care products and air-blasting, and (2) secondary MPs produced due to fragmentation of larger plastic pieces or as a result of removed fibers through washing of clothes (Browne et al., 2011). Raju et al. (2018) described the primary MPs as PE, PP, PS granules and PPS from personal care products, scrubbing agents, air blasting media, and injection molding powders, and the secondary MPs as nylon, PP, PVA fibers from polyester, rayon, and nylon fabrics, carpets, ropes, fishing gear, and industrial wastewater. The primary MPs mainly are represented by microbeads, usually less than 1 mm in diameter, which are widely used in personal care products, such as toothpaste and face and body scrubs. The plastic particles presented in wastewater can be classified by size as big MP (1–5 mm), small MP (100 nm–1 mm), and nanoplastic (1–100 nm) (Herbort et al., 2018). However, the coronavirus outbreak added a new, unexpected source of plastic and MP aquatic pollution – the improper disposal of face masks (Abuwatfa et al., 2021). It was estimated that from 0.15–0.39 million tons of plastic debris from face masks entered the oceans within a year of the pandemic (Chowdhury et al., 2021).

To decrease MP accumulation in nature and protect the fragile fresh and marine ecosystems, the methods for its removal from streams discharging in water bodies should be developed and implemented in real life in order to diminish the load and protect the fragile marine ecosystems and aquatic environments (Stabnikov et al., 2023; Vaseashta et al. 2021). There are some ways to diminish the negative effect of MPs on the aquatic environment. It is known that MPs are not removed during the treatment of wastewater in municipal wastewater treatment plants (MWWP), instead being released in great amounts with treated water into natural aquatic bodies. It was estimated that 1 m^3 of effluent from municipal wastewater treatment plant in Germany contains up to 9,000 pieces of MPs sized >0.5 mm, with polyethylene predominating (Mintenig et al., 2017). MPs in effluent from MMWT serve as carriers for sewage-treated microorganisms as well as for heavy metals and different recalcitrant organic compounds into aquatic environments. MP particles can also be covered with pathogenic microorganisms and thus can serve as carriers of human pathogens. Therefore, the need to develop approaches for MP removal from wastewaters is vital. The microplastics entering with treated wastewater into freshwater natural reservoirs are also a carrier of microorganisms that are alien to natural microbial communities and of antibiotic resistance genes that pose a threat to environmental safety (Eckert et al., 2018).

The conventional activated sludge strategy is also used to treat wastewater in water treatment plants but shows lower efficiency. The removal of microplastics from wastewater depends on the technological processes used for treatment. It has been shown that the treatment of wastewater in WWTPs remove a major part of MPs – from 83–95% (Dris et al., 2015). Preliminary and primary treatment (screening, grit and oil removal, settling) removes from 40% (Dris et al., 2015) or 45% (Habib et al., 2021; Murphy et al., 2016) to 77% of total MPs (Tagg et al., 2020). Secondary treatment (biological treatment and sedimentation) resulted in further removal of approximately 50% (Enfrin et al., 2019). The concentrations of MPs in the influents of 7 municipal WWTPs in the Netherlands ranged from 68,000–910,000 MPs/m^3, while in the effluents they were from 51,000 81,000 MPs/m^3 (Leslie et al., 2017). The efficiency of removal of MPs from wastewater after treatment in a secondary wastewater treatment plant (population equivalent 650,000) was 98.41%; the concentration of MPs in the final effluent was not high – 250 pieces/m^3; meanwhile, the estimated annual amount of MP particles released in the environment from this wastewater treatment enterprise is 23,812 million (Murphy et al., 2016). Furthermore, Carr with co-authors (2016) found that secondary treatment processes (WWTP, Southern California, USA) removed more than 99.9% of MPs; their content was only 1 particle in 1.4 m^3 of effluent (in case of tertiary treatment, it was 0.02 particle in 1 m^3). They also found that MP particles were removed mainly during primary treatment via the solids skimming and sludge settling processes. According to the authors of this publication, the existing methods for wastewater treatment are quite effective for removal of MPs. As the MPs in treated wastewater represented mainly polyethylene particles found in toothpaste, there is a real way to diminish their loads on WWTPs by the replacement of these additives in cosmetic products with more environmentally friendly materials. Tertiary treatment using a membrane bioreactor removed from 97% (Mintenig et al., 2017) to 99.9% of MPs (Talvitie et al., 2017); effluents after the tertiary treatment contain less than 2% of MPs (Sun et al., 2019) (Table 4).

A study of application of different advanced final-stage water treatment technologies after conventional wastewater treatment in WWTPs in Finland gave the follow results: membrane bioreactor removed 99.9% of MPs during the treatment (from 6.9–0.005 MP/L), rapid sand filter – 97% (from 0.7–0.02 MP/L), dissolved air flotation – 95% (from 2.0–0.1 MP/L), and disc filter – 40 to 98.5% (from 0.5-2.0–0.03-0.3 MP/L) (Talvitie et al. 2017).

The effectiveness of the MPs removal was studied in three WWTPs, which used primary treatment (grift and primary setting), secondary treatment (bioreactor and secondary setting), and tertiary treatment including coagulation

Table 4: Removal of microplastics from wastewater using different treatment methods

Treatment Method	WW Load	MPs per m^3 of		WWTP Location	Reference
		Influent	Effluent		
Preliminary and Primary Treatment					
Screening (mesh size of 5 mm), grit removal, and sedimentation	N/S	N/S	1,500	Sydney, Australia	Ziajahromi et al, 2017
Screening, grit and oil removal, settling	240,000 m^3/day	260000–320000	50,000–120,000	Paris, France	Dris et al. 2015
Secondary					
Aeration, sedimentation, and UV disinfection	N/S	N/S	480	Sydney, Australia	Ziajahromi et al, 2017
Aeration, sedimentation	1,510,000 m^3/day	N/S	1	Southern California, USA	Carr et al., 2016
Biofiltering	240,000 m^3/day	50000– 120000	14,000– 50,000	Paris, France	Dris et al., 2015
Aaeration, clarifier	650,000 PE	15700	250	Glasgow, UK	Murphy et al., 2016
Mechanically, chemically, and biologically	12,000 PE	15100	8	Lysekil, Sweden	Magnusson & Norén, 2014
Activated biological anoxic treatment, activated biological aerobic treatment	200,000 PE	n/s	1,500	East Midlands, UK	Tagg et al., 2020
Tertiary					
Flocculation, disinfection/de-chlorination, ultrafiltration	n/s	n/s	280	Sydney, Australia	Ziajahromi et al., 2017

Reverse osmosis and de-carbonation	n/s	n/s	210	Sydney, Australia	Ziajahromi et al., 2017
Membrane microfiltration, continuous backwash upflow dual sand microfiltration, rapid sand filters	n/s	n/s	0	New York, USA	Schneiderman, 2015
Gravity filters	n/s	n/s	0.02	Southern California, USA	Carr et al., 2016
Tertiary-level	n/s	n/s	1,000	NSW, Australia	Browne et al., 2011
Post-filtration	35,616 PE	50	0 (MP > 500 μm)	Oldenburg, Germany	Mintenig et al., 2017
Membrane bioreactor	n/s	6900	5	Mikkeli, Finland	Talvitie et al., 2017
Rapid sand filter	n/s	700	20	Turku, Finland	
Dissolved air flotation	n/s	2000	100	Hameenlinna, Finland	
Disc filter, pore size 10 μm 20 μm	n/s	 500 2000	 300 30	Helsinki, Finland	
Membrane bioreactor	n/s	4400	920	Murcia, Spain	Bayo et al., 2020
Rapid sand filtration	n/s	4400	1,080		
Ozone	n/s	164000	33000	Daegu, South Korea	Hidayaturrahman & Lee, 2019
Membrane disc-filter	n/s	1444000	297000		
Rapid sand filter	n/s	215000	66000		

Abbreviations: n/s – Not specified; PE – Population equivalent; NSW – New South Wales.

with ozone: (1) membrane disc-filter, (2) or rapid sand filter, (3) technologies (Hidayaturrahman & Lee, 2019). It was shown that the efficiency of MP removal using primary and secondary treatment ranged from 75–91.9% and exceeded 98% after tertiary treatment. The total removal of MPs with the application of ozone was 99.2%, membrane disc filter – 99.1%, and rapid sand filter – 98.9%. However, due to the high initial concentration of MPs in wastewater, its final concentrations in effluents also remained high, and these effluents constantly discharging in large volumes into rivers cause irreparable harm to the environment (Hidayaturrahman & Lee, 2019).

One of the methods to diminish the quantity of MPs released with effluent from WWTPs is its additional treatment. It is possible to increase the efficiency of MP removal from effluent from WWTPs by the introduction of additional steps into the tertiary stage of wastewater treatment. Several approaches have been suggested to reduce MP release with treated wastewater into the environment (Lastovina & Budnyk, 2021; Padervand et al., 2020). The removal methods include physical sorption and filtration, biological removal and ingestion, and chemical treatments (Padervand et al., 2020). It was found that the introduction of a post-filtration step into the tertiary treatment of wastewater removed 97% of MPs from the effluent, and MPs sized >0.5 mm were removed completely (Mintenig et al., 2017). However, not-so-high efficiencies of MP removal were reported for rapid sand filtration – 79.01%, and MBR – 75.49% (Bayo et al., 2020). Despite the fact that the application of membrane technologies as a tertiary treatment has shown good results for MP removal, their utilization, especially in large-scale operations, is challenging, because even low concentrations of MPs in wastewater cause an increase in the extent of membrane fouling (LaRue et al., 2022). The dynamic membrane technology was proposed for use as a possible technique to remove MPs from wastewater. A dynamic membrane is formed by a semi-permeable layer from suspended particles presented in wastewater on the surface of a conventional membrane so that these deposited particles serve as a composite membrane, which is in dynamic equilibrium with the treated solution. The advantages of such self-forming dynamic membranes (DM) are their formation directly in the membrane reactor, non-use of chemicals, no by-products produced, reduction of energy consumption, and cost of treatment. The DMs formed on a 90 μm supporting mesh were effective in removing particles with sizes ranging from 1.65–516 μm from synthetic wastewater (Li et al., 2018).

Analysis of the content of MPs in the effluents of three WWTPs in Sydney, Australia, that use different technologies for wastewater treatment showed the following numbers in particles per m^3: 1,500 (after primary treatment – screening [mesh size of 5 mm], grit removal, and sedimentation) with deep ocean discharge at first; 480 (after primary treatment – screening [mesh size of 3 mm], grit removal, and sedimentation; secondary treatment – aeration, sedimentation, and

UV disinfection) with the treated effluent discharged to the ocean at second; 280 (in a highly treated effluent by applying primary treatment – screening [mesh size of 3 mm] and sedimentation; secondary treatment – biological, and tertiary treatment – flocculation, disinfection/de-chlorination, ultrafiltration), or 210 (after reverse osmosis and de-carbonation) with the treated effluent discharged into an urban river (Ziajahromi et al., 2017).

It is necessary to be aware that the MPs removed from wastewater during treatment in WWTPs are accumulated in sewage sludge, which will become a new source of MPs threatening the environment. The mean retention efficiency of MPs in the sewage sludge at seven WWTPs in the Netherlands was 72% (Leslie et al., 2017). The sludge after aerobic treatment at Sari WWTP, Iran, contained 238 pcs of MPs/g dry weight, which amounts to 601 million pcs of MPs/day (Petroody et al., 2021). In other studies, sewage sludge from WWTPs contained the following quantities of MPs (pcs/kg of dw): 16,700 (Magnusson & Norén, 2014); 1,000 (Mintenig et al., 2017); 4,196–15,385 (Mahon et al., 2017).

The cost-effectiveness of using a combination of wastewater treatment methods such as activated sludge, rapid sand filtration, or membrane bioreactor, and sewage sludge treatment methods such as anaerobic digestion or incineration to reduce the impact of MPs on aquatic and terrestrial ecosystems has been assessed (Vuori and Ollikainen, 2022). When conducting the assessment, such environmental issues were taken into account as the external costs related to climate and water pollution from treated wastewater and sludge management. The best option turned out to be wastewater treatment in a membrane bioreactor combined with sludge incineration to prevent the accumulation of MPs in the soil. The second option was conventional activated sludge (CAS) technology enhanced with rapid sand filtration, which is more suitable for large-scale applications. CAS and sludge incineration ranked third.

All of the studies described above showed that the modern wastewater treatment process is effective enough for the removal of MPs, but, taking into account the large volumes of effluents, WWTPs still continue to be significant sources of MPs entering the environment. The use of tertiary treatment methods can significantly reduce the effect of pollution, but this requires expensive equipment and high operation and maintenance costs given the volume of water that needs to be treated. Analysis of the effluents from 17 WWTPs across the USA showed a small amount of MPs – an average of 50±24 MPs/m^3; at the same time, over 4 million microparticles were released into the environment from each wastewater treatment plant daily (Mason et al., 2017). Hence, the search for new non-traditional methods for removing MPs from treated wastewater is an urgent task that can help reduce the quantities of MPs entering the aquatic environment. By the same token, less costly and simpler methods to be used for the large-scale removal of MPs from wastewater should be proposed and implemented in practice. One such method could be the adsorption of MPs on biosorbents.

3.2. Application of Biosorbents for Removal of Microplastics from Wastewater

The application of cheap natural materials such as zeolites and bentonites to remove MPs from wastewater with an efficiency of 91–93% was shown (Spacilova et al., 2023). A more prospective approach is to use biomass sorbents (agricultural, forestry biomass waste) for the removal of MPs, as this approach can be considered sustainable, and it has demonstrated its efficiency for the simultaneous removal of other wastewater contaminants. Biosorbents more often are obtained from raw waste materials by pyrolysis, which is the thermal treatment of biomass at a temperature above 500°C in a limited oxygen environment. Pyrolysis ensures the decomposition of organic materials and the production of volatile components, and biochar – a carbon-rich, charcoal-like substance – is left after emitting these components. The temperature at which pyrolysis occurs and the nature of the raw material largely determine the properties of the resulting biochar. The high temperature of pyrolysis ensures the production of biochar with a developed surface area, high porosity, and high content of carbon and mineral residues due to the increased formation of volatile substances compared to the product obtained by pyrolysis at lower temperatures (Tomczyk et al., 2020).

The preparation of biochar from various organic materials and the development of methods for its modification in order to increase the efficiency of use for the removal of various water and soil pollutants comprise one of the priority areas of current research (Qiu et al., 2022). The possibility of using biochar to remove MPs from effluent after WWTPs is also attracting the attention of researchers.

The exact mechanism of adsorption of MPs by biochar is unknown. It is supposed that MP particles can be retained among biochar particles or within biosorbent pores. Thus, the porosity of the adsorbent should play a key role in the retention of micro- and NPs. This assumption is supported by experiments that found that biochar with a relatively low surface area and macro-scale porosity showed a high retention capability for MP particles (Siipola et al., 2020). Wang et al. (2020), using the environmental scanning electron microscope (ESEM) images, showed that microbeads were retained on the surfaces and pores of the examined types of biochar from corn straw and hardwood. The authors described the retention mechanisms as 'sticking' (microbeads get stuck among biochar particles that are smaller than the particle size), 'trapping' (microbeads get trapped inside the pores of biochar), and 'entanglement' (microbeads get entangled by small biochar chips, increasing their size and decreasing mobility) (Figure 2) (Wang & Sedighi, 2023).

Biochar produced from pine and spruce bark by slow pyrolysis at 475°C was steam-activated at 800°C and used for wastewater treatment. It was shown that the steam-activated biochar could serve as a suitable adsorbent for MPs from stormwater and wastewater. It also adsorbs cations and residual organics from aqueous media. However, the removal of small particles sized 10 μm was not

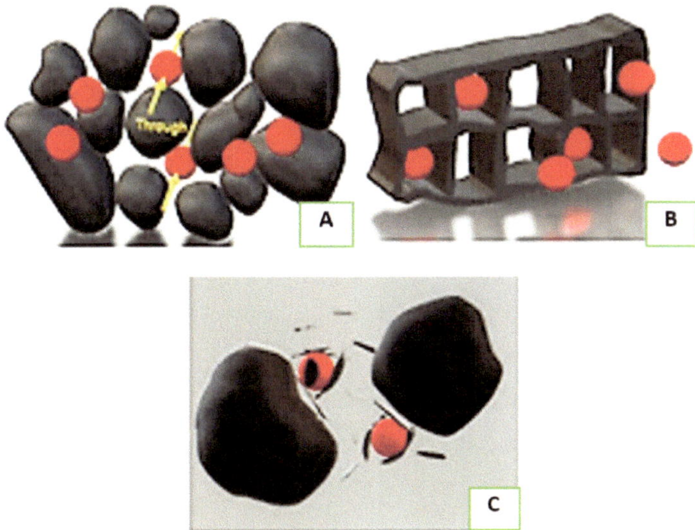

Figure 2: Three mechanisms of interaction between biochar and microplastic spheres observed in ESEM: A – MPs get stuck among biochar particles; B – MPs get trapped inside the pores of biochar; C – MPs get entangled by biochar particles (adopted from Wang & Sedighi, 2023).

as effective as the removal of MP particles of large sizes (Siipola et al., 2020) (Table 5).

Efficiencies of applying the biochar types produced from corn straw and hardwood for the removal of microbeads as a step to increase the efficiency of the conventional sand filter technology were investigated (Wang et al., 2020). Microbeads with a 10 μm diameter, which usually are present in high amounts in effluents from different tertiary wastewater treatment systems, had a low adsorption capacity in a sand filter due to their smooth surface and round shape. Straw biochar obtained by the slow pyrolysis process at 400 or 500°C and hardwood biochar obtained by the conventional open-fire stove method showed microbead removal efficiency higher than 95%; meanwhile a biochar filter combining the bottom and top layers of silica sand and the intermediate layer of corn straw biochar obtained by pyrolysis at 500°C showed a 100% microbead removal efficiency (Wang et al., 2020).

Furthermore, the efficiency of the application of an iron-modified, biochar-based nanocomposite for the removal of NPs has been shown (Singh et al., 2021). Carboxylate (NP1)- and amine (NP2)-functionalized NPs sized 1,000 nm were used in this study. The proposed adsorbent showed a higher removal efficiency

Table 5: Biosorbents used for removal of microplastics from wastewater

Biosorbent	Process of Biosorbent Production	Removal Capability	Reference
Pine and spruce bark biochar	Slow pyrolysis at 475°C, and activation by steam at 800°C	165–293 mg MPs/g of biochar	Siipola et al., 2020
Iron-modified biochar-based nanocomposite	Pyrolysis at 850°C	225 mg NP1/g of biochar	Singh et al., 2021
	Pyrolysis at 550°C	290 mg NP2/g of biochar	
Biochar from corn straw	Slow pyrolysis process at 400–500°C	Microbead removal ≥95%	Wang et al., 2020
Fly ash	Iron-modified	PS nanoplastics 82.8–89.9 mg/g of biochar	Zhao et al., 2022
Sawdust treated with Fe^{3+} and	$Fe(NO_3)_3 \cdot 9H_2O$, 550°C	PS microspheres (100 mg/mL) 94.81%	Wang et al., 2021
Modified with Mg^{2+}	$Mg(NO_3)_2 \cdot 6H_2O$	98.75%	
Modified with Zn^{2+}	$Zn(NO_3)_2 \cdot 6H_2O$	99.46%	
Magnetic biochar from rice husks		99.96%	Wu et al., 2023

Notes: NP1 (carboxylate-functionalized) and NP2 (amine-functionalized) – nanoparticles size 1,000 nm; PS – Polystyrene.

(100%) for NPs in comparison with non-modified biochar (75%), due to surface complexation and electrostatic interactions between the NPs and the nanoparticles.

There have been numerous studies proposing the use of magnetic biochar for MP removal from wastewater. Fly ash modified with Fe ions resulted in a new magnetic material, which strongly interacts with polystyrene NPs in aqueous solutions through the attachment of them to the surface and pores of the modified biochar. It was shown that electrostatic attraction, complexation, and π–π interactions participated in the adsorption process, and the adsorption amounts of polystyrene NPs varied from 82.8–89.9 mg/g of magnetic biochar at pH 5-7 (Figure 3) (Zhao et al., 2022).

To increase the retention of MP particles sized 0.2–2 μm from groundwater in porous media, for example, quartz sand, the addition of biochar or magnetic biochar (Fe_3O_4-biochar) was recommended (Tong et al., 2020). Addition of Mg- or Zn-modified biochar adsorbents (sawdust soaked in a solution containing $Fe(NO_3)_3 \cdot 9H_2O$, dried at 105°C, heated at 550°C for 2 h, dried at 105°C, and then treated with $Mg(NO_3)_2 \cdot 6H_2O$ or $Zn(NO_3)_2 \cdot 6H_2O$) to an aqueous solution containing polystyrene microspheres (1 μm, 100 mg/mL) resulted in their

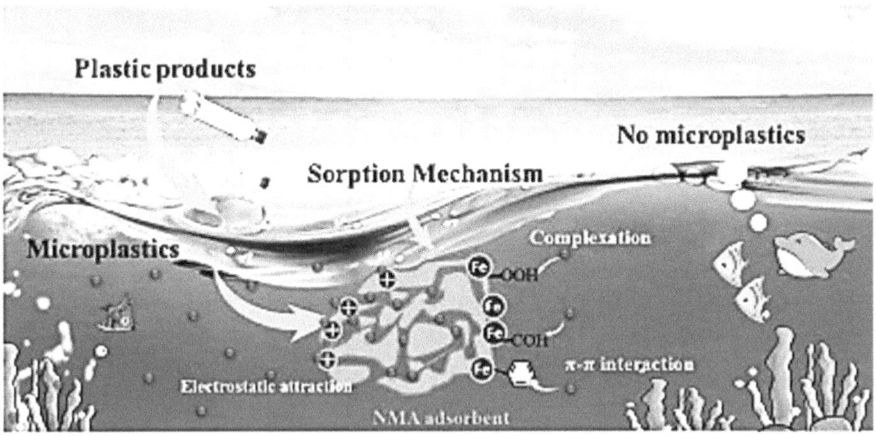

Figure 3: Sorption mechanism of microplastic particles on magnetic biochar (adopted from Zhao et al., 2022).

removal with the efficiencies of 94.81%, 98.75%, and 99.46%, respectively (Wang et al., 2021). Thermal treatment of biochar ensured simultaneous regeneration and degradation of adsorbed MPs, and the biochar maintained high MP removal efficiency even after 5 adsorption-pyrolysis cycles. However, if the temperature regime of pyrolysis is not observed and it does not reach at least 450°C, MP particles with a rougher surface and increased adsorption capacity with respect to contaminants may appear (Ni et al., 2020).

Application of magnetic biochar prepared from agricultural waste rice husks allowed for removal by adsorption under acidic conditions up to 99.96% of MPs present in treated water due to electrostatic attraction, surface complexation, hydrogen bonding, and π-π interactions (Wu et al., 2023). Recovered by thermal treatment (500°C), biochar also showed high efficiency of MP removal up to 90%. The addition of a thin layer of biochar (1% w/w), produced via pyrolysis at a temperature 700°C from woodchips, into a sand filtration column allowed the complete removal of MP particles with a diameter of 1 μm from real river or sewage water (Hsieh et al., 2022).

3.3. Agglomeration as a Way to Remove Microplastic from Wastewater

Since MP particles are not removed by such conventional wastewater treatment plant technologies as flocculation or sedimentation, there is an intensive search for new approaches to solve this problem. Agglomerate formation is one of these new approaches. The sol-gel-induced agglomeration of MP particles, which is aimed at forming larger particle agglomerates that could be easily separated from wastewater using a simple separation system, such as a sand trap, has been proposed (Herbort et al., 2018a, b). The possibility of increasing the volume of the original MP particles by 666 times using alkoxy-silyl compounds has been shown. Application of such coagulants as ferric chloride or polyaluminum chloride in doses that are conventionally used for tertiary treatment of wastewater resulted in the 99.4% removal efficiency of polystyrene spheres sized 1 μm and 6.3 μm (Rajala et al., 2020).

Electrocoagulation is considered to be used to separate MPs from wastewater. During electrocoagulation, metal ions produced at the sacrificial anode (Fe or Al) and hydroxide ions produced at the cathode due to water electrolysis are combined to form the metal hydroxide, which serves as a coagulant for the formation of flocs with MPs. Electrocoagulation increased the size and density of MPs and converted them into separable flocs by centrifugation. The best MP removal rates by electrocoagulation with aluminum anode at pH 7.2 were shown to be 93.2% for polyethylene, 91.7% for poly(methyl methacrylate), and 98.4% for polypropylene (Shen et al., 2022). Electrocoagulation treatment using aluminum electrodes removed 99.24% spherical microbeads from the model wastewater at a pH of 7.5 (Perren et al., 2018). The removal efficiency of MPs from real wastewater after chlorination using electrocoagulation with aluminum electrodes at an initial pH of 4 and a current density of 2.88 mA/cm^2 was 96.5% (Elkhatib et al., 2021). To ensure effective MP removal (97.3%), combined treatment of wastewater including conventional biological treatment, electrocoagulation, and treatment with granular activated carbon (with thermal regeneration) was proposed (Kim & Park, 2021). To concede, studies of agglomeration and coagulation applications to remove MPs from aquatic environments currently take place in laboratories only.

4. New Nontraditional Methods to Remove Microplastics from the Aquatic Environment

New approaches are being developed to extract microplastic particles from polluted water. For example, methods based on the change of surface properties of MPs by hydrophobic Fe nanoparticles, which bind to plastic, allowing magnetic recovery. 92% of 10–20 μm polyethylene and polystyrene beads and 93% of >1 mm MPs (polyethylene, polyethylene terephthalate, polystyrene, polyurethane, polyvinyl chloride, and polypropylene) were recovered from seawater (Grbic et al., 2019). The addition of 1.3 g/L nano-Fe_3O_4 and treatment for 150 min resulted in the adsorption of these particles on the surface of MPs, causing their magnetization. The magnetized MPs (polyethylene, polypropylene, and polystyrene in sizes of approximately 200–900 μm) were removed by magnetic force from the aquatic environment with an average removal rate of over 80% (Shi et al., 2022). To remove polystyrene from aqueous solutions, a novel material with a metal-organic framework capable of adsorbing MP particles was proposed. Polystyrene MP removal efficiency at their concentration of 5 mg/L was up to 92% at pH 8 and 25°C due to hydrogen bond interactions, π-π stacking, and electrostatic interactions between MPs and the new adsorbent (Wan et al., 2022).

Freezing and thawing of plastic in aquatic environments can lead to its mechanical degradation (Zhang et al., 2023). Rapid degradation of MPs, namely polystyrene, was found in frozen solution (–20°C) due to the selective generation of singlet oxygen in the narrow space of the liquid layer when particles are trapped and squeezed between ice crystals (Tian et al., 2022). Recently, photocatalytic treatment has been proposed to remove MPs from wastewater (Mahmud et al., 2022; Sacco et al., 2023; Wang et al., 2019) and degrade them (Ariza-Tarazona, 2019), though these methods have so far only been investigated at a laboratory and their application in a large scale currently is limited due to the high cost and the complexity of execution.

5. Bioplastics as a Way to Diminish the Microplastic Load on the Marine Environment

The accumulation of MPs in the environment is due to the fact that plastic does not biodegrade under natural conditions. Therefore, the development and use of biodegradable polymers has attracted attention as a promising solution to the pollution problem caused by the widespread use of conventional polymers. It was proposed to use: (a) based on fossil resources but biodegradable polybutylene adipate terephthalate (PBAT), polybutylene succinate (PBS), polycaprolactone (PCL), and (b) bio-based and biodegradable polyhydroxyalkanoates (PHAs), particularly polyhydroxybutyrate (PHB) (Ivanov & Stabnikov, 2020; McAdam et al., 2020), and polylactic acid (PLA) that is considered to be the most promising material (Tao et al., 2024). It was believed that the use of biodegradable, bio-based plastics would be a desirable environmental solution reducing the negative impact on the environment. However, a number of new studies indicate that degradation products of bioplastics, in particular PLA, may have the same impact as conventional plastics on marine ecosystems. For example, PLA microparticles lowered fertilization rates of the solitary ascidian *Microcosmus exasperates* (Anderson & Shenkar, 2021). It was shown that biodegradable polylactic acid MPs and conventional polyethylene and polyvinyl chloride MPs in sandy sediments had the same negative impact on the primary productivity of lugworms, *Arenicola marina* (Green et al., 2016). At the same time it was shown that the duration of wastewater treatment in the activated sludge system is not enough for PLA degradation, and PLA MPs enter natural water bodies with effluent from WWTPs (Rom et al., 2017). González-Pleiter et al. (2019) found that secondary NPs released from polyhydroxybutyric acid microplastics significantly decreased the growth rate of cyanobacteria and green algae, which constitute an important part of phytoplankton. All these observations indicate that the application of biodegradable plastics does not ensure environmental safety. Biodegradable polymers undergo fragmentation into small particles during the biodegradation process, resulting in the formation of large quantities of small plastic particles. These MPs, originating from biodegradable plastic products, can be present for a long time in natural conditions posing a risk to the environment. Even polylactic acid, which degrades quickly in laboratory scale tests, undergoes a much slower process of biodegradation in real conditions. As a result, depending on the type of biopolymer and environmental factors, MPs formed from biodegradable materials can persist in the natural environment for decades. Under certain conditions, bio-MPs have even been shown to cause greater negative effects than conventional microplastics, as they may have a stronger affinity for environmental pollutants and may serve as a more significant vector for the transfer of both microorganisms and chemical toxicants. For example, it was shown that biodegradable polybutylene succinate adsorbed pesticides by 2 and 7 times more than polyethylene and polyvinyl chloride (Jiang et al., 2020). However, there are also studies showing that biodegradable MPs have a smaller negative impact on marine organisms compared with conventional polymers, and the application of biodegradable materials may be the way to diminish stress on marine ecosystems (Hawke et al., 2024). Therefore, further systematic studies of biodegradable MPs and their role in environmental safety are necessary.

6. Conclusions

MPs, as a widespread and ever-increasing pollutant of the world's water basin, entering the global ocean are of concern. Notably, the estimated number of MPs floating in the ocean is much less than the estimated input. Different ways of removing MPs from the ocean have been found. Natural processes leading to their reduction in the aquatic environment include its removal: (a) with marine snow; (b) due to incorporation into iron-containing flocs; (c) by coral activity; (d) with phytoplankton aggregates; (e) by aggregation of MPs: (i) due to covering with biofilms and biofouling, followed by the precipitation of aggregates and their accumulation in the ocean sediments; (ii) due to consumption by marine inhabitants; (iii) due to degradation. However, all these processes can remove only a negligible part of the total MP waste. Therefore, it is necessary to find ways to reduce the level of this pollutant entering the water basins.

The largest permanent source of MP pollution is treated wastewater, which, even with a small residual MP content, supplies huge amounts of it, taking into account the volumes of input. Recently, the improvement of traditional methods of tertiary wastewater treatment and the development of new, modern technological methods for post-treatment of water for almost complete removal of MPs have been given great attention, and the number of publications on this topic constantly increases. However, the existing cleaning technologies based on the separation of MPs are associated with the need for periodic cleaning of the equipment and also have a problem of disposal of the MPs removed from the water. In this regard, technologies based on the destruction of MPs have advantages, although the disadvantages in this case are the high cost of these cleaning methods and also the possibility of obtaining even more harmful degradation products than the source MPs. Among the newly proposed technologies, attention is drawn to the use of cheap and renewable biosorbents with modified surface properties, the regeneration of which by the thermal method simultaneously ensures the complete utilization of the isolated MP. The production and use of biodegradable plastics may not protect the environment from existing MP pollution. Recent research suggests the need to further study the production, and existence of biodegradable plastics and their impact on ecosystems to prevent possible negative consequences for the global environment.

References

Abuwatfa, W.H., Al-Muqbel, D., Al-Othman, A., Halalsheh, N., & Tawalbeh, M. (2021). Insights into the removal of microplastics from water using biochar in the era of COVID-19: A mini review. *Case Studies in Chemical and Environmental Engineering*, 100151. https://doi.org/10.1016/j.cscee.2021.100151.

Andrady, A.L. (2011). Microplastics in the marine environment. *Marine Pollution Bulletin*, 62(8), 1596–1605. https://doi.org/10.1016/j.marpolbul.2011.05.030.

Amaral-Zettler, L.A., Zettler, E.R., Mincer, T.J., Klaassen, M.A., & Gallagher, S.M. (2021). Biofouling impacts on polyethylene density and sinking in coastal waters: A macro/micro tipping point? *Water Research*, 201, 117289. https://doi.org/10.1016/j.watres.2021.117289.

Anderson G., & Shenkar, N. (2021). Potential effects of biodegradable single-use items in the sea: Polylactic acid (PLA) and solitary ascidians. *Environmental Pollution*, 268, 115364. https://doi.org/10.1016/j.envpol.2020.115364.

Artham, T., Sudhakar, M., Venkatesan, R., Madhavan Nair, C., Murty, K.V.G.K., & Doble, M. (2009). Biofouling and stability of synthetic polymers in sea water. *International Biodeterioration & Biodegradation*, 63, 884–890. https://doi.org/10.1016/j.ibiod.2009.03.003.

Ariza-Tarazona, M.C., Villarreal-Chiu, J.F., Barbieri, V., Siligardi, C., & Cedillo-González, E.I. (2019). New strategy for microplastic degradation: Green photocatalysis using a protein-based porous N-TiO_2 semiconductor. *Ceramics International*, 45(7), 9618–9624. https://doi.org/10.1016/j.ceramint.2018.10.208.

Auta, H.S., Emenike, C.U., & Fauziah, S.H. (2017). Screening of *Bacillus* strains isolated from mangrove ecosystems in Peninsular Malaysia for microplastic degradation. *Environmental Pollution*, 231(Pt 2), 1552–1559. https://doi.org/ 10.1016/j.envpol.2017.09.043.

Bayo, J., López-Castellanos, J., & Olmos, S. (2020). Membrane bioreactor and rapid sand filtration for the removal of microplastics in an urban wastewater treatment plant. *Marine Pollution Bulletin*, 156, 111211. https://doi.org/10.1016/j.marpolbul.2020.111211.

Bellasi, A., Binda, G., Pozzi, A., Galafassi, S., Volta, P., & Bettinetti, R. (2020). Microplastic contamination in freshwater environments: A review, focusing on interactions with sediments and benthic organisms. *Environments*, 7, 30. https://doi.org/10.3390/environments7040030.

Bessa, F., Barría, P., Neto, J.M., Frias, J.P.G.L., Otero, V., Sobral, P., & Marques, J.C. (2018). Occurrence of microplastics in commercial fish from a natural estuarine environment. *Marine Pollution Bulletin,* 128, 575–584. https://doi.org/10.1016/j.marpolbul.2018.01.044.

Bhattacharya, P., Lin, S., Turner, J.P., & Ke, P.C. (2010). Physical adsorption of charged plastic nanoparticles affects algal photosynthesis. *The Journal of Physical Chemistry C,* 114(39), 16556–16561. https://doi.org/10.1021/jp1054759.

Brandon, J., Goldstein, M., & Ohman, M.D. (2016). Long-term aging and degradation of microplastic particles: Comparing *in situ* oceanic and experimental weathering patterns. *Marine Pollution Bulletin, 110*(1), 299–308. https://doi.org/10.1016/j.marpolbul.2016.06.048.

Browne, M.A., Crump, P., Niven, S.J., Teuten, E., Tonkin, A., Galloway, T., & Thompson, R. (2011). Accumulation of microplastic on shorelines worldwide: Sources and sinks. *Environmental & Science Technology,* 45(21), 9175–9179. https://doi.org/10.1021/es201811s.

Bryant, J.A., Clemente, T.M., Viviani, D.A., Fong, A.A., Thomas, K.A., Kemp, P. et al. (2016). Diversity and activity of communities inhabiting plastic debris in the North Pacific Gyre. *mSystems,* 1(3), e00024–16. https://doi.org/10.1128/msystems.00024-16.

Cai, Z., Li, M., Zhu, Z., Wang, X., Huang, Y., Li, T. et al. (2023). Biological degradation of plastics and microplastics: A recent perspective on associated mechanisms and influencing factors. *Microorganisms,* 11(7), 1661. https://doi.org/10.3390/microorganisms11071661.

Carr, S.A., Liu, J., & Tesoro, A.G. (2016). Transport and fate of microplastic particles in wastewater treatment plants. *Water Research,* 91, 174–182. https://doi.org/10.1016/j.watres.2016.01.002.

Catarino, A.I., Macchia, V., Sanderson, W.G., Thompson, R.C., & Henry, T.B. (2018). Low levels of microplastics (MP) in wild mussels indicate that MP ingestion by humans is minimal compared to exposure via household fibres fallout during a meal. *Environmental Pollution,* 237, 675–684. https://doi.org/10.1016/j.envpol.2018.02.069.

Chen, X., Xiong, X., Jiang, X., Shi, H., & Wu, C. (2019). Sinking of floating plastic debris caused by biofilm development in a freshwater lake. *Chemosphere,* 222, 856–864. https://doi.org/10.1128/mSystems.00024-16.

Chenillat, F., Huck, T., Maes, C., Grima, N., & Blanke, B. (2021). Fate of floating plastic debris released along the coasts in a global ocean model. *Marine Pollution Bulletin,* 165, 112116. https://doi.org/10.1016/j.marpolbul.2021.112116.

Chow, J.S., Lee, C., & Engel, A. (2015). The influence of extracellular polysaccharides, growth rate, and free coccoliths on the coagulation efficiency of *Emiliania huxleyi*. *Marine Chemistry*, 175, 5–17. https://doi.org/10.1016/j.marchem.2015.04.010.

Chowdhury, H., Chowdhury, T., & Sait, S.M. (2021). Estimating marine plastic pollution from COVID-19 face masks in coastal regions. *Marine Pollution Bulletin*, 168, 112419. https://doi.org/10.1016/J.MARPOLBUL.2021.112419.

Cole, M., Lindeque, P., Fileman, E., Halsband, C., Goodhead, R., Moger, J., & Galloway, T.S. (2013). Microplastic ingestion by zooplankton. *Environmental Science & Technology,* 47(12), 6646–6655. https://doi.org/10.1021/es400663f.

Collicutt, B., Juanes, F., & Dudas, S.E. (2019). Microplastics in juvenile Chinook salmon and their nearshore environments on the east coast of Vancouver Island. *Environmental Pollution,* 244, 135–142. https://doi.org/10.1016/j.envpol.2018.09.137.

Costa, E., Piazza, V., Lavorano, S., Faimali, M., Garaventa, F., & Gambardella, C. (2020). Trophic transfer of microplastics from copepods to jellyfish in the marine environment. *Frontiers in Environmental Science*, 4, 158. https://doi.org/10.3389/fenvs.2020.571732.

Cózar, A., Echevarría, F., González-Gordillo, J.I., Irigoien, X., Ubeda, B., Hernández-León, S. et al. (2014). Plastic debris in the open ocean. *Proceedings of the National Academy of Sciences*, USA, 111(28), 10239-10244. https://doi.org/10.1073/pnas.1314705111.

Cunha, C., Faria, M., Nogueira, N., Ferreira, A., & Cordeiro, N. (2019). Marine vs freshwater microalgae exopolymers as biosolutions to microplastics pollution. *Environmental Pollution*,249, 372–380. https://doi.org/10.1016/j.envpol.2019.03.046.

Danso, D., Chow, J., & Streit, W.R. (2019). Plastics: Microbial degradation, environmental and biotechnological perspectives. *Applied and Environmental Microbiology,* 85(19). https://doi.org/10.1128/AEM.01095-19.

De Monte, C., Locritani, M., Merlino, S., Ricci, L., Pistolesi, A., & Bronco, S. (2022). An *in situ* experiment to evaluate the aging and degradation phenomena induced by marine environment conditions on commercial plastic granules. *Polymers,* 14, 1111. https://doi.org/10.3390/polym14061111.

Dris, R., Gasperi, J., Rocher, V., Saad, M., Renault, N., & Tassin, B. (2015). Microplastic contamination in an urban area: A case study in Greater Paris. *Environmental Chemistry,* 12(5), 592–599. https://doi.org/10.1071/EN14167.

Du, Y., Liu, X., Dong, X., & Yin, Z. (2022). A review on marine plastisphere: Biodiversity, formation, and role in degradation. *Computational and Structural Biotechnology Journal,* 20, 975–988. https://doi.org/10.1016/j.csbj.2022.02.008.

Dudek, K.L., Cruz, B.N., Polidoro, B., & Neuer, S. (2020). Microbial colonization of microplastics in the Caribbean Sea. *Limnology and Oceanography Letters*, 5, 5–17. https://doi.org/10.1002/lol2.10141.

Eckert, E.M., Di Cesare, A., Kettner, M.T., Arias-Andres, M., Fontaneto, D., Grossart, H.P., & Corno, G. (2018). Microplastics increase impact of treated wastewater on freshwater microbial community. *Environmental Pollution,* 234, 495–502. https://doi.org/10.1016/j.envpol.2017.11.070.

Elkhatib, D., Oyanedel-Craver, V., & Carissimi, E. (2021). Electrocoagulation applied for the removal of microplastics from wastewater treatment facilities. *Separation and Purification Technology,* 276, 118877. https://doi.org/10.1016/j.seppur.2021.118877.

Enfrin, M., Dumée, L., & Lee, J. (2019). Nano/microplastics in water and wastewater treatment processes: Origin, impact, and potential solutions. *Water Research,* 161, 621–638. https://doi.org/10.1016/j.watres.2019.06.049.

Engel, A. (2000). The role of transparent exopolymer particles (TEP) in the increase in apparent particle stickiness (alpha) during the decline of a diatom bloom. *Journal of Plankton Research,* 22(3), 485–497. https://doi.org/10.1093/plankt/22.3.485.

Eriksen, M., Lebreton, L.C., Carson, H.S., Thiel, M., Moore, C.J., Borrero, J.C. et al. (2014). Plastic pollution in the world's oceans: More than 5 trillion plastic pieces weighing over 250,000 tons afloat at sea. *PLoS One,* 9(12), e111913. https://doi.org/10.1371/journal.pone.0111913.

Eryaşar, A.R., Gedik, K., & Mutlu, T. (2022). Ingestion of microplastics by commercial fish species from the southern Black Sea coast. *Marine Pollution Bulletin,* 177, 113535. https://doi.org/10.1016/j.marpolbul.2022.113535.

Fan, T., Zhao, J., Chen, Y., Wang, M., Wang, X., Wang, S. et al. (2021). Coexistence and adsorption properties of heavy metals by polypropylene microplastics. *Adsorption Science and Technology*, 4938749. https://doi.org/10.1155/2021/4938749.

FAO. (2022). *The State of World Fisheries and Aquaculture 2022.* Available from: https://www.fao.org/3/cc0461en/online/sofia/2022/world-fisheries-aquaculture.html. Accessed on January 06, 2024.

Fazey, F.M.C., & Ryan, P.G. (2016). Biofouling on buoyant marine plastic: An experimental study into the effect of size on surface longevity. Environmental Pollution, 210, 254–363. https://doi.org/10.1016/j.envpol.2016.01.026

Fischer, R., Lobelle, D., Kooi, M., Koelmans, A., Onink, V., Laufkötter, C. et al. (2022). Modelling submerged biofouled microplastics and their vertical trajectories. *Biogeosciences,* 19, 2211–2234, https://doi.org/10.5194/bg-19-2211-2022, 2022.

Flemming, H.C., & Wingender, J. (2010). The biofilm matrix. *Nature Reviews Microbiology,* 8(9), 623–633. https://doi.org/10.1038/nrmicro2415.

Garnier, Y., Jacob, H., Guerra, A.S., Bertucci, F., & Lecchini, D. (2019). Evaluation of microplastic ingestion by tropical fish from Moorea Island, French Polynesia. *Marine Pollution Bulletin,* 140, 165–170. https://doi.org/10.1016/j.marpolbul.2019.01.038.

Gao, R., Liu, R., & Sun, C. (2022). A marine fungus *Alternaria alternata* FB1 efficiently degrades polyethylene. *Journal of Hazardous Materials*, 431, 128617. https://doi.org/10.1016/j.jhazmat.2022.128617.

Gerritse, J., Leslie, H.A., de Tender, C.A., Devriese, L.I., & Vethaak, A.D. (2020). Fragmentation of plastic objects in a laboratory seawater microcosm. *Scientific Reports,* 10, 10945. https://doi.org/10.1038/s41598-020-67927-1.

Gedik, K., Eryaşar, A.R., & Gözler, A.M. (2022). The microplastic pattern of wild-caught Mediterranean mussels from the Marmara Sea. *Marine Pollution Bulletin,* 175, 113331. https://doi.org/10.1016/j.marpolbul.2022.113331.

Geyer, R. (2020). Production, use, and fate of synthetic polymers. *In:* T.M. Letcher (Ed.), *Plastic Waste and Recycling: Environmental Impact, Societal Issues, Prevention, and Solutions* (pp. 13–32). Academic Press, London. https://doi.org/10.1016/B978-0-12-817880-5.00002-5.

Ghatge, S., Yang, Y., Ahn, J.H., & Hur H.G. (2020). Biodegradation of polyethylene: A brief review. *Applied Biological Chemistry,* 63, 27. https://doi.org/10.1186/s13765-020-00511-3.

Ghosh, G.C., Akter, S.M., Islam, R.M., Habib, A., Chakraborty, T.K., Zaman, S. et al. (2021). Microplastics contamination in commercial marine fish from the Bay of Bengal. *Regional Studies in Marine Science,* 44, 101728. https://doi.org/10.1016/j.rsma.2021.101728.

Gong, M., Yang, G., Zhuang, L., & Zeng, E.Y. (2019). Microbial biofilm formation and community structure on low-density polyethylene microparticles in lake water microcosms. *Environmental Pollution,* 252(Part A), 94–102. https://doi.org/10.1016/j.envpol.2019.05.090.

González-Pleiter, M., Tamayo-Belda, M., Pulido-Reyes, G., Amariei, G., Leganés, F., Rosal, R. & Fernández-Piñas, F. (2019). Secondary nanoplastics released from a biodegradable microplastic severely impact freshwater environments. *Environmental Science: Nano,* 6, 1382–1392. https://doi.org/10.1039/C8EN01427B.

Grbic, J., Nguyen, B., Guo, E., You, J.B., Sinton, D., & Rochman, C.M. (2019). Magnetic extraction of microplastics from environmental samples. *Environmental Science & Technology Letters,* 6(2), 68–72. https://doi.org/ 10.1021/acs.estlett.8b00671.

Green, D.S., Boots, B., Sigwart, J., Jiang, S., & Rocha, C. (2016). Effects of conventional and biodegradable microplastics on a marine ecosystem engineer (*Arenicola marina*) and sediment nutrient cycling. *Environmental Pollution,* 208, 426–434. https://doi.org/10.1016/j.envpol.2015.10.010.

Gutow, L., Eckerlebe, A., Gimenez, L., & Saborowski, R. (2015). Experimental evaluation of seaweeds as a vector for microplastics into marine food webs. *Environmental Science & Technology,* 50(2), 915–923. https://doi.org/10.1021/acs.est.5b02431.

Habib, R., Al Kendi, R., & Thiemann, T. (2021). The effect of wastewater treatment plants on retainment of plastic microparticles to enhance water quality: A review. *Journal of Environmental Protection,* 12, 161– 195. https://doi.org/10.4236/jep.2021.123011.

Halstead, J.E., Smith, J.A., Carter, E.A., Lay, P.A., & Johnston, E.L. (2018). Assessment tools for microplastics and natural fibres ingested by fish in an urbanised estuary. *Environmental Pollution,* 234, 552–561. https://doi.org/10.1016/j.envpol.2017.11.085.

Harrison, P., Sapp, M., Schratzberger, M., & Osborn, A.M. (2011). Interactions between microorganisms and marine microplastics: A call for research. *Marine Technology Society Journal,* 45(2), 12–20. https://doi.org/10.4031/MTSJ.45.2.2.

Hawke, A.M., Trujillo, J.E., Oey, I., Giteru, S.G., & Allan, B.J.M. (2024). Exposure to petroleum-derived and biopolymer microplastics affect fast start escape performance and aerobic metabolism in a marine fish. *Science of The Total Environment,* 906, 167423. https://doi.org/10.1016/j.scitotenv.2023.167423.

Herbort, A.F., Sturm, M.T., Fiedler, S., Abkai G., & Schuhen K. (2018a). Alkoxy-silyl induced agglomeration: A new approach for the sustainable removal of microplastic from aquatic systems. *Journal of Polymers and the Environment,* 26, 4258–4270. https://doi.org/10.1007/s10924-018-1287-3.

Herbort, A.F., Sturm, M.T., & Schuhen, K. (2018b). A new approach for the agglomeration and subsequent removal of polyethylene, polypropylene, and mixtures of both from freshwater systems: A case study. *Environmental Science and Pollution Research,* 25(1–3), 15226–15234. https://doi.org/10.1007/s11356-018-1981-7.

Hidayaturrahman, H., & Lee, T. (2019). A study on characteristics of microplastic in wastewater of South Korea: Identification, quantification, and fate of microplastics during treatment process. *Marine Pollution Bulletin,* 146, 696–702. https://doi.org/10.1016/ j.marpolbul.2019.06.071.

Hochberg, E.J., & Gierach, M. (2021). Missing the reef for the corals: Unexpected trends between coral reef condition and the environment at the ecosystem scale. *Frontiers in Marine Science,* 8, 727038. https://doi.org/10.3389/fmars.2021.727038.

Hsieh, L., He, L., Zhang, M., Lv, W., Yang, K., & Tong, M. (2022). Addition of biochar as thin preamble layer into sand filtration columns could improve the microplastics removal from water. *Water Research,* 21, 118783. https://doi.org/ 10.1016/j.watres.2022.118783.

Huang, J.S., Koongolla, J.B., Li, H.X., Lin, L., Pan, Y.F., Liu, S. et al. (2020). Microplastic accumulation in fish from Zhanjiang mangrove wetland, South China. *Science of The Total Environment,* 708, 134839. https://doi.org/10.1016/j.scitotenv.2019.134839.

Huang, W., Chen, M., Song, B., Deng, J., Shen, M., Chen, Q. et al. (2021). Microplastics in the coral reefs and their potential impacts on corals: A mini-review. *Science of The Total Environment,* 762, 143112. https://doi.org/10.1016/j.scitotenv.2020.143112.

Iñiguez, M.E., Conesa, J.A., & Fullana, A. (2017). Microplastics in Spanish table salt. *Scientific Reports,* 7, 8620. https://doi.org/10.1038/s41598-017-09128-x.

Issac, M.N., & Kandasubramanian, B. (2021). Effect of microplastics in water and aquatic systems. *Environmental Science and Pollution Research,* 28, 19544–19562. https://doi.org/10.1007/s11356-021-13184-2.

Ivanov, V., & Stabnikov, V. (2020). Environmental safety of biotechnological materials and processes. *In:* F. Pacheco-Torgal, V. Ivanov, & D.C.W. Tsang (Eds.), *Bio-based Materials and Biotechnologies for Eco-efficient Construction* (pp. 359–376). Woodhead Publishing Series in Civil and Structural Engineering (1st Edn.). Elsevier, United States. https://doi.org/10.1016/B978-0-12-819481-2.00017-9.

Jalón-Rojas, I., Romero-Ramírez, A., Fauquembergue, K., Rossignol, L., Cachot, J., Sous,

D., & Morin, B. (2022). Effects of biofilms and particle physical properties on the rising and settling velocities of microplastic fibers and sheets. *Environmental Science & Technology,* 56(12), 8114-8123. https://doi.org/10.1021/acs.est.2c01302.

Jiang, M., Hu, L., Lu, A., Liang, G., Lin, Z., Zhang, T., & Gong, W. (2020). Strong sorption of two fungicides onto biodegradable microplastics with emphasis on the negligible role of environmental factors. *Environmental Pollution,* 267, 115496. https://doi.org/10.1016/j.envpol.2020.115496.

Kaiser, D., Estelmann, A., Kowalski, N., Glockzin, M., & Waniek, J.J. (2019). Sinking velocity of sub-millimeter microplastic. *Marine Pollution Bulletin,* 139, 214–220. https://doi.org/10.1016/j.marpolbul.2018.12.0.

Kaiser, D., Kowalski, N., & Waniek, J.J. (2017). Effects of biofouling on the sinking behavior of microplastics. *Environmental Research Letters,* 12(12), 124003. https://doi.org/10.1088/1748-9326/aa8e8b.

Kalčíková, G. (2023). Beyond ingestion: Adhesion of microplastics to aquatic organisms. *Aquatic Toxicology,* 258, 106480. https://doi.org/ 10.1016/j.aquatox.2023.106480.

Kane, I.A., Clare, M.A., Miramontes, E., Wogelius, R., Rothwell, J.J., Garreau, P., & Pohl, F. (2020). Seafloor microplastic hotspots controlled by deep-sea circulation. *Science,* 368(6495), 1140-1145. https://doi.org/10.1126/science.aba5899.

Klavins, M., Klavins, L., Stabnikova, O., Stabnikov, V., Marynin, A., Ansone-Bertina, L. et al. (2022). Interaction between microplastics and pharmaceuticals depending on the composition of aquatic environment. *Microplastics,* 1, 520–535. https://doi.org/10.3390/microplastics1030037.

Kim, K.T., & Park, S. (2021). Enhancing microplastics removal from wastewater using and granule-activated carbon with thermal regeneration. *Processes,* 9(4), 617. https://doi.org/10.3390/pr9040617.

Kirstein, I.V., Wichels, A., Krohne, G., & Gerdts, G. (2018). Mature biofilm communities on synthetic polymers in seawater: Specific or general? *Marine Environmental Research,* 142, 147–154. https://doi.org/10.1016/j.marenvres.2018.09.028.

Kolandhasamy, P., Su, L., Li, J., Qu, X., Jabeen, K., & Shi, H. (2018). Adherence of microplastics to soft tissue of mussels: A novel way to uptake microplastics beyond ingestion. *Science of The Total Environment,* 610–611, 635–640. https://doi.org/10.1016/j.scitotenv.2017.08.053.

Kooi, M., Van Nes, E.H., Schefer, M. & Koelmans, A.A. (2017). Ups and downs in the ocean: Effects of biofouling on vertical transport of microplastics. *Environmental Science and Technology,* 51(14), 7963–7971. https://doi.org/ 10.1021/acs.est.6b04702.

Kouzuma, A., & Watanabe, K. (2015). Exploring the potential of algae/bacteria interactions. *Current Opinion in Biotechnology,* 33, 125–129. https://doi.org/10.1016/j.copbio.2015.02.007.

Krasowska, K., Heimowska, A., & Rutkowska, M. (2015). Environmental degradability of polyurethanes. *In:* C.K. Das (Ed.), *Thermoplastic Elastomers: Synthesis and Applications*. InTech. https://doi.org/10.5772/60925.

Kvale, K., Prowe, A.E.F., Chien, CT. Landolfi, A., & Oschlies, A. (2020). The global biological microplastic particle sink. *Scientific Reports,* 10(1), 16670. https://doi.org/10.1038/s41598-020-72898-4.

Kumari, A., Chaudhary, D.R. & Jha, B. (2019). Destabilization of polyethylene and polyvinyl chloride structure by marine bacterial strain. *Environmental Science and Pollution Research,* 26, 1507–1516. https://doi.org/10.1007/s11356-018-3465-1.

Lagarde, F., Olivier, O., Zanella, M., Daniel, P., Hiard, S., & Caruso, A. (2016). Microplastic interactions with freshwater microalgae: Hetero-aggregation and changes in plastic density appear strongly dependent on polymer type. *Environmental Pollution,* 215, 331–339. https://doi.org/10.1016/j.envpol.2016.05.006.

LaRue, R.J., Patterson, B., O'Brien, S., & Latulippe, D.R. (2022). Evaluation of membrane fouling by microplastic particles in tertiary wastewater treatment processes. *ACS EST Water,* 2(6), 955–966. https://doi.org/10.1021/acsestwater.1c00430.

Latva, M., Dedman, C.J., Wright, R.J., Polin, M., & Christie-Oleza, J.A. (2022). Microbial pioneers of plastic colonisation in coastal seawaters. *Marine Pollution Bulletin*, 179, 113701. https://doi.org/10.1016/j.marpolbul.2022.113701.

Lebreton, L.C.M., van der Zwet, J., Damsteeg, J.W., Slat, B., Andrady, A., & Reisser, J. (2017). River plastic emissions to the world's oceans. *Nature Communications,* 8, 15611. https://doi.org/10.1038/ncomms15611.

Leiser, R., Schumann, M., Dadi, T., & Wendt-Potthoff, K. (2021a). Burial of microplastics in freshwater sediments facilitated by iron-organo flocs. *Scientific Reports,* 11(1), 24072. https://doi.org/10.1038/s41598-021-02748-4.

Leiser, R., Jongsma, R., Bakenhus, I., Möckel, R., Philipp, B., Neu, T.R., & Wendt-Potthoff, K. (2021b). Interaction of cyanobacteria with calcium facilitates the sedimentation of microplastics in a eutrophic reservoir. *Water Research,* 189, 116582. https://doi.org/10.1016/j.watres.2020.116582.

Leslie, H.A., Brandsma, S.H., van Velzen, M.J.M., & Vethaak, A.D. (2017). Microplastics en route: Field measurements in the Dutch river delta and Amsterdam canals, wastewater treatment plants, North Sea sediments and biota. *Environment International,* 101, 33–142. https://doi.org/10.1016/j.envint.2017.01.018.

Lestari, P., Trihadiningrum, Y., & Warmadewanthi, I.D.A.A. (2022). Simulated degradation of low-density polyethylene and polypropylene due to ultraviolet radiation and water velocity in the aquatic environment. *Journal of Environmental Chemical Engineering,* 10(3), 107553. https://doi.org/10.1016/j.jece.2022.107553.

Li, J., Yang, D., Li, L., Jabeen, K., & Shi, H. (2015). Microplastics in commercial bivalves from China. *Environmental Pollution,* 207, 190–195. https://doi.org/10.1016/j.envpol.2015.09.018.

Li, L., Xu, G., Yu, H. & Xing, J. (2018). Dynamic membrane for micro-particle removal in wastewater treatment: Performance and influencing factors. *The Science of The Total Environment,* 627, 332–340. https://doi.org/10.1016/j.scitotenv.2018.01.239.

Li, N., Wang, P., Wang, S., Wang, C., Zhou, H., Kapur, S. et al. (2022). Electrostatic charges on microalgae surface: Mechanism and applications. *Journal of Environmental Chemical Engineering,* 10(3), 107516. https://doi.org/10.1016/j.jece.2022.107516.

Li, S., Keenan, J.I., Shaw, I.C., & Frizelle, F.A. (2023). Could microplastics be a driver for early onset colorectal cancer? *Cancers*, 15(13), 3323. https://doi.org/10.3390/cancers15133323.

Liu, S., Huang, Y., Luo, D., Wang, X., Wang, Z., Ji, X. et al. (2022). Integrated effects of polymer type, size, and shape on the sinking dynamics of biofouled microplastics. *Water Research,* 220, 118656. https://doi.org/10.1016/j.watres.2022.118656.

Long, M., Moriceau, B., Gallinari, M., Lambert, C., Huvet, A., Raffray, J. & Soudant, P. (2015). Interactions between microplastics and phytoplankton aggregates: Impact on their respective fates. *Marine Chemistry,* 175, 39-46. https://doi.org/10.1016/j.marchem.2015.04.003.

Lopez, A.G., Najjar, R.G., Friedrichs, M.A.M., Hickner, M.A., & Wardrop, D.H. (2021). Estuaries as filters for riverine microplastics: Simulations in a large, coastal-plain estuary. *Frontiers in Marine Science,* 26, 715924. https://doi.org/10.3389/fmars.2021.715924.

Lusher, A., Mchugh, M., & Thompson, R. (2013). Occurrence of microplastics in the gastrointestinal tract of pelagic and demersal fish from the English Channel. *Marine Pollution Bulletin,* 67(1–2), 94–99. https://doi.org/10.1016/j.marpolbul.2012.11.028.

Ma, B., Xue, W., Ding, Y., Hu, C., Liu, H., & Qu, J. (2019). Removal characteristics of microplastics by Fe-based coagulants during drinking water treatment. *Journal of Environmental Science,* 78, 267–275. https://doi.org/10.1016/j.jes.2018.10.006.

Magnusson, K., & Noren, F. (2014). *Screening of Microplastic Particles in and Downstream a Wastewater Treatment Plant. Report C55.* Swedish Environmental Research Institute: Stockholm. https://www.diva-portal.org/smash/get/diva2:773505/FULLTEXT01.pdf.

Mahmud, A., Wasif, M.M., Roy, H., Mehnaz, F., Ahmed, T., Pervez, M.N. et al. (2022). Aquatic microplastic pollution control strategies: Sustainable degradation techniques, resource recovery, and recommendations for Bangladesh. *Water,* 14(23), 3968. https://doi.org/10.3390/w14233968.

Mahon, A.M., O'Connell, B., Healy, M.G., O'Connor, I., Officer, R., Nash, R., & Morrison, L. (2017). Microplastics in sewage sludge: Effects of treatment. *Environmental Science and Technology,* 51(2), 810–818. https://doi.org/10.1021/acs.est.6b04048.

Martin, C., Corona, E., Mahadik, G.A., & Duarte, C.M. (2019). Adhesion to coral surface as a potential sink for marine microplastics. *Environmental Pollution,* 255(Part 2), 113281. https://doi.org/10.1016/j.envpol.2019.113281.

Mason, S.A., Garneau, D., Sutton, R., Chu, Y., Ehmann, K., Barnes, J. et al. (2016). Microplastic pollution is widely detected in US municipal wastewater treatment plant effluent. *Environmental Pollution,* 218, 1045–1054. https://doi.org/10.1016/j.envpol.2016.08.056.

Mateos-Cárdenas, A., Scott, D.T., Seitmaganbetova, G., van Pelt, F.N.A.M., O'Halloran, J., & Jansen, M.A.K. (2019). Polyethylene microplastics adhere to *Lemna minor* (L.), yet have no effects on plant growth or feeding by *Gammarus duebeni* (Lillj.). *Science of The Total Environment,* 689, 413–421. https://doi.org/ 10.1016/j.scitotenv.2019.06.359.

Matsuguma, Y., Takada, H., Kumata, H., Kanke, H., Sakurai, S., Suzuki, T. et al. (2017). Microplastics in sediment cores from Asia and Africa as indicators of temporal trends in plastic pollution. *Archives of Environmental Contamination and Toxicology,* 73, 230–239. https://doi.org/10.1007/s00244-017-0414-9.

Mayali, X. (2018). Editorial: Metabolic interactions between bacteria and phytoplankton. *Frontiers in Microbiology,* 9, 727. https://doi.org/10.3389/fmicb.2018.00727.

McAdam, B., Brennan Fournet, M., McDonald, P., & Mojicevic, M. (2020). Production of polyhydroxybutyrate (PHB) and factors impacting its chemical and mechanical characteristics. *Polymers*, 12(12), 2908. https://doi.org/10.3390/polym12122908.

Menicagli, V., Castiglione, M.R., Balestri, E., Giorgetti, L., Bottega, S., Sorce, C. et al. (2022). Early evidence of the impacts of microplastic and nanoplastic pollution on the growth and physiology of the seagrass *Cymodocea nodosa. Science of the Total Environment,* 838(Pt 3), 156514. https://doi.org/ 10.1016/j.scitotenv.2022.156514.

Miao, L., Gao, Y., Adyel, T. M., Huo, Z., Liu, Z., Wu, J., & Hou, J. (2021). Effects of biofilm colonization on the sinking of microplastics in three freshwater environments. *Journal of Hazardous Materials,* 413, 125370. https://doi.org/10.1016/j.jhazmat.2021.125370.

Michels, J., Stippkugel, A., Lenz, M., Wirtz, K., & Engel, A. (2018). Rapid aggregation of biofilm-covered microplastics with marine biogenic particles. *Proceedings of the Royal Society B: Biological Sciences*, 285(1885), 20181203. https://doi.org/10.1098/rspb.2018.1203.

Mintenig, S.M., Int-Veen, I., Löder, M.G.J., Primpke, S., & Gerdts, G. (2017). Identification of microplastic in effluents of waste water treatment plants using focal plane array-based micro-Fourier-transform infrared imaging. *Water Research,* 108(1), 365–372. https://doi.org/10.1016/j.watres.2016.11.015.

Mohanrasu, K., Premnath, N., Prakash, G.S., Sudhakar, M., Boobalan, T., & Arun, A. (2018). Exploring multi potential uses of marine bacteria: An integrated approach for PHB production, PAHs, and polyethylene biodegradation. *Journal of Photochemistry and Photobiology, B: Biology,* 185, 55–65. https://doi.org/10.1016/j.jphotobiol.2018.05.014.

Murphy, F., Ewins, C., Carbonnier, F., & Quinn, B. (2016). Wastewater treatment works (WwTW) as a source of microplastics in the aquatic environment. *Environmental Science and Technology,* 50(11), 5800–5808. https://doi.org/10.1021/acs.est.5b05416.

Muthukumar, T., Aravinthan, A., Lakshmi, K., Venkatesan, R., Vedaprakash, L., & Doble, M. (2011). Fouling and stability of polymers and composites in marine environment. *International Biodeterioration & Biodegradation,* 65, 276–284. https://doi.org/10.1016/j.ibiod.2010.11.012.

Ni, B.J., Zhu, Z.R., Li, W.H., Yan, X., Wei, W., Xu, Q. et al. (2020). Microplastics mitigation in sewage sludge through pyrolysis: The role of pyrolysis temperature. *Environmental Science and Technology Letters,* 7(12), 961–967. http://hdl.handle.net/10453/146730.

Nolte, T.M., Hartmann, N.B., Kleijn, J.M., Garnæs, J., van de Meent, D., Hendriks, J.A., & Baun, A. (2017). The toxicity of plastic nanoparticles to green algae as influenced by surface modification, medium hardness, and cellular adsorption. *Aquatic Toxicology,* 183, 11–20. https://doi.org/10.1016/j.aquatox.2016.12.005.

Oberbeckmann, S., Loeder, M.G.J., Gerdts, G., & Osborn, A.M. (2014). Spatial and seasonal variation in diversity and structure of microbial biofilms on marine plastics in Northern European waters. *FEMS Microbiology Ecology,* 90(2), 478–492. https://doi.org/10.1111/1574-6941.12409.

Oberbeckmann, S., Osborn, A.M., & Duhaime, M.B. (2016). Microbes on a bottle: Substrate, season, and geography influence community composition of microbes colonizing marine plastic debris. *PLoS ONE,* 11(8), e0159289. https://doi.org/10.1371/journal.

Oberbeckmann, S., & Labrenz, M. (2020). Marine microbial assemblages on microplastics: Diversity, adaptation, and role in degradation. *Annual Review of Marine Science*, 12, 209–232. https://www.annualreviews.org/doi/10.1146/annurev-marine-010419-010633.

Ogunola, S.O., Reis-Santos, P., Wootton, N., & Gillanders, B.M. (2022). Microplastics in decapod crustaceans sourced from Australian seafood markets. *Marine Pollution Bulletin,* 179, 113706. https://doi.org/10.1016/j.marpolbul.2022.113706.

Paço, A., Duarte, K., da Costa, J.P., Santos, P.S.M., Pereira, R., Pereira, M.E. et al. (2017). Biodegradation of polyethylene microplastics by the marine fungus *Zalerion maritimum*. *Science of The Total Environment,* 586, 10–15. https://doi.org/10.1016/j.scitotenv.2017.02.017.

Padervand, M., Lichtfouse, E., Robert D., & Wang, C. (2020). Removal of microplastics from the environment: A review. *Environmental Chemistry Letters,* 18(3), 807–828. https://hal.archives-ouvertes.fr/hal-02562545.

Perren, W., Wojtasik, A., & Cai, Q. (2018). Removal of microbeads from wastewater using electrocoagulation. *ACS Omega,* 3(3), 3357–3364. https://pubs.acs.org/doi/10.1021/acsomega.7b02037.

Petroody, S.S.A., Hashemi, S.H., & van Gestel, C.A.M. (2021). Transport and accumulation of microplastics through wastewater treatment sludge processes. *Chemosphere,* 278, 130471. https://doi.org/10.1016/j.chemosphere.2021.130471.

Porter, A., Lyons, B.P., Galloway, T.S., & Lewis, C. (2018). Role of marine snows in microplastic fate and bioavailability. *Environmental Science & Technology,* 52(12), 7111–7119. https://doi.org/10.1021/acs.est.8b01000.

Piarulli, S., & Airoldi, L. (2020). Mussels facilitate the sinking of microplastics to bottom sediments and their subsequent uptake by detritus-feeders. *Environmental Pollution,* 266, 115151. https://doi.org/10.1016/j.envpol.2020.115151.

Pivokonsky, M., Cermakova, L., Novotna, K., Peer, P., Cajthaml, T., & Janda, V. (2018). Occurrence of microplastics in raw and treated drinking water. *Science of the Total Environment,* 643, 644–651. https://doi.org/10.1016/j.scitotenv.2018.08.102.

Qiu, M., Liu, L., Ling, Q., Cai, Y., Yu, S., Wang, S. et al. (2022). Biochar for the removal of contaminants from soil and water: A review. *Biochar,* 4, 19. https://doi.org/10.1007/s42773-022-00146-1.

Rajala, K., Grönfors, O., Hesampour, M., & Mikola, A. (2020). Removal of microplastics from secondary wastewater treatment plant effluent by coagulation/flocculation with iron, aluminum, and polyamine-based chemicals. *Water Research,* 183, 116045. https://doi.org/10.1016/j.watres.2020.116045.

Raju, S., Carbery, M., Kuttykattil, A., Senathirajah, K., Subashchandrabose, S.R., Evans, G.M., & Thavamani, P. (2018). Transport and fate of microplastics in wastewater treatment plants: Implications to environmental health. *Reviews in Environmental Science and Bio/Technology,* 17, 637–653. https://doi.org/10.1007/s11157-018-9480-3.

Reichert, J., Arnold, A.L., Hammer, N., Miller, I.B., Rades, M., Schubert, P. et al. (2021). Reef-building corals act as long-term sink for microplastic. *Global Change Biology,* 28, 33–45. https://doi.org/10.1111/gcb.15920.

Rochman, C.M., Tahir, A., Williams, S.L., Baxa, D.V., Lam, R., Miller, J.T. et al. (2015). Anthropogenic debris in seafood: Plastic debris and fibers from textiles in fish and bivalves sold for human consumption. *Scientific Reports,* 5, 14340. https://doi.org/10.1038/srep14340.

Rom, M., Fabia, J., Grübel, K., Sarna, E., Graczyk, T., & Janicki, J. (2017). Study of the biodegradability of polylactide fibers in wastewater treatment processes. Polimery, 62(11–12), 834–840. https://doi.org/10.14314/polimery.2017.834.

Rotjan, R.D., Sharp, K.H., Gauthier, A.E., Yelton, R., Lopez, E.M.B., Carilli, J., & Urban-Rich, J. (2019). Patterns, dynamics and consequences of microplastic ingestion by the temperate coral, *Astrangia poculata. Proceedings of the Royal Society B: Biological Sciences,* 286(1905), 20190726. https://doi.org/10.1098/rspb.2019.0726.

Rozman, U., Blažič, A., & Kalčíková, G. (2023). Phytoremediation: A promising approach to remove microplastics from the aquatic environment. *Environmental Pollution,* 338, 122690. https://doi.org/10.1016/j.envpol.2023.122690.

Rummel, C.D., Jahnke, A., Gorokhova, E., Kühnel, D., & Schmitt-Jansen, M. (2017). Impacts of biofilm formation on the fate and potential effects of microplastic in the aquatic environment. *Environmental Science & Technology Letters,* 4, 258–267. https://doi.org/10.1021/acs.estlett.7b00164.

Rutkowska, M., Heimowska, A., Krasowska, K., & Janik, H. (2002). Biodegradability of polyethylene starch blends in sea water. *Polish Journal of Environmental Studies,* 11(3), 267–274. https://www.academia.edu/19029210/Biodegradability_of_Polyethylene_Starch_Blends_in_Sea_Water

Sacco, N.A., Zoppas, F.M., Devard, A., González-Muñoz, M.P., García, G., & Marchesini, F.A. (2023). Recent advances in microplastics removal from water with special attention given to photocatalytic degradation: Review of scientific research. *Microplastics,* 2(3), 278–303. https://doi.org/10.3390/microplastics2030023.

Schirinzi, G.F., Pérez-Pomeda, I., Sanchís, J., Rossini, C., Farré, M., & Barceló, D. (2017). Cytotoxic effects of commonly used nanomaterials and microplastics on cerebral and epithelial human cells. *Environmental Research,* 159, 579–587. https://doi.org/10.1016/j.envres.2017.08.043.

Schneiderman, E.T. (2015). Discharging microbeads to our waters: An examination of wastewater treatment plants in New York. http://www.ag.ny.gov/pdfs/2015_Microbeads_Report_FINAL.pdf.

Shen, M., Zhang, Y., Almatrafi, E., Hu, T., Zhou, C., Song, B. et al. (2022). Efficient removal of microplastics from wastewater by an electrocoagulation process. *Chemical Engineering Journal,* 428, 131161. https://doi.org/10.1016/j.cej.2021.131161.

Shi, X., Zhang, X., Gao, W., Zhang, Y., & He, D. (2022). Removal of microplastics from water by magnetic nano-Fe_3O_4. *Science of The Total Environment,* 802, 149838. https://doi.org/10.1016/j.scitotenv.2021.149838.

Siipola, V., Pflugmacher, S., Romar, H., Wendling, L., & Koukkari, P. (2020). Low-cost biochar adsorbents for water purification including microplastics removal. *Applied Sciences,* 10(3), 788. https://doi.org/10.3390/app10030788.

Singh, N., Khandelwal, N., Ganie, Z.A., Tiwari, E., & Darbha, G.K. (2021). Eco-friendly magnetic biochar: An effective trap for nanoplastics of varying surface functionality and size in the aqueous environment. *Chemical Engineering Journal,* 418, 129405. https://doi.org/10.1016/J.CEJ.2021.129405.

Smith, M., Love, D.C., Rochman, C.M., & Neff, R.A. (2018). Microplastics in seafood and the implications for human health. *Current Environmental Health Reports,* 5(3), 375-386. https://doi.org/ 10.1007/s40572-018-0206-z

Spacilova, M., Dytrych, P., Lexa, M., Wimmerova, L., Masin, P., Kvacek, R., & Solcova, O. (2023). An innovative sorption technology for removing microplastics from wastewater. *Water,* 15(5), 892. https://doi.org/10.3390/w15050892.

Stabnikov, V., Mikolayenko, R., & Stabnikov, D. (2023). Removal of microplastics from wastewater for world water pool protection. *Scientific Works of NUFT,* 29(6), 23–45. https://doi.org/10.24263/2225-2924-2023-29-6-4.

Stabnikova, O., Stabnikov, V., Marinin, A., Klavins, M., Klavins, L., & Vaseashta, A. (2021). Microbial life on the surface of microplastics in natural waters. *Applied Sciences,* 11(24), 11692. https://doi.org/10.3390/app112411692.

Stabnikova, O., Stabnikov, V., Marinin, A., Klavins, M., & Vaseashta, A. (2022). The role of microplastics biofilm in accumulation of trace metals in aquatic environment. *World Journal of Microbiology and Biotechnology,* 38, 117. https://doi.org/10.1007/s11274-022-03293-6.

Sudhakar, M., Trishul, A., Doble, M., Suresh Kumar, K., Syed Jahan, S., Inbakandan, D. et al. (2007). Biofouling and biodegradation of polyolefins in ocean waters. *Polymer Degradation and Stability,* 92(9), 1743–1752. https://doi.org/10.1016/j.polymdegradstab.2007.03.029.

Sudhakar, M., Doble, M., Murthy, P.S., & Venkatesan, R. (2008). Marine microbe-mediated biodegradation of low- and high-density polyethylenes. *International Biodeterioration & Biodegradation,* 61, 203–213. https://doi.org/10.1016/j.ibiod.

Sun, J., Dai, X., Wang, Q., van Loosdrecht, M.C.M., & Ni, B.J. (2019). Microplastics in wastewater treatment plants: Detection, occurrence, and removal. *Water Research,* 152, 21–37. https://doi.org/10.1016/j.watres.2018.12.050.

Sundbæk, K.B., Koch, I.D.W., Villaro, C.G., Rasmussen, N.S., Holdt, S.L., & Hartmann, N.B. (2018). Sorption of fluorescent polystyrene microplastic particles to edible seaweed *Fucus vesiculosus*. *Journal of Applied Phycology,* 30, 2923–2927. https://doi.org/10.1007/s10811-018-1472-8.

Sutherland, I.W. (2001). Biofilm exopolysaccharides: A strong and sticky framework. *Microbiology,* 147(1), 3–9 https://doi.org/10.1099/00221287-147-1-3.

Syranidou, E., Karkanorachaki, K., Amorotti, F., Avgeropoulos, A., Kolvenbach, B., Zhou, N. et al. (2019). Biodegradation of mixture of plastic films by tailored marine consortia. *Journal of Hazardous Materials,* 375, 33–42. https://doi.org/10.1016/j.jhazmat.2019.04.078.

Tagg, A.S., Sapp, M., Harrison, J.P., Sinclair, C.J., Bradley, E., Ju-Nam, Y. & Ojeda, J.J. (2020). Microplastic monitoring at different stages in a wastewater treatment plant using reflectance micro-FTIR imaging. *Frontiers in Environmental Science,* 8, 145. https://doi.org/10.3389/fenvs.2020.00145.

Talvitie, J., Mikola, A., Koistinen, A., & Setälä, O. (2017). Solutions to microplastic pollution: Removal of microplastics from wastewater effluent with advanced wastewater treatment technologies. *Water Research,* 123, 401–407. https://doi.org/10.1016/j.watres.2017.07.005.

Tao, S., Li, T., Li, M., Yang, S., Shen, M., & Liu, H. (2024). Research advances on the toxicity of biodegradable plastics derived micro/nanoplastics in the environment: A review. *Science of The Total Environment,* 916, 170299. https://doi.org/10.1016/j.scitotenv.2024.170299.

Thushari, G.G.N., & Senevirathna, J.D.M. (2020). Plastic pollution in the marine environment. *Heliyon,* 6, e04709. https://doi.org/10.1016/j.heliyon.2020.e04709.

Tian, C., Lv, J., Zhang, W., Wang, H., Chao, J., Chai, L., & Lin, Z. (2022). Accelerated degradation of microplastics at the liquid interface of ice crystals in frozen aqueous solutions. *Angewandte Chemie International Edition,* 61(31), e202206947. https://doi.org/10.1002/anie.202206947.

Tipping, E., Woof, C. & Cooke, D. (1981). Iron oxide from a seasonally anoxic lake. *Geochimica et Cosmochimica Acta,* 45(9), 1411–1419. https://doi.org/10.1016/0016-7037(81)90275-1.

Tomczyk, A., Sokołowska, Z. & Boguta, P. (2020). Biochar physicochemical properties: Pyrolysis temperature and feedstock kind effects. *Reviews in Environmental Science and Bio/Technology,* 19, 191–215. https://doi.org/10.1007/s11157-020-09523-3.

Tong, H., Jiang, Q., Hu, X., & Zhong, X. (2020). Occurrence and identification of microplastics in tap water from China. *Chemosphere*, 252, 126493. https://doi.org/10.1016/j.chemosphere.2020.126493.

Tong, M., He, L., Rong, H., Li, M., & Kim, H. (2020). Transport behaviors of plastic particles in saturated quartz sand without and with biochar/Fe_3O_4-biochar amendment. *Water Research,* 169, 115284. https://doi.org/10.1016/j.watres.2019.115284.

Triebskorn, R., Braunbeck, T., Grummt, T., Hanslik, L., Huppertsberg, S., Jekel, M. et al. (2018). Relevance of nano- and microplastics for freshwater ecosystems: A critical review. *TrAC Trends in Analytical Chemistry*, 110, 375–392. https://doi.org/10.1016/j.trac.2018.11.02.

Urbanek, A.K., Rymowicz, W. & Mirończuk, A.M. (2018). Degradation of plastics and plastic-degrading bacteria in cold marine habitats. *Applied Microbiology and Biotechnology*, 102, 7669–7678. https://doi.org/10.1007/s00253-018-9195-y.

Vaksmaa, A., Knittel, K., Abdala Asbun, A., Goudriaan, M., Ellrott, A., Witte, H.J. et al. (2021). Microbial communities on plastic polymers in the Mediterranean Sea. *Frontiers in Microbiology,* 12, 673553. https://doi.org/10.3389/fmicb.2021.673553.

Vaksmaa, A., Polerecky, L., Dombrowski, N., Kienhuis, M.V.M., Posthuma, I., Gerritse, J. et al. (2023). Polyethylene degradation and assimilation by the marine yeast *Rhodotorula mucilaginosa. ISME Communications,* 3(1), 68. https://doi.org/ 10.1038/s43705-023-00267-z.

Van Cauwenberghe, L., Vanreusel, A., Mees, J., & Janssen, C.R. (2013). Microplastic pollution in deep-sea sediments. *Environmental Pollution,* 182, 495–499. https://doi.org/10.1016/j.envpol.2013.08.013.

Vaseashta, A., Ivanov, V., Stabnikov, V., & Marinin, A. (2021). Environmental safety and security investigations of neustonic microplastic aggregates near water-air interphase. *Polish Journal of Environmental Studies,* 30(4), 3457–3469. https://doi.org/10.15244/pjoes/131947.

Vaseashta, A., Gevorgyan, G., Kavaz, D., Ivanov, O., Jawaid, M., & Vasović, D. (2021). Exposome, biomonitoring, assessment and data analytics to quantify universal water quality. *In:* Vaseashta, A., Maftei, C. (Eds.), *Water Safety, Security, and Sustainability: Advanced Sciences and Technologies for Security Applications*. Springer, Cham. https://doi.org/10.1007/978-3-030-76008-3_4.

Vaseashta, A., Duca, G., & Travin, S. (Eds.). (2022). *Handbook of Research on Water Sciences and Society*. IGI Global. https://doi.org/10.4018/978-1-7998-7356-3.

Vázquez-Rowe, I., Ita-Nagy, D., & Kahhat, R. (2021). Microplastics in fisheries and aquaculture: Implications to food sustainability and safety. *Current Opinion in Green and Sustainable Chemistry*, 29, 100464. https://doi.org/10.1016/j.cogsc.2021.100464.

Vidyasakar, A., Krishnakumar, S., Suresh Kumar, K., Neelavannan, K., Anbalagan, S., Kasilingam, K. et al. (2021). Microplastic contamination in edible sea salt from the largest salt-producing states of India. *Marine Pollution Bulletin,* 171, 112728. https://doi.org/10.1016/j.marpolbul.2021.112728.

Vuori, L., & Ollikainen, M. (2022). How to remove microplastics in wastewater? A cost-effectiveness analysis. *Ecological Economics,* 192, 107246. https://doi.org/10.1016/j.ecolecon.2021.107.

Waite, H.R., Donnelly, M.J., & Walters, L.J. (2018). Quantity and types of microplastics in the organic tissues of the eastern oyster *Crassostrea virginica* and Atlantic mud crab *Panopeus herbstii* from a Florida estuary. *Marine Pollution Bulletin,* 129(1), 179–185. https://doi.org/10.1016/j.marpolbul.2018.02.026.

Walkinshaw, C., Lindeque, P.K., Thompson, R., Tolhurst, T., & Cole, M. (2020). Microplastics and seafood: Lower trophic organisms at highest risk of contamination. *Ecotoxicology and Environmental Safety*, 190, 110066. https://doi.org/10.1016/j.ecoenv.2019.110066.

Wan, H., Wang, J., Sheng, X., Yan, J., Zhang, W., & Xu, Y. (2022). Removal of polystyrene microplastics from aqueous solution using the metal–organic framework material of ZIF-67. *Toxics,* 10(2), 70. https://doi.org/10.3390/toxics10020070.

Wang, J., Sun, C., Huang, Q.X., Chi, Y., & Yan, J.H. (2021). Adsorption and thermal degradation of microplastics from aqueous solutions by Mg/Zn modified magnetic biochars. *Journal of Hazardous Materials,* 419, 126486. https://doi.org/10.1016/j.jhazmat.2021.126486.

Wang, L., Käppler, A., Fischer, D., & Simmchen, J. (2019). Photocatalytic TiO_2 micromotors for removal of microplastics and suspended matter. *ACS Applied Materials & Interfaces,* 11(36), 32937–32944. https://doi.org/10.1021/acsami.9b06128.

Wang, Z., Sedghii, M., & Lea-Langton, A. (2020). Filtration of microplastic spheres by biochar: Removal efficiency and immobilisation mechanisms. *Water Research,* 184, 116165. https://doi.org/10.1016/j.watres.2020.116165.

Wang, Z., & Sedighi, M. (2023). Transport and retardation of microplastic in biochar. *In: Proceedings of the 9ICEG 9th International Congress on Environmental Geotechnics,* 25–28 June, 2023, Chania, Greece. https://doi.org/10.53243/ICEG2023-250.

Wayman, C., & Niemann, H. (2021). The fate of plastic in the ocean environment: A mini review. *Environmental Science: Processes & Impacts,* 23, 198–212. https://doi.org/10.1039/D0EM00446D

Woodall, L.C., Sanchez-Vidal, A., Canals, M., Paterson, G.L.J., Coppock, R., Sleight, V. et al. (2014). The deep sea is a major sink for microplastic debris. *Royal Society Open Science,* 1, 140317. https://doi.org/10.1098/rsos.140317.

Wootton, N., Ferreira, M., Reis-Santos, P., & Gillanders, B.M. (2021). A comparison of microplastic in fish from Australia and Fiji. *Frontiers in Marine Science,* 8, 690991. https://doi.org/10.3389/fmars.2021.690991.

Wu, J., Yang, C., Zhao, H., Shi, J., Liu, Z., Li, C., & Song, F. (2023). Efficient removal of microplastics from aqueous solution by a novel magnetic biochar: Performance, mechanism, and reusability. *Environmental Science and Pollution Research,* 30, 26914–26928. https://doi.org/10.1007/s11356-022-24130-1.

Zettler, E.R., Mincer, T.J., & Amaral Zettler, L.A. (2013). Life in the "plastisphere": Microbial communities on plastic marine debris. Environmental Science and Technology, 47(13), 7137–7146. https://doi.org/10.1021/es401288x

Zhang, K., & Wu, C. (2023). Airborne microplastics: Analysis, fate, and human health effects. *Comprehensive Analytical Chemistry,* 100, 1–16. https://doi.org/10.1016/bs.coac.2022.07.001.

Zhao, H., Huang, X., Wang, L., Zhao, X., Yan, F., Yang, Y. et al. (2022). Removal of polystyrene nanoplastics from aqueous solutions using a novel magnetic material: Adsorbability, mechanism, and reusability. *Chemical Engineering Journal,* 430(Part 4), 133122. https://doi.org/10.1016/j.cej.2021.133122.

Zhao, S., Danley, M., Ward, J.E., Li, D., & Mincer, T.J. (2017). An approach for extraction, characterization and quantitation of microplastic in natural marine snow using Raman microscopy. *Analytical Methods,* 9(9), 1470–1478. https://doi.org/ 10.1039/C6AY02302A.

Zhao, S., Zettler, E.R., Bos, R.P., Lin, P., Amaral-Zettler, L.A., & Mincer, T.J. (2022). Large quantities of small microplastics permeate the surface ocean to abyssal depths in the South Atlantic Gyre. *Global Change Biology,* 28(9), 2991–3006. https://doi.org/10.1111/gcb.16089.

Ziajahromi, S., Neale, P.A., Rintoul, L., & Leusch, F.D.L. (2017). Wastewater treatment plants as a pathway for microplastics: Development of a new approach to sample wastewater-based microplastics. *Water Research,* 112, 93–99. https://doi.org/10.1016/j.watres.2017.01.042.

3 | Properties and Related Effects of Microplastics in the Aquatic Environment: From the Organismic to Cellular Level

Michael Schwarzer, Julian Brehm, David Gitschier, Luca Liprandi, Anja FRM Ramsperger, Sven Ritschar, Jona Schmitt, Reetha Thomas, Simon Wieland, Martin GJ Löder, Holger Kress and Christian Laforsch*

Universität Bayreuth, Universitätsstraße 30, 95447 Bayreuth, Germany

Abstract: Plastics have become an essential part of our everyday life. Over the past decades, increasing plastic production has resulted in a substantial rise in plastic waste, which entered the environment through littering and inadequate waste management. Although the effects of larger plastic items in the environment have been studied for quite a while, research has been focusing on microplastics (MPs) – plastic particles smaller than 1 mm, originating from the fragmentation of larger plastic items, abrasion, or direct release from industrial applications – only since the past two decades. MP particles are a very diverse suite of particles with varying physicochemical properties, which differ in size, shape, polymer type, density, unintentionally or intentionally added chemical compounds or residual monomer content, aging state, surface charge, surface roughness, environmental coating, and biofouling as well as other physicochemical parameters. Once released, this complexity increases even further due to interactions with biotic and abiotic environmental factors. Abiotic interactions are primarily related to exposure to physical or chemical factors that lead to aging such as UV-radiation. Interactions with biotic factors range from the formation of an, depending on the particle size, eco-corona or biofilm on the surfaces of an MP particle to uptake by various organisms, which potentially also modifies their physicochemical properties. On the other hand, MPs have been shown to affect organisms at different biological levels, from the base of

*Corresponding author: christian.laforsch@uni-bayreuth.de

the food web up to top predators. The reported effects of MPs range from highly toxic to negligible, making it challenging to draw general conclusions about the hazards of MPs. To enlighten the variable results in the ecotoxicity of MPs, this chapter explores the wide range of MP physicochemical properties and their potential influence on interactions with and effects on organisms and cells.

1. Introduction

1.1. From the Success Story of Plastic Materials to the Dilemma of Plastic Waste

Modern society greatly benefits from plastic materials, and our modern life would not be imaginable without their invention (Andrady & Neal, 2009; Pilapitiya & Ratnayake, 2024). Plastic is relatively cheap, durable, mechanically robust, moldable in different applications, lightweight, and producible in constant quality. Combining those properties makes plastic suitable for various applications. It is, therefore, intensively used in all aspects of our everyday lives and in every sector of industry, clearly showing the success story of plastics.

In 2022, the global plastic production reached 400.3 Mt. In Europe, the production amount of plastics covers different sectors such as houseware, leisure, and sports activities (4.1%), agriculture and gardening (4.4%), electrical and electronic (5.7%), automotive (8.3%), 'others' like medical devices or furniture (15.9%), followed by building and construction (22.9%), and the highest share is contributed to packaging (39.0%) (PlasticsEurope, 2024). This widespread use of plastic materials results from advantageous properties compared to other materials. For instance, packaging materials made of plastic enable transporting goods while reducing transportation costs and emissions due to lower their weight compared to other materials. Further, its good barrier properties prolong the shelf lives of food items (Andrady & Neal, 2009; Robertson, 2016; White & Lockyer, 2020). Since plastics are used in various sectors and each sector demands a vast variety of material properties, a wide range of polymer types are produced. However, only a few polymers account for the largest share of plastic produced. In Europe, for instance, almost half (46.6%) of the numerous polymer types produced are shared between three polymer types: polyethylene (PE, in low and high density, 22.1%), polypropylene (PP, 15.4%), and polyvinylchloride (PVC, 9.1%). These are followed by polystyrene (PS, 5.4%), polyurethane (PUR, 5.2%), and polyethylene terephthalate (PET, 5.0%) (PlasticsEurope, 2024).

With the increasing production and use of plastic materials, particularly single-use plastics such as packaging, plastic waste is constantly increasing. Although the share of plastics captured by circular economy increased by 30% in 5 years (2018–2022) (PlasticsEurope 2024), it is assumed that since the 1960s, 6.3 billion tonnes of plastic waste have been produced, 79% of which has been released into the environment (Geyer et al., 2017). Either the waste ends up in landfills, and from there enters other environmental compartments, or it is intentionally or unintentionally introduced into terrestrial or aquatic ecosystems (Fagnani et al., 2021).

1.2. What are Microplastics, and How are They Distributed in the Environment?

Since the last two decades, small plastic particles, so-called MPs have gained awareness from science and the general public, and it has become evident that MPs abundantly and ubiquitously occur in the environment, exceeding the number of larger plastic items by far (Eriksen et al., 2014; Geyer et al., 2017). MPs are any plastic particles insoluble in water within a size range between 1–1000 μm in diameter, and plastic particles smaller than 1 μm are defined as nanoplastics (Hartmann et al., 2019). MPs can be intentionally produced in the respective size as the so-called primary MP and find their applications in various products like personal care and cosmetic products (Lin et al., 2023). However, the more significant fraction of MP found in the environment originates from the fragmentation of larger plastic debris or by abrasion of particles from plastic items during daily usage and is defined as secondary MP (Kasmuri et al., 2022; Wang et al., 2021). Meanwhile, the occurrence of MPs has been reported in marine, freshwater, and terrestrial systems (Coyle et al., 2020; Dris et al., 2020; Kernchen et al., 2022, 2024; Kokalj et al., 2024; Schrank et al., 2022; Wang et al., 2023). The presence of MPs in the environment does not necessarily correlate with local anthropogenic activities since MPs were found in remote areas like uninhabited islands, the poles, and even the deep sea (Imhof et al., 2017; Lacerda et al., 2019; Laforsch et al., 2021; Ross et al., 2021). Thus, over the last two decades, many efforts have been made to analyze the concentrations of MPs and understand their transportation routes and sinks (Laforsch et al., 2021). Here, atmospheric transport and aerial deposition are key to understanding the worldwide distribution of MPs and the contamination of remote places (Evangeliou et al., 2020). Regardless of whether MP particles are transported by wind or water currents, their transport route, deposition, accumulation, and resuspension strongly depend on their physicochemical properties (Dichgans et al., 2023; Elagami et al., 2023; Esders et al., 2024; Kernchen et al., 2024; Lehmann et al., 2023; Oehlschlägel et al., 2024). The transport and fate of MPs in the environment further determine whether and to which type of MP organisms are exposed in different ecosystems.

MPs in the aquatic environment are not homogeneously dispersed, and reported concentrations, size classes, polymer types, and shapes can vary tremendously between sampling sites (Nava et al., 2023; Xu et al., 2020; Gupta et al., 2024). These high variations in reported concentrations and properties of MPs in water bodies around the world are also due to methodological restrictions, as there are no standardized methods for sampling, sample processing, and particle characterization, which impedes the comparability of studies (Li et al., 2024; Möller et al., 2022; Moses et al., 2023; Sharma et al., 2024). Further, most studies provide only a snapshot in time, not considering the spatial and temporal variability of this particulate contaminant. Nevertheless, a thorough characterization of the plethora of physicochemical properties of MPs derived from environmental samples would be vital for hazard assessment.

1.3. Interactions of Microplastics and Biota: Microplastics ≠ Microplastics

Given that a wide variety of organisms from terrestrial and aquatic ecosystems are exposed to MPs and are reported to ingest or inhale them (Alfaro-Núñez et al., 2021; Kahane-Rapport et al., 2022; Thompson et al., 2024), the question arises: are MPs harmful? To answer this question, we must consider the above-mentioned complexity of MPs to be able to elucidate their effects on environmental and human health. it is essential to note that MP is not just a single and defined entity but a highly complex group of particulate contaminants. MPs differ in size, shape, polymer type, density, unintentionally or intentionally added chemical compounds or residual monomer content, aging state, surface charge, surface roughness, environmental coating, biofouling, and other physicochemical parameters. All these factors come in different combinations, change constantly in the environment, and potentially determine the interactions and effects of MPs with biota in the aquatic environment. Some MP properties, however, are more prominent and, thus, are used in most MP ecotoxicological studies to distinguish between the effects of different types of MPs. These properties include geometric ones, such as the size and shape of MP particles.

2. Physicochemical Properties of MP

2.1. Geometric Properties of MP: Size, Shape, and Surface

One prominent physical property determining the ecotoxicity of MPs is particle size. By definition, the size of MP particles varies across three orders of magnitude (Aragaw, 2021; Haleem et al., 2024) (1–1000 μm, ISO 24187: 2023; Hartmann et al., 2019). However, the analysis of MPs in the aquatic environment is often size-biased due to sampling methods and analytical restrictions in particle detection (Huppertsberg & Knepper, 2018; Löder et al., 2015; Oßmann et al., 2018). For instance, when water is sampled using filtration, the pore sizes of the filters used determine the size of the collected particles (Covernton et al., 2019). Furthermore, the standard particle-based analytical methods used to detect MPs in environmental samples have a detection limit for particles of a specific size. Here, Fourier-Transform Infrared- (FT-IR) and Raman-spectroscopy have lower size limits of 10–20 μm and 1 μm, respectively (Kniese et al., 2023; Löder et al., 2015).

Another relevant geometric property is the shape of the MPs. The main shapes of MPs are spheres, fragments, and fibers (Figure 1) (Koelmans et al.,

2022). However, larger MP particles can sometimes also be found to be described as foams, films, or other shapes (Aragaw, 2021; Chen et al., 2021; Hartmann et al., 2019; Hu et al., 2022; Koelmans et al., 2022). The shape of MP determines properties like packing density in agglomerates, mechanical strength, and surface reactivity (Ali et al., 2024). MP size and shape can affect its transport, like the sinking speed in the water column (Ali et al., 2024; Chen et al., 2021; Du et al., 2021; Haleem et al., 2024; Hu et al., 2022), which strongly determines their availability and uptake probability for organisms.

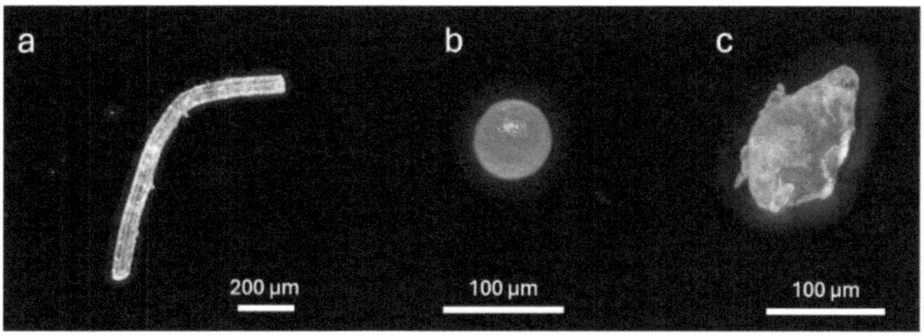

Figure 1: Three main shapes of MPs: (a) polystyrene (PS) fiber, (b) polyethylene (PE) sphere, and (c) polystyrene (PS) fragment.

Closely connected to the size and shape of MPs are parameters related to the particle's surface. Here, the particle's surface area and surface roughness are most relevant. Surface roughness, defined in the literature as the deviation of a surface from being perfectly flat (Burrows et al., 2020; Mao et al., 2020), thereby influences, e.g., the sorption behavior of chemicals onto the surface of MP particles (Burrows et al., 2020). The surface area of a particle is determined by the particle's roughness, size, and shape. For instance, irregularly shaped fragments have a larger surface area than fibers or spheres (Abadi et al., 2024). Reactivity and dissolution rate increase with surface area, influencing the particle's stability (Ali et al., 2024). The surface area is often seen in relation to the volume of MP particles and is described as the surface-to-volume ratio. The surface-to-volume ratio of a particle increases with decreasing size of the particle (Ziani et al., 2023). A higher surface-to-volume ratio allows for more interactions with the surrounding substrate or medium and, therefore, determines how conducive the particle is to the adsorption of (heavy) metals or organic pollutants (Ahmed et al., 2023; Ali et al., 2024; Chen et al., 2023; Du et al., 2021; Hamidian et al., 2021; Xu et al., 2020). Further, more possibilities for interaction between particles and their surroundings also typically means increased potential for chemical leaching (Haleem et al., 2024). It has already been shown that both parameters, surface area, and surface roughness, influence the interactions and toxicity of MPs in the environment (Ateia et al., 2020; Burrows et al., 2020).

2.2. Polymer Type

The most common types of MP found in the aquatic environment reflect the most important polymers in production, including PE, PP, PVC, PS, and PET (Rossatto et al., 2023). The stability, flexibility, and reactivity of MPs are greatly influenced by the chemical composition of the polymer backbone (Lambert et al., 2017). For instance, the simple hydrocarbon backbones of PE and PP make them hydrophobic and largely chemically inert. In contrast, the addition of chlorine atoms in PVC makes it more reactive and less stable (Costa et al., 2024). A polymer type can also vary in its level of crystallinity, which influences its physicochemical properties (Lambert et al., 2017) and determines its density (Kholodovych & Welsh, 2007). For example, amorphous polymers, like low-density polyethylene (LDPE), show less crystallinity, contrary to their high-density polyethylene (HDPE) counterpart. The crystallinity of polymers also determines how rigid they are and how brittle they can become (Maddison et al., 2023; Zhang et al., 2024). Hence, the polymer type of MPs, like their geometric properties, strongly determines their environmental behavior and their interactions with organisms (Ali et al., 2024).

2.3. Additives, Residual Monomers, and Plastic-associated Chemicals

Different types of additives are being added to the polymers to enhance their properties such as flexibility, durability, and thermal stability (Marturano et al., 2019). These additives comprise antistatic agents (e.g., ethoxylated amines), nucleating agents (e.g., sodium benzoate), adhesives (e.g., epoxy resins), lubricants (e.g., stearic acid), plasticizers (e.g., Diisodecyl-phthalate), antioxidants (e.g., butylhydroxytoluene), pigment agents (e.g., titanium dioxide), flame retardants (e.g., tris-chloropropylphosphate), heat stabilizers (e.g., calcium-zinc stabilizers), and more (Hahladakis et al., 2018), leading to more than 30,000 commercially available additives (https://polymer-additives.specialchem.com; search category 'additives'). Residual monomers can originate from incomplete polymerization reactions during the production of plastics (da Costa et al., 2023). For instance, the monomer styrene originates from the polymerization processes of PS (Wang et al., 2023). Residual monomers can leach out of the polymer matrix into the surrounding media, potentially making them available for organismic uptake (Kwan & Takada, 2016). Besides intentionally added additives, a vast array of non-intentionally added substances (NIAS) can be associated with plastics (Brander et al., 2024; Wagner et al., 2024). These NIAS can originate from different processes including polymer and additive degradation, impurities from raw materials or additives, and other contaminants (Nerin et al., 2013). Plastic chemicals pose a potential threat to the organisms exposed to MPs as those may leach from the material (da Costa et al., 2023).

2.4. Artificial and Environmental Alterations of Physicochemical MP Properties

Besides the surface area and surface roughness of an MP particle, the surface characteristics of MPs are also influenced by the surroundings and the physicochemical composition of plastics. The impact of physicochemical alterations on MP particles due to photooxidation, mechanical stress, or biotic interactions and, hence, the level of degradation of MPs is often expressed using the terms of weathering or aging, i.e., the sum of biotic and abiotic processes that MP particles undergo in the environment or artificially in laboratories (Figure 2).

Figure 2: Visualization of an artificial aging process of a PS plastic particle: (a) pristine and (b) aged due to artificial UV weathering for 1400 h.

The level of degradation by weathering or aging processes on plastics within the aquatic environment depends on multiple factors including physical abrasion, microbial activity, chemical oxidation, and UV radiation (Folino et al., 2020; Liu et al., 2020; Reineccius et al., 2023). These processes lead to drastic changes in the physical properties and chemical composition of MP particles (Ahmed et al., 2023; Chen et al., 2023; da Costa et al., 2023; Haleem et al., 2024; Hu et al., 2022; Luo et al., 2022). This results in, e.g., a lower molecular weight, a more hydrophilic character (Menzel et al., 2022), or changes in crystallinity and sorption behavior (Liu et al., 2020). After aging, further surface alterations of MP particles have been observed, changing the particle's surface roughness (Pinlova & Nowack, 2024), surface porosity (Bhagat et al., 2022), and surface charges (Hao et al., 2024).

The surface charge is another crucial surface characteristic of MP particles, primarily measured as the zeta potential. The zeta potential is a quantitative value of the electrical potential of a particle in a medium (Ramsperger et al., 2020), which is typically positive at low pH, negative at high pH, or neutral at the isoelectric point (Rahman et al., 2024). The zeta potential of an MP particle is influenced by the introduction of functional groups on the particles surface by, e.g., surface oxidation through artificial UV aging (Bhagat et al., 2022). These functional groups substantially alter the particle's properties by, e.g., making them hydrophobic due to the polar nature of oxygen-containing groups (i.e., hydroxyl, ether, and carbonyl groups) (Bhagat et al., 2022). Functional groups, therefore, substantially determine the particle's sorption behavior (Yao et al., 2022; Zhao et al., 2022), which depends on, e.g., particle hydrophobicity (Fu et al., 2021) or crystallinity (Prajapati et al., 2022). Thus, the physicochemical surface characteristics of MP determine the potential of MP particles to be coated with biomolecules in the environment. These coatings on MP particles are referred to as eco-corona.

An eco-corona is a complex matrix that forms on the surface of MPs upon environmental exposure. Extracellular polymeric substances (EPS, originating from microbial activity), other naturally occurring organic matter, and inorganic materials such as minerals and ions form the eco-corona (Junaid & Wang, 2021; Ramsperger et al., 2020). The eco-corona thereby changes the physicochemical properties on the surface of MP particles. For instance, the eco-corona makes a particle rougher (Wieland et al., 2024) but softer (Witzmann et al., 2022), introduces functional groups such as hydroxyl and carbonyl groups (Kim et al., 2023; Liu et al., 2023; Ramsperger et al., 2020; Wieland et al., 2024), and changes the particle's surface charge (Wieland et al., 2024).

In the aquatic environment, MP particles large enough to be colonized by microorganisms, such as bacteria, algae, fungi, and protozoa (Weig et al., 2021; Zhai et al., 2023) provide a substrate where a biofilm, the so-called plastisphere, is formed (Huang et al., 2024). Biofilm formation and composition depend on environmental factors and MP properties (Jia et al., 2024) such as size, shape, surface roughness, hydrophobicity, and the particles' surface charge (Agboola & Benson, 2021; Du et al., 2021; Hamidian et al., 2021; Luo et al., 2022; Parrish & Fahrenfeld, 2019; Rahman et al., 2024; Xia et al., 2023; Yao et al., 2023). Both eco-corona and biofilm modify the particle's properties (Koelmans et al., 2022), also changing the behavior of the particles in the aquatic environment. For example, they can alter the density of the particles, which in turn affects their transport and sedimentation behavior in water (Hamidian et al., 2021; Parrish & Fahrenfeld, 2019), and their location in the water column, and thus the probability of interaction with organisms.

3. Physicochemical Properties of MP Determining Uptake and Effects on Organisms

Due to the widespread occurrence of MPs, interactions with organisms are likely to occur in the aquatic environment (Kibria et al., 2022). From the organismic perspective, ecological factors, such as feeding behavior and habitat, are essential in determining the uptake probability of MP particles. Filter, suspension, and deposit feeders with relatively indiscriminate feeding behaviors are most prone to ingest MP particles unintentionally (Figure 3) (Setälä et al., 2016). Once organisms have taken up MP particles, these particles may provoke detrimental effects while interacting with the body's natural barriers, such as skin tissues, respiratory organs, or the digestive tract (Zhu et al., 2023).

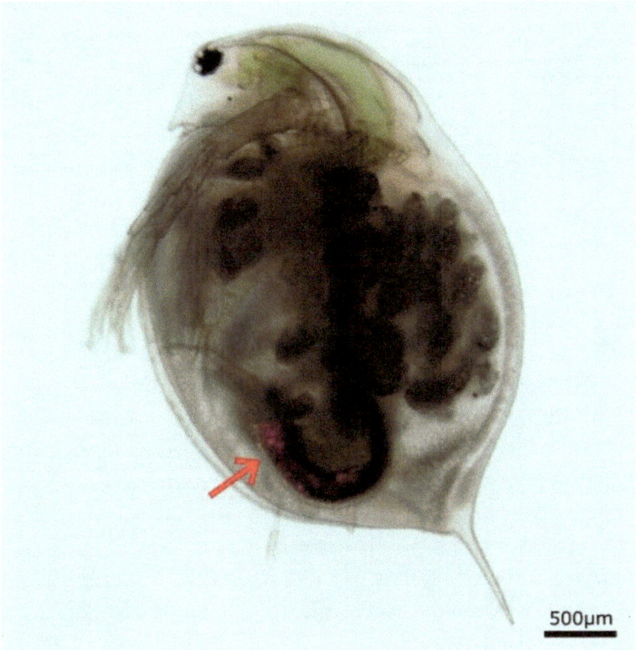

Figure 3: Female *Daphnia magna* adult after uptake of fluorescently labeled PS MPs. MP particles (fragments <20 µm) are depicted in red in the lower part of the gut (abdominal region, red arrow).

Laboratory experiments under controlled conditions are often performed to investigate the impact of MPs on aquatic organisms. These studies frequently use model organisms (Ankeny & Leonelli, 2011). Here, Mollusca (e.g., *Mytilus* spp. (marine); *Dreissena* spp. (limnic)) and Arthropoda (*Carcinus maenas* (marine); *Daphnia* spp. (limnic)) are among the most studied invertebrates, while fish are used (e.g., *Danio rerio* (limnic)) as a model system for vertebrates (Prokić et al., 2021). Commonly investigated endpoints are mortality and immobilization, life history parameters (e.g., days until maturity), oxidative stress levels, and behavioral aspects (de Sá et al., 2018). Distinguishing between the effects of the different MP properties is crucial for identifying potential toxicity mechanisms and improving MP's ecological risk assessment (Moura et al., 2023). Consequently, many efforts have already been made to investigate and address the multidimensionality of MP particles in effect studies with organisms.

3.1. Geometric MP Properties Determining MP Toxicity on Aquatic Organisms

The size and shape of MP particles are critical determinants of how they interact with aquatic organisms and their potential impact on ecosystems. Particle size significantly influences whether MPs enter the food web, how they are transferred across trophic levels, and their potential for egestion (Ahmed et al., 2023; Brehm et al., 2023; Chen et al., 2021; Du et al., 2021; Hoang & Felix-Kim, 2020; Kinjo et al., 2019; Mondellini et al., 2024; Xu et al., 2020). Small-sized MPs, particularly in the low micron range, are accessible to a broader range of organisms, including unicellular species, and generally exhibit more substantial adverse effects on both vertebrate and invertebrate species (Gambardella et al., 2019).

One reason for the increased toxicity of smaller MPs may be their higher surface-to-volume ratio, which increases their interactions with biological tissues. This increased ratio allows smaller MPs to cross biological barriers more efficiently, potentially entering the circulatory system or neighboring tissues (Lu et al., 2016; Qiao et al., 2019; Thornton Hampton et al., 2022). As a result, smaller MPs often pose a higher toxicity than larger particles that are less capable of traversing biological boundaries. Studies focusing on the size of MPs have demonstrated that once ingested, MPs can induce various size-dependent effects including inflammatory reactions, oxidative stress, metabolic changes, behavioral alterations, and tissue damage (Abarghouei et al., 2021; Ahmed et al., 2023). For example, vertebrates like goldfish showed size-dependent histopathological effects for particles of 8 μm (strongest effect) in size (Abarghouei et al., 2021), and invertebrates like shrimps showed size-dependent histopathological effects of smaller MPs (sizes 0.1 μm, 1 μm, and 5 μm) and enrichment of pathogenic bacteria in the gut microbiome (size 1 μm and 5 μm) (Zhou et al., 2023). For daphnids, size-dependent differences for smaller MP particles (size 6 μm) in comparison to larger particles (size ~20 μm) on the number of offspring were detected (Schwarzer et al., 2022).

Beyond size, laboratory experiments have also shown that MP shapes influence the effects on organisms. First, MP shape can potentially determine the uptake rates of organisms, depending on their feeding behavior (Porter et al., 2023). Second, different MP shapes elicit differing effects after ingestion. For example, particles with various shapes made of the same material (PS) show shape-dependent adverse effects on the reproductive success (offspring per brood) of *Daphnia.* Strongest effects have been shown for irregular PS fragments and spheres, whereas only slight effects were observed for PS fibers (length ~50 μm) (Schwarzer et al., 2022). However, MP fibers longer than 50 μm have been found to elicit adverse effects on aquatic organisms, such as the pacific mole crab *Emerita analoga* (Horn et al., 2020) or the goldfish *Carassius auratus* (Jabeen et al., 2018). Several conceivable mechanisms may account for differences in effects between the different shapes of MP particles. For example, it is hypothesized that fibers can adhere to the epithelium of the digestive tract, leading to gut clogging and, consequently, less nutrient uptake. Moreover, irregular particles (size range 1–400 μm) can have sharp edges, leading to a higher probability of tissue fissures, which can increase the translocation of MPs to other tissues of the organisms (Bacchetta et al., 2018). Further, it is discussed that the potential physical damage inflicted by irregularly shaped MPs might also lead to different inflammatory responses within the tissues (Franzellitti et al., 2019). Next to the physical stress, the differing surface-to-volume ratio might also be important in determining the toxicity of the different MP shapes (Deng et al., 2017; Lin et al., 2024). In this context, fragments with larger surface-to-volume areas could be more toxic than similarly sized fibers or spheres (Ahmed et al., 2023).

The surface-to-volume ratio emerges as a critical parameter in assessing the toxicity of MPs, as it directly influences their interactions with biological systems (Ahmed et al., 2023; Xu et al., 2020). Consequently, both size and shape must be carefully considered when evaluating the toxicological impact of MPs on aquatic life (Koelmans et al., 2022).

3.2. Effects of the Polymer Type on Aquatic Organisms

In addition to the physical properties discussed above, the chemical properties of MPs also play a critical role in determining their toxicity and availability to aquatic organisms. Among these chemical characteristics, the polymer type of MPs is essential. Different polymers have varying densities, directly affecting their buoyancy and distribution within aquatic ecosystems. For instance, polymers like PP and PE are buoyant and tend to float in the upper water layers. In contrast, polymers like PET and polyamide (PA) are denser and often accumulate in sediment layers (He et al., 2021). This polymer-dependent distribution influences the likelihood of exposure, i.e., inhalation or ingestion by certain organisms, with floating MPs posing a greater ingestion risk to surface-dwelling species (Du et al., 2021). Research indicates that the adverse effects of MPs on organisms may vary depending on the type of polymer (Enyoh et al., 2022; Lagarde et al., 2016). In daphnids and mussels, polymer-dependent MP effects on life history traits, enzyme activity, behavior, and proteomic responses have been observed, demonstrating that toxicity varies by polymer type (Brehm et al., 2022; Zimmermann et al., 2020). In these examples, recycled PET MP elicited more pronounced effects on mussels than polylactic acid (PLA), PA, and PS (Brehm et al., 2022). In daphnids, PVC MP elicited more pronounced effects than PLA and PUR MP (Zimmermann et al., 2020). These differences in the polymer type-dependent effects might also be attributed to variations in their physicochemical properties, such as, in the case of Brehm et al. (2022), residual monomers and plastic-associated chemicals. Tissue-level effects show similar polymer dependency, as evidenced in fish and mussels exposed to various types of MP (e.g., PA, PE, PET, PP, PS, and PVC) (Lei et al., 2018; Teng et al., 2021). Different polymers induce distinct histopathological changes in particular tissues (Espinosa et al., 2019), such as epithelial cell coagulation, digestive tract necrosis, and intestinal villi cracking (Lei et al., 2018; Teng et al., 2021). Contrary to popular belief, biodegradable polymer MP particles do not necessarily pose a lower risk than conventional MPs. Biodegradable MPs have been found to cause similar or even more significant harm to organisms and their tissues (An et al., 2024; Lara-Topete et al., 2024). This harm may be due to their rougher surfaces, which increases physical tissue damage and facilitates the leaching of toxic chemicals, such as plastic additives (Tao et al., 2024). Notably, ecotoxicological data on the effects of different MP polymer types still needs to be expanded, as laboratory studies often focus on spherical PS particles that do not resemble the particles found in nature (Enyoh et al., 2022). Further research is essential to clarify the diverse impacts of polymer types on aquatic organisms and ecosystems since the polymer type determines many physicochemical properties of MPs, such as their crystallinity and, hence, their degradability (Sutkar et al., 2023) and the content of plastic associated chemicals (Wagner et al., 2024).

3.3. Effects of Additives, Residual Monomers, and Plastic-associated Chemicals on Aquatic Organisms

Properties closely related to the polymer type include additional chemicals associated with the polymer such as additives, monomers, and NIAS. For example, plastic additives are polymer-type dependent and are used to enhance the processability and functionality of plastic products (Höfer & Hinrichs, 2009). These chemicals are gaining increased attention in MP effect research, as they are prone to leach from the MP particle into the surrounding water or biological compartment, probably causing harm to organisms (da Costa et al., 2023). For additives, the most pertinent effects include neurotoxicity (Maric et al., 2024), inflammation, disturbance of lipid metabolism, and cancer pathways (Yu et al., 2024). In laboratory tests, additives, including phthalates and brominated flame retardants, have been demonstrated to interfere with several biological functions, including metabolism and skeletal development (Jeong & Choi, 2020). Further, the plasticizer diisononyl phthalate incorporated in different concentrations into PVC MP evoked fewer offspring in exposed daphnids. Pure exposure to PVC MP particles without additives evoked no effects (Schrank et al., 2019). While the data availability for additives is robust, very few studies deal with the toxicity of residual monomers on organisms. However, there are first hints that the toxicity of MP increases with increasing residual monomer content (Brehm et al., 2022; Zhang et al., 2023). However, residual monomers might be less toxic than some additives (Meng et al., 2023). Consequently, as plastics are a mixture of one polymer plus various additives and chemicals (Andrady & Rajapakse, 2016), in the context of hazard assessment, it is crucial to characterize the MP particles used in laboratory studies in-depth (Brehm et al., 2022; Koelmans et al., 2022). Studies conducted with in-depth characterized compositions of the used MPs emphasize this by showing that in the presence of a higher number of chemicals like additives or NIAS, the toxicity on aquatic organisms increases (Brehm et al., 2022; Capolupo et al., 2023; Uribe-Echeverría & Beiras, 2022). Plastic-associated chemicals are also considered one of the primary sources affecting the gut microbiome in fish (Adamovsky et al., 2021) and other organisms, resulting in adverse health implications (Fackelmann & Sommer, 2019).

3.4. Effects of Artificial and Environmental Alterations of Physicochemical MP Properties on Aquatic Organisms

MP particles in aquatic environments can undergo surface modifications via various factors, affecting their biogeochemical behavior and toxicity (Liu et al., 2023). Among these factors are the level of aging (also described as photooxidation, weathering, and degradation) of MPs, artificial surface modifications, and the attachment of biomolecules (eco-corona) or tiny organisms like bacteria or fungi forming a biofilm. The surface characteristics are next to the surface-to-volume ratio and potentially toxic MP-associated chemicals, the main determinants regarding organism or tissue interactions and drivers of MP toxicity. For instance, aged MP particles have been observed to be more prone to heterogeneous aggregation with algae cells (Ni et al., 2023). Consequently, aged MP particles are more frequently observed to be ingested by zooplankton (Ni et al., 2023; Vroom et al., 2017). This enhanced ingestion and interaction may account for the observed rise in toxicity compared to pristine MP particles, including increased oxidative stress, heightened immune responses, and a more significant disruption of endocrine functions (Alimba & Faggio, 2019).

Closely related to the surface changes caused by the aging of MPs, are properties often intentionally altered in laboratory experiments. These are properties like, e.g., purposely added functional groups such as amino or carboxyl groups. Functional groups on MPs affect their surface charge, changing their electrostatic properties (Zhu et al., 2020). These changes influence their affinity for pollutants like heavy metals (Adeleye et al., 2024; Yu et al., 2024). Studies also showed that charged functional groups on MP particles can inhibit the sorption of organic contaminants like naphthalene, potentially due to increased polarity and weakened π-π interactions (Yu et al., 2020). Apart from the altered adsorption behavior of MPs due to functional groups, amine, and carboxy functionalized spherical PS MPs were also shown to induce an increase in metabolism and detoxification enzymes in *D. magna* (Panagiotidis et al., 2023). However, besides data on nanoplastics, where an increased toxicity for similarly functionalized spherical PS particles was found (Zhang et al., 2022), the number of studies on the direct effects of functionalized MPs on organisms is very limited. Nonetheless, the trend of increased toxicity due to surface modifications of MPs highlights the importance of considering functional groups when assessing the impacts and risks associated with MPs. These controlled surface modifications in laboratories (e.g., intentionally added functional groups by UV-aging) may have valuable implications for MP ecotoxicity by providing model MP particles that exhibit properties closer to the properties of environmentally exposed ones. However, these particles cannot reflect the entirety of physicochemical alterations MP particles undergo in nature.

Therefore, MP surface alterations happening in the environment are referred to as uncontrolled surface modifications (i.e., naturally occurring) (Brehm et al., 2023). These uncontrolled surface modifications of MP particles are, e.g., coating with an eco-corona or biofilm, which changes their surface characteristics, hence their biogeochemical behavior and toxicity (Liu et al., 2023; Rummel et al., 2017). The ecological consequences of eco-corona/biofilm-MP interactions, such as trophic transfer and adverse effects, remain largely unknown and require further investigation (Moyal et al., 2023; Rummel et al., 2017). Nonetheless, MP particles may carry antibiotic-resistance genes and pathogenic microorganisms over long distances, increasing their potential hazard to the biota (Bowley et al., 2021; Tuvo et al., 2023; Zhao et al., 2024; Zhong et al., 2023). Further, studies using MP particles coated with an eco-corona have demonstrated that the eco-corona mediates uptake, retention, and toxicity, affecting feeding behavior and energy storage (Nasser & Lynch, 2016; Galloway et al., 2017).

Since most aquatic organisms are constantly exposed to particulates suspended in the water column (Ogonowski et al., 2023), they can most likely egest MP particles (Frydkjær et al., 2017; Xiong et al., 2019) as they do with other naturally occurring particulates. However, the pure gut passage of MPs can induce changes in the gut microbiome of organisms (Wang et al., 2024), which is important for, e.g., the immune system (Luan et al., 2023). Further, as mentioned before, MP particles can interact with tissues or translocate into them. Once these particles surpass biological barriers, they will probably interact with cells. In aquatic organisms, studies with vertebrate cell lines from the rainbow trout (Boháčková et al., 2023; Heinrich & Braunbeck, 2020) or the atlantic salmon (Solhaug et al., 2024) were conducted. However, data on cell lines from aquatic organisms is still limited. Therefore, in the following section, cell lines from terrestrial vertebrates, like, e.g., mammals, will also be included to cover the whole spectrum of known MP effects on cells.

4. Physicochemical Properties of MP in the Aquatic Environment Determining Uptake and Effects on Cells

The primary intake pathways for MP particles in organisms are respiratory pathways and ingestion. Therefore, among the first cells to encounter MP particles are cells of the respiratory and gastrointestinal tract (Prata, 2018; Revel et al., 2018; Wieland et al., 2022). There, cell types such as epithelial cells, alveolar macrophages, goblet cells, and M cells may interact with the MP particles (Figure 4) (Ramsperger et al., 2023; Wieland et al., 2022).

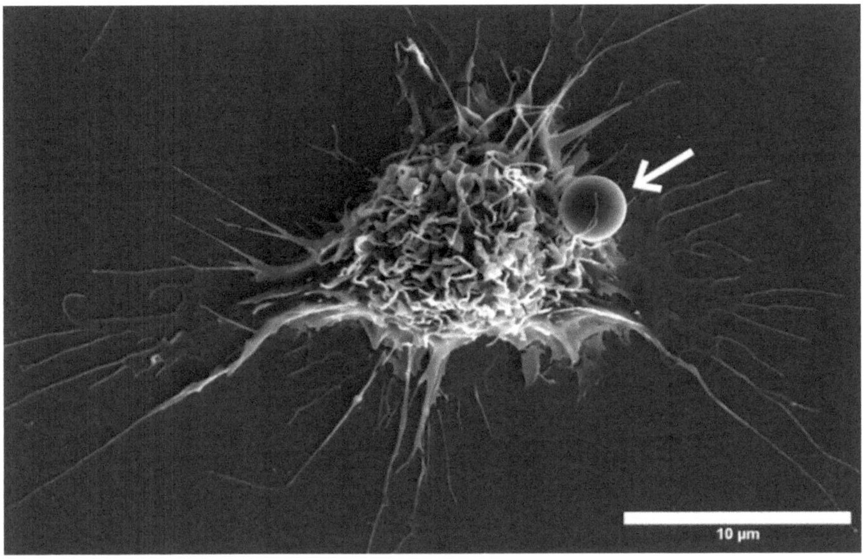

Figure 4: Scanning electron microscope (SEM) image of a J774A.1-macrophage cell, interacting with a PS MP particle (sphere, white arrow). The particle was incubated in freshwater two weeks before the image was taken.

In this context, immune cells such as macrophages and neutrophils may be of particular interest due to their increased ability to take up particulate matter and their ability to move within the organism (Ramsperger et al., 2023; Wieland et al., 2022).

The physicochemical properties of MPs can affect particle-cell interactions, adhesion strength, internalization, and their cellular fate. Here, due to the size of MPs in the micrometer range, two internalization pathways are important: (1) phagocytosis, which is triggered by specific receptor-ligand interactions (Freeman & Grinstein, 2014), and (2) macropinocytosis, which is constitutive in many cell types and enables the non-selective internalization of surrounding media and particles (Swanson et al., 1995). After the cellular internalization of an MP particle is complete, a membrane-bound organelle containing the particle is formed, the so-called phagosome.

Once internalized, MPs can negatively impact the cells. Cytotoxicity, reactive oxygen species (ROS) generation, and inflammatory responses are toxicological mechanisms by which MP particles may affect cells and, consequently, organisms (Ramsperger et al., 2023; Wieland et al., 2022). Oxidative stress in cells can have two potential causes: first, MP particles may disrupt the balance of oxidants and antioxidants in the cells' metabolism (Pérez-Albaladejo et al., 2020). Second, oxidative stress may also be induced by ROS carried by the MPs, e.g., due to the polymerization process or aging processes in the environment (Bhuyan, 2022; Pérez-Albaladejo et al., 2020). Oxidative stress can additionally affect inflammatory responses and lead to cytotoxicity (Bhuyan, 2022; Brehm et al., 2022; Halappanavar & Mallach, 2021). Furthermore, the adverse effects of MP particles on cellular metabolism can mediate their cytotoxicity (Bhuyan, 2022; Brehm et al., 2022).

However, although adverse effects of MP particles on cells were observed in various cell types from different organisms such as fish (Boháčková et al., 2023), humans (Wu et al., 2019), or mice (Rudolph et al., 2021), the current results seem to indicate that the cellular responses are highly dose-dependent and usually only occur at high concentrations of MPs (Danopoulos et al., 2022; Ramsperger et al., 2022). Also, *in vitro* experiments examining the potential adverse effects of MP particles on cells are typically conducted using PS model MPs that reflect the MP particles found in the environment only to a certain extent (Lim, 2021; Wieland et al., 2024).

4.1. Geometric Properties of MPs Determining Cellular Uptake and Effects

The size of particles determines their surface-to-volume ratio, which is vital for their adhesion to cells. The adhesion strength of a particle is a crucial determinant for its internalization because particles that interact stronger with cells are more likely to be internalized (Lesniak et al., 2013; Wieland et al., 2024). It is expected that smaller particles interact stronger with cells and tissues than larger particles due to their increased surface-to-volume ratio relative to their size, leading to stronger adhesive forces for the same total particle mass (Otto et al., 2015; Schmid & Stoeger, 2016). The size of MP particles also affects the probability of internalization. For macrophages, the internalization probability is highest for particles with a size of 2–3 μm, and particles outside that range are usually internalized less frequently (Baranov et al., 2021; Rudolph et al., 2021, Champion et al., 2008). Ultimately, the maximal size of particles that macrophages can successfully internalize is limited by the available membrane to engulf the particle (Cannon & Swanson, 1992). In contrast, the internalization probability increases with decreasing particle size in the case of epithelial cells (Rudolph et al., 2021). Similarly, a trend is observable in unicellular freshwater organisms, where the number [...] of ingested MPs particles increases with the decreasing size of MPs (Mondellini et al., 2024). Furthermore, there is evidence that the duration of the maturation process is affected by particle size (Koval et al., 1998) and that smaller MPs induce an increased lysosomal activity (Boháčková et al., 2023). Due to these factors, the size of the MPs determines their effects on cells. For example, PS MPs were observed to affect the metabolic activity of murine macrophage and epithelial cells and ROS generation in a concentration-, cell-line, and size-dependent way (Rudolph et al., 2021). Similarly, in cells from rainbow trout (gill, ovary, and liver), smaller MP particles had a stronger effect on ROS production (Boháčková et al., 2023). Likewise, size can influence the impact of MPs on inflammatory processes, such as the secretion of the inflammatory cytokine IL-1β by human immune cells (Beijer et al., 2022).

The shape of MP particles can affect the internalization of MPs in a similar way. The speed with which a particle is engulfed by a cell depends on the local geometry of the particle and, therefore, differs between spherical and ellipsoidal particles (Champion & Mitragotri, 2006; Paul et al., 2013). Furthermore, in the case of elongated particles and fibers, maturation may not progress uniformly but rather be distributed in a gradient along the particle's length, which can also affect the maturation kinetics (Baranov et al., 2021). Notably, the prominent example of fibers demonstrates how shape can even lead to toxicity: for bio-persistent microfibers that exceed the typical length of the macrophages, internalization often can't be completed. This phenomenon can lead to chronic inflammation, which can cause pneumoconiosis and cancer (Donaldson et al., 2010).

4.2. Effects of the Polymer Type on Cellular Uptake and Effects

The polymer type of MPs is an essential parameter for the cellular interactions and effects of MPs because it determines underlying physical and chemical properties, such as the surface charge and hydrophobicity, their capability to adsorb environmental pollutants, or the chemical nature of potential residual monomers (Wieland et al., 2022). For example, one study observed that the polymer type of similar-sized MPs spheres determined the interactions and uptake efficiency of MP in different murine cell lines (Jasinski et al., 2024). While macrophages generally internalized MPs of various polymer types (PE, PS, PVC, PLA), hepatic epithelial cells only internalized PS MPs, and intestinal epithelial cells did not internalize any of these MPs. Consequently, the effects on cells differed between polymer types and cell lines, significantly increasing ROS generation in macrophages upon exposure to PLA MPs, which was not observed for other polymer types. Similarly, in cells from rainbow trout, PVC particles appeared to induce increased ROS generation, whereas PET particles had no any effect (Boháčková et al., 2023). The responses of human immune cells were similarly affected by the polymer type. For example, the secretion of IL-1β was higher when the cells were exposed to PET or PP, compared to PS and PE of the same size (Beijer et al., 2022).

4.3. Effects of Additives, Residual Monomers, and Plastic-associated Chemicals on Cellular Uptake and Effects

Similar to the polymer type, the content of additives, residual monomers, and plastic-associated chemicals of MPs can have different effects on cells. Once internalized by cells, the MPs may release these chemical compounds, potentially producing chemical toxicity in cells (Rozman & Kalčíková, 2022). For example, one study demonstrated that the viability of human fibroblast cells decreased upon exposure to PS MPs with increasing residual monomer content, indicating that leaching residual monomers significantly contributed to MP toxicity (Zhang et al., 2023). In addition to that, leachates from environmental MPs have been found to cause DNA damage in rainbow trout liver cells (Pannetier et al., 2019). In general, many monomers and additives used in the production of plastics are known to cause oxidative stress and cytotoxicity (Pérez-Albaladejo et al., 2020). Therefore, the leaching of additives, monomers, and other plastics-associated chemicals may play a substantial role in the cellular toxicity of MPs.

4.4. Effects of Artificial and Environmental Alterations of Physicochemical MP Properties on Cellular Uptake and Effects

The various surface characteristics of MPs, which are also highly variable, are essential parameters of cellular uptake and effects. For instance, environmental or artificial aging is expected to substantially affect the cellular interactions, internalization, and impacts of MPs. Consistently, one study observed that aging significantly altered the interactions and internalization of PS MPs and human placental cells (Dusza et al., 2022). Aging also causes effects in murine macrophages. When exposed to increasingly aged PS MPs, the metabolic activity of the cells decreased, and ROS generation increased, indicating oxidative stress. Also, the cytokine secretion of the cells changed upon exposure to aged MPs compared to pristine MPs (Völkl et al., 2022). Furthermore, another study observed that the aging of PUR MPs promoted the release of organic compounds, which caused cytotoxic effects on human colon cells (Xu et al., 2022).

Another critical parameter determining the particles' interactions with cells is their zeta potential (Jeon et al., 2018; Ramsperger et al., 2022; Tabata & Ikada, 1988). PS MP particles with a more negative zeta potential generally get bound more quickly and adhere stronger to cells, likely due to electrostatic interactions between particles and cells (Wieland et al., 2024). Due to these stronger particle-cell interactions, the absolute internalization probability also increased for particles with a more negative zeta potential. Consequently, particles with a more negative zeta potential showed higher cytotoxicity, likely due to the increased interactions and internalization (Ramsperger et al., 2022). Although these studies focused on otherwise identical commercial model PS spheres, their findings are of high environmental relevance since many parameters of an MP particle affect its surface charge and, thus, its zeta potential. For example, different polymer types may carry different surface charges, which may be modified by adding functional groups (Johansson, 2017). Additionally, aging processes such as photooxidation can generate functional surface groups on the MP particle, resulting in a more negative zeta potential (Menzel et al., 2022; Meides et al., 2021). Likewise, due to the formation of an eco-corona, the zeta potential of an environmental MP particle can change because biomolecules such as proteins carry functional groups that may introduce charges to the MP surface (Wieland et al., 2024).

Biomolecules adsorbed on the surface of MP particles can significantly alter their binding and adhesion to cells (Ramsperger et al., 2020; Wieland et al., 2024). It was observed that environmentally exposed MP particles, covered with an eco-corona, bound to cells more quickly and had stronger adhesion forces than pristine MP particles (Wieland et al., 2024). Two factors might be important here: First, biomolecules in an eco-corona often carry charges (Jayalath et al., 2018; Kretzschmar et al., 1997), affecting the zeta potential and, consequently, the cellular interactions of MP particles (Wieland et al., 2024). Second, these biomolecules might also form specific receptor-ligand bonds, for example, with pattern recognition receptors of macrophages such as Toll-like receptors (Barton & Medzhitov, 2002) or scavenger receptors (Platt & Gordon, 2001). These receptor-ligand interactions might contribute to the binding and adhesion of environmentally exposed MP particles to cells (Wieland et al., 2022). Binding and adhesion also affect the cellular internalization of eco-corona-coated MPs. On the one hand, as discussed above, the adhesion strength is a crucial determinant for the internalization of MPs. On the other hand, the biomolecules of the eco-corona are an essential factor for their internalization , facilitating the activation of phagocytic receptors (Freeman & Grinstein, 2014). Considering the consequences of an eco-corona for cellular interactions and internalization, we expect cellular effects to be substantially altered for eco-corona-coated MPs as well. However, so far, most studies have not investigated these effects for MPs in cell culture studies, which makes it difficult to draw any conclusions. Future studies focusing on cell cultures will be necessary to achieve a better understanding of the role of eco-coronae and biofilms in the cytotoxicity of MPs.

5. Conclusions

Many aquatic organisms are affected by the presence of MPs in their habitats. Various studies showed adverse effects of MPs on these organisms, such as increased mortality, life history changes, growth parameters, and molecular changes. Of these studies, only a few reported them in relation to the physicochemical properties that are decisive for MP toxicity. In most cases, only properties like the particles' shape and size were reported. However, reliable explanations for the adverse effects of MPs on organisms are still lacking. Many of the measured effects most likely originate from cellular processes post-ingestion. Therefore, potential adverse effects of MPs on the cellular level of organisms have to be considered in MP ecotoxicology. At the cellular level, diverse toxic effects were found to be influenced by the size, shape, or surface charge of MPs. In particular, smaller, more negatively charged particles adhere more strongly to cells, increasing internalization and potential cytotoxicity. Environmental aging further enhances these effects by altering the surface properties of MPs. However, the complex properties of MPs that are assumed to determine MP toxicity in organisms and cells, thus far, remain understudied. These properties include, for instance, specific surface properties of MP particles (including eco-corona and biofilm) or plastic-associated chemicals. As most studies work with high MP concentrations in experiments, it is difficult to entangle the specific threshold for toxicity, which might also be dependent on the physicochemical properties of MPs. It is, therefore, crucial to understand how the various physicochemical properties of MPs contribute to their toxicity on both the organismic and cellular levels. To improve the current understanding of the potential toxicity of MPs for organisms and cells, studies working with both well-defined model MPs as well as environmentally realistic MPs (complex polymers containing, e.g., additives, biofilms, and eco-coronae, or weathered MPs) will be necessary. In either case, a thorough characterization of the physicochemical properties of the MP particles is imperative. This will help to establish a mechanistic understanding of the role of the individual and combined physicochemical particle properties for the hazard potential of MP particles. Additionally, elaborate experimental designs must use equally characterized control particles that closely resemble natural particles occurring in the aquatic environment. These control particles would enable scientists to draw deliberate conclusions about MP toxicity by ruling out effects that could be caused by particle exposure per se. However, it remains a grand challenge to the scientific community to keep an appropriate perspective on the causes of MP toxicity since some adverse effects might be attributed to, e.g., the size of MPs while in reality, the measured effect might have been the response to other, more harmful physicochemical properties of the same particle such as, e.g., specific surface characteristics. Therefore, a thorough characterization of employed MPs combined with the well-established use of control particles and, thus, a reliable identification of the causes of MP toxicity in ecotoxicological experiments would create profound knowledge that could pave the way for a reliable hazard assessment of MPs and for the development of less harmful plastic materials.

References

Abadi, Z.T.R., Abtahi, B., Fathi, M.B. et al. (2024). Size, shape, and elemental composition as predictors of microplastic surface erosion. *Journal of Hazardous Materials*, 476, 134961. https://doi.org/10.1016/j.jhazmat.2024.134961.

Abarghouei, S., Hedayati, A., Raeisi, M., Hadavand, B.S., Rezaei, H., & Abed-Elmdoust, A. (2021). Size-dependent effects of microplastic on uptake, immune system, related gene expression, and histopathology of goldfish (*Carassius auratus*). *Chemosphere*, 276, 129977. https://doi.org/10.1016/j.chemosphere.2021.129977.

Adamovsky, O., Bisesi, J.H. & Martyniuk, C.J. (2021). Plastics in our water: Fish microbiomes at risk. *Comparative Biochemistry and Physiology – Part D: Genomics and Proteomics*, 39(September 2020), 100834. https://doi.org/10.1016/j.cbd.2021.100834.

Adeleye, A.T., Bahar, M.M., Megharaj, M. et al. (2024). The unseen threat of the synergistic effects of microplastics and heavy metals in aquatic environments: A critical review. *Current Pollution Reports*, 10(3), 478–497. https://doi.org/10.1007/s40726-024-00298-7.

Agboola, O.D., & Benson, N.U. (2021). Physisorption and chemisorption mechanisms influencing micro (nano) plastics-organic chemical contaminants interactions:

A review. *Frontiers in Environmental Science*, 9, 1–27. https://doi.org/10.3389/fenvs.2021.678574.

Ahmed, A.S., Billah, M.M., Ali, M.M. et al. (2023). Microplastics in aquatic environments: A comprehensive review of toxicity, removal, and remediation strategies. *Science of The Total Environment*, 876. https://doi.org/10.1016/j.scitotenv.2023.162414.

Alfaro-Núñez, A., Astorga, D., Cáceres-Farías, L. et al. (2021). Microplastic pollution in seawater and marine organisms across the tropical eastern pacific and galápagos. *Scientific Reports*, 11(1), 1–8. https://doi.org/10.1038/s41598-021-85939-3.

Ali, N., Khan, M.H., Ali, M. et al. (2024). Insight into microplastics in the aquatic ecosystem: Properties, sources, threats, and mitigation strategies. *Science of The Total Environment*, 913: 169489. https://doi.org/10.1016/j.scitotenv.2023.169489.

Alimba, C.G., & Faggio, C. (2019). Microplastics in the marine environment: Current trends in environmental pollution and mechanisms of toxicological profile. *Environmental Toxicology and Pharmacology*, 68, 61–74. https://doi.org/10.1016/j.etap.2019.03.001.

An, G., Na, J., Song, J., & Jung, J. (2024). Chronic toxicity of biodegradable microplastic (polylactic acid) to *Daphnia magna*: A comparison with polyethylene terephthalate. *Aquatic Toxicology*, 266, 106790. https://doi.org/10.1016/j.aquatox.2023.106790.

Andrady, A.L., & Neal, M.A. (2009). Applications and societal benefits of plastics. *Philosophical Transactions of the Royal Society B: Biological Sciences*, 364(1526), 1977–1984. https://doi.org/10.1098/rstb.2008.0304.

Andrady, A.L., & Rajapakse, N. (2016). Additives and chemicals in plastics. *In:* Takada, H., Karapanagioti, H.K. (Eds.), *Hazardous Chemicals Associated with Plastics in the Marine Environment. The Handbook of Environmental Chemistry* (Vol. 78). Springer, Cham. https://doi.org/10.1007/698_2016_124.

Ankeny, R.A., & Leonelli, S. (2011). What's so special about model organisms? *Studies in History and Philosophy of Science Part A*, 42(2), 313–323. https://doi.org/10.1016/j.shpsa.2010.11.039.

Aragaw, T.A. (2021). Microplastic pollution in African countries' water systems: A review on findings, applied methods, characteristics, impacts, and managements. *SN Applied Sciences*, 3(6). https://doi.org/10.1007/s42452-021-04619-z.

Ateia, M., Zheng, T., Calace, S. et al. (2020). Sorption behavior of real microplastics (MPs): Insights for organic micropollutants adsorption on a large set of well-characterized MPs. *Science of The Total Environment*, 720. https://doi.org/10.1016/j.scitotenv.2020.137634.

Bacchetta, R., Santo, N., Valenti, I. et al. (2018). Comparative toxicity of three differently shaped carbon nanomaterials on *Daphnia magna*: Does a shape effect exist? *Nanotoxicology*, 12(3), 201–223. https://doi.org/10.1080/17435390.2018.1430258.

Baranov, M.V., Kumar, M., Sacanna, S. et al. (2021). Modulation of immune responses by particle size and shape. *Frontiers in Immunology*, 11, 1–23. https://doi.org/10.3389/fimmu.2020.607945.

Barton, G.M., & Medzhitov, R. (2002). Toll-like receptors and their ligands. *In: Current Topics in Microbiology and Immunology* (Vol. 270). Springer, Berlin, Heidelberg. https://doi.org/10.1007/978-3-642-59430-4_5.

Beijer, N.R., Dehaut, A., Carlier, M.P. et al. (2022). Relationship between particle properties and immunotoxicological effects of environmentally-sourced microplastics. *Frontiers in Water*, 4, 1–16. https://doi.org/10.3389/frwa.2022.866732.

Bhagat, K., Barrios, A.C., Rajwade, K. et al. (2022). Aging of microplastics increases their adsorption affinity towards organic contaminants. *Chemosphere*, 298, 134238. https://doi.org/10.1016/j.chemosphere.2022.134238.

Bhuyan, M.S. (2022). Effects of microplastics on fish and in human health. *Frontiers in Environmental Science*, 10, 1–17. https://doi.org/10.3389/fenvs.2022.827289.

Boháčková, J., Havlíčková, L., Semerád, J. et al. (2023). *In vitro* toxicity assessment of polyethylene terephthalate and polyvinyl chloride microplastics using three cell lines from rainbow trout (*Oncorhynchus Mykiss*). *Chemosphere*, 312. https://doi.org/10.1016/j.chemosphere.2022.136996.

Bowley, J., Baker-Austin, C., Porter, A. et al. (2021). Oceanic hitchhikers: Assessing pathogen risks from marine microplastic. *Trends in Microbiology*, 29(2), 107–116. https://doi.org/10.1016/j.tim.2020.06.011.

Brander, S.M., Senathirajah, K., Fernandez, M.O. et al. (2024). The time for ambitious action is now: Science-based recommendations for plastic chemicals to inform an effective global plastic treaty. *Science of The Total Environment*, 949. https://doi.org/10.1016/j.scitotenv.2024.174881.

Brehm, J., Wilde, M.V., Reiche, L. et al. (2022). In-depth characterization revealed polymer type and chemical content specific effects of microplastic on *Dreissena bugensis*. *Journal of Hazardous Materials*, 437. https://doi.org/10.1016/j.jhazmat.2022.129351.

Brehm, J., Ritschard, S., Laforsch, C. et al. (2023). The complexity of micro- and nanoplastic research in the Genus daphnia: A systematic review of study variability and a meta-analysis of immobilization rates. *Journal of Hazardous Materials*, 458. https://doi.org/10.1016/j.jhazmat.2023.131839.

Burrows, S.D., Frustaci, S., Thomas, K.V. et al. (2020). Expanding exploration of dynamic microplastic surface characteristics and interactions. *TrAC – Trends in Analytical Chemistry*, 130, 115993. https://doi.org/10.1016/j.trac.2020.115993.

Cannon, G.J., & Swanson, J.A. (1992). The macrophage capacity for phagocytosis. *Journal of Cell Science*, 101(4), 907–913. https://doi.org/10.1242/jcs.101.4.907.

Capolupo, M., Rafiq, A., Coralli, I. et al. (2023). Bioplastic leachates characterization and impacts on early larval stages and adult mussel cellular, biochemical, and physiological responses. *Environmental Pollution*, 319, 120951. https://doi.org/10.1016/j.envpol.2022.120951.

Champion, J.A., & Mitragotri, S. (2006). Role of target geometry in phagocytosis. *In: Proceedings of the National Academy of Sciences of the United States of America*, 103(13), 4930–4934. https://doi.org/10.1073/pnas.0600997103.

Champion, J.A., Walker, A. & Mitragotri, S. (2008). Role of particle size in phagocytosis of polymeric microspheres. *Pharm Res.*, 25: 1815–1821. https://doi.org/10.1007/s11095-008-9562-y

Chen, J., Wang, W., Liu, H. et al. (2021). A review on the occurrence, distribution, characteristics, and analysis methods of microplastic pollution in ecosystems. *Environmental Pollutants and Bioavailability*, 33(1), 227–246. https://doi.org/10.1080/26395940.2021.1960198.

Chen, Y.N., Rani, A., Chiang, C.Y. et al. (2023). Monitoring, control, and assessment of microplastics in bioenvironmental systems. *Environmental Technology and Innovation*, 32, 103250. https://doi.org/10.1016/j.eti.2023.103250.

Costa, C.Q., Afonso, I.I., Cruz, J. et al. (2024). Environmental markers of plastics and microplastics. *Environmental Science and Technology*, 58(20), 8889–8898. https://doi.org/10.1021/acs.est.3c09662.

Covernton, G.A., Pearce, C.M., Gurney-Smith, H.J. et al. (2019). Size and shape matter: A preliminary analysis of microplastic sampling technique in seawater studies with implications for ecological risk assessment. *Science of The Total Environment*, 667, 124–132. https://doi.org/10.1016/j.scitotenv.2019.02.346.

Coyle, R., Hardiman, G., & O'Driscoll, K. (2020). Microplastics in the marine environment: A review of their sources, distribution processes, uptake, and exchange in ecosystems. *Case Studies in Chemical and Environmental Engineering*, 2. https://doi.org/10.1016/j.cscee.2020.100010.

Da Costa, J.P., Avellan, A., Mouneyrac, C. et al. (2023). Plastic additives and microplastics as emerging contaminants: Mechanisms and analytical assessment. *TrAC – Trends in Analytical Chemistry*, 158. https://doi.org/10.1016/j.trac.2022.116898.

Danopoulos, E., Twiddy, M., West, R. et al. (2022). A rapid review and meta-regression analyses of the toxicological impacts of microplastic exposure in human cells. *Journal of Hazardous Materials*, 427, 127861. https://doi.org/10.1016/j.jhazmat.2021.127861.

De Sá, L.C., Oliveira, M., Ribeiro, F. et al. (2018). Studies of the effects of microplastics on aquatic organisms: What do we know and where should we focus our efforts in the future? *Science of The Total Environment*, 645, 1029–1039. https://doi.org/10.1016/j.scitotenv.2018.07.207.

Deng, Y., Zhang, Y., Lemos, B. et al. (2017). Tissue accumulation of microplastics in mice and biomarker responses suggest widespread health risks of exposure. *Scientific Reports*, 7, 1–10. https://doi.org/10.1038/srep46687.

Dichgans, F., Boos, J.P., Ahmadi, P. et al. (2023). Integrated numerical modeling to quantify transport and fate of microplastics in the hyporheic zone. *Water Research*, 243, 120349. https://doi.org/10.1016/j.watres.2023.120349.

Donaldson, K., Murphy, F.A., Duffin, R. et al. (2010). Asbestos, carbon nanotubes and the pleural mesothelium: A review of the hypothesis regarding the role of long fibre retention in the parietal pleura, inflammation, and mesothelioma. *Particle and Fibre Toxicology*, 7, 1–17. https://doi.org/10.1186/1743-8977-7-5.

Dris, R., Agarwal, S., & Laforsch, C. (2020). Plastics: From a success story to an environmental problem and a global challenge. *Global Challenges*, 4(6), 1–2. https://doi.org/10.1002/gch2.202000026.

Du, S., Zhu, R., Cai, Y. et al. (2021). Environmental fate and impacts of microplastics in aquatic ecosystems: A review. *RSC Advances*, 11(26), 15762–15784. https://doi.org/10.1039/D1RA00880C.

Dusza, H.M., Katrukha, E.A., Nijmeijer, S.M. et al. (2022). Uptake, transport, and toxicity of pristine and weathered micro- and nanoplastics in human placenta cells. *Environmental Health Perspectives*, 130(9). https://doi.org/10.1289/EHP10873.

Elagami, H., Frei, S., Boos, J.P. et al. (2023). Quantifying microplastic residence times in lakes using mesocosm experiments and transport modelling. *Water Research*, 229, 119463. https://doi.org/10.1016/j.watres.2022.119463.

Enyoh, C.E., Wang, Q., Ovuoraye, P.E. et al. (2022). Toxicity evaluation of microplastics to aquatic organisms through molecular simulations and fractional factorial designs. *Chemosphere*, 308(P2), 136342. https://doi.org/10.1016/j.chemosphere.2017.06.048 and https://doi.org/10.1016/j.scitotenv.2021.147670

Eriksen, M., Lebreton, L.C., Carson, H.S., Thiel, M., Moore, C.J., Borerro, J.C., Galgani, F., Ryan, P.G., Reisser, J. (2014). Plastic pollution in the world's oceans: More than 5 Trillion Plastic Pieces Weighing over 250,000 Tons Afloat at Sea. PLoS One. 9(12), e111913. doi: 10.1371/journal.pone.0111913

Esders, E.M., Georgi, C., Babel, W. et al. (2024). Vertical concentrations gradients and transport of airborne microplastics in wind tunnel experiments. *Aerosol Research*, 2(2), 235–243. https://doi.org/10.5194/ar-2-235-2024.

Espinosa, C., Esteban, M.Á., & Cuesta, A. (2019). Dietary administration of PVC and PE microplastics produces histological damage, oxidative stress and immunoregulation in European sea bass (*Dicentrarchus labrax* L.). *Fish & Shellfish Immunology*, 95, 574–583. https://doi.org/10.1016/j.fsi.2019.10.072.

Evangeliou, N., Grythe, H., Klimont, Z. et al. (2020). Atmospheric transport is a major pathway of microplastics to remote regions. *Nature Communications*, 11(1). http://dx.doi.org/10.1038/s41467-020-17201-9.

Fackelmann, G., & Sommer, S. (2019). Microplastics and the gut microbiome: How chronically exposed species may suffer from gut dysbiosis. *Marine Pollution Bulletin*, 143, 193–203. https://doi.org/10.1016/j.marpolbul.2019.04.030.

Fagnani, D.E., Tami, J.L., Copley, G. et al. (2021). 100th anniversary of macromolecular science viewpoint: Redefining sustainable polymers. *ACS Macro Letters*, 10(1), 41–53. https://doi.org/10.1021/acsmacrolett.0c00789.

Folino, A., Karageorgiou, A., Calabrò, P.S. et al. (2020). Biodegradation of wasted bioplastics in natural and industrial environments: A review. *Sustainability*, 12(15), 6030. https://doi.org/10.3390/su12156030.

Franzellitti, S., Canesi, L., Auguste, M. et al. (2019). Microplastic exposure and effects in aquatic organisms: A physiological perspective. *Environmental Toxicology and Pharmacology*, 68, 37–51. https://doi.org/10.1016/j.etap.2019.03.009.

Freeman, S.A., & Grinstein, S. (2014). Phagocytosis: Receptors, signal integration, and the cytoskeleton. *Immunological Reviews*, 262(1), 193–215. https://doi.org/10.1111/imr.12212.

Frydkjær, C.K., Iversen, N., & Roslev, P. (2017). Ingestion and egestion of microplastics by the Cladoceran Daphnia magna: Effects of regular and irregular shaped plastic and sorbed phenanthrene. *Bulletin of Environmental Contamination and Toxicology*, 99(6), 655–661. http://dx.doi.org/10.1007/s00128-017-2186-3.

Fu, L., Li, J., Wang, G. et al. (2021). Adsorption behavior of organic pollutants on microplastics. *Ecotoxicology and Environmental Safety*, 217, 112207. https://doi.org/10.1016/j.ecoenv.2021.112207.

Gambardella, C., Piazza, V., Albentosa, M. et al. (2019). Microplastics do not affect standard ecotoxicological endpoints in marine unicellular organisms. *Marine Pollution Bulletin*, 143, 140–143. https://doi.org/10.1016/j.marpolbul.2019.04.055.

Galloway, T., Cole, M., & Lewis, C. (2017). Interactions of microplastic debris throughout the marine ecosystem. *Nat. Ecol. Evol.*, 1, 0116. https://doi.org/10.1038/s41559-017-0116.

Geyer, R., Jambeck, J.R., & Law, K.L. (2017). Production, use, and fate of all plastics ever made. *Science Advances*, 3(7), 25–29. https://www.science.org/doi/10.1126/sciadv.1700782.

Gupta, N., Parsai, T., & Kulkarni, H.V. (2024). A review on the fate of micro and nano plastics (MNPs) and their implication in regulating nutrient cycling in constructed wetland systems. *Journal of Environmental Management*, 350, 119559. https://doi.org/10.1016/j.jenvman.2023.119559.

Hahladakis, J.N., Velis, C.A., Weber, R. et al. (2018). An overview of chemical additives present in plastics: Migration, release, fate, and environmental impact during their use, disposal, and recycling. *Journal of Hazardous Materials*, 344, 179–199. http://dx.doi.org/10.1016/j.jhazmat.2017.10.014.

Halappanavar, S., & Mallach, G. (2021). Adverse outcome pathways and *in vitro* toxicology strategies for microplastics hazard testing. *Current Opinion in Toxicology*, 28, 52–61. https://doi.org/10.1016/j.cotox.2021.09.002.

Haleem, N., Kumar, P., Zhang, C. et al. (2024). Microplastics and associated chemicals in drinking water: A review of their occurrence and human health implications. *Science of the Total Environment*, 912, 169594. https://doi.org/10.1016/j.scitotenv.2023.169594.

Hamidian, A.H., Ozumchelouei, E.J., Feizi, F. et al. (2021). A review on the characteristics of microplastics in wastewater treatment plants: A source for toxic chemicals. *Journal of Cleaner Production*, 295, 126480. https://doi.org/10.1016/j.jclepro.2021.126480.

Hao, L., Ma, H., & Xing, B. (2024). Surface characteristics and adsorption properties of polypropylene microplastics by ultraviolet irradiation and natural aging. *Science of The Total Environment*, 944, 173962. https://doi.org/10.1016/j.scitotenv.2024.173962.

Hartmann, N.B., Huffer, T., Thompson, R.C. et al. (2019). Are we speaking the same language? Recommendations for a definition and categorization framework for plastic debris. *Environmental Science and Technology*, 53(3), 1039–1047. https://doi.org/10.1021/acs.est.8b05297.

He, B., Smith, M., Egodawatta, P. et al. (2021). Dispersal and transport of microplastics in river sediments, *Environmental Pollution*, 279, 116884. https://doi.org/10.1016/j.envpol.2021.116884.

Heinrich, P., & Braunbeck, T. (2020). Microplastic particles reduce EROD-induction specifically by highly lipophilic compounds in RTL-W1 cells. *Ecotoxicology and Environmental Safety*, 189, 110041. https://doi.org/10.1016/j.ecoenv.2019.110041.

Hoang, T.C., & Felix-Kim, M. (2020). Microplastic consumption and excretion by fathead minnows (*Pimephales Promelas*): Influence of particles size and body shape of fish. *Science of The Total Environment*, 704, 135433. https://doi.org/10.1016/j.scitotenv.2019.135433.

Höfer, R., & Hinrichs, K. (2009). Additives for the manufacture and processing of polymers. *In:* Eyerer, P., Weller, M., Hübner, C. (Eds.), *Polymers – Opportunities and Risks II: The Handbook of Environmental Chemistry* (Vol 12). Springer, Berlin, Heidelberg. https://doi.org/10.1007/698_2009_12.

Horn, D.A., Granek, E.F., & Steele, C.L. (2020). Effects of environmentally relevant concentrations of microplastic fibers on pacific mole crab (*Emerita Analoga*) mortality and reproduction. *Limnology and Oceanography Letters*, 5(1), 74–83. https://doi.org/10.1002/lol2.10137.

Hu, K., Yang, Y., Zuo, J. et al. (2022). Emerging microplastics in the environment: Properties, distributions, and impacts. *Chemosphere*, 297, 134118. https://doi.org/10.1016/j.chemosphere.2022.134118.

Huang, H., Wang, F., Ma, S. et al. (2024). 262 reviews of environmental contamination and toxicology. *In: The Aquatic Plastisphere: Methodology, Biofilm Formation Mechanism, and Microbial Diversity*. Springer International Publishing. https://doi.org/10.1007/s44169-024-00063-3.

Huppertsberg, S., & Knepper, T.P. (2018). Instrumental analysis of microplastics—Benefits and challenges. *Analytical and Bioanalytical Chemistry*, 410(25), 6343–6352. https://doi.org/10.1007/s00216-018-1210-8.

Imhof, H.K., Sigl, R., Brauer, E. et al. (2017). Spatial and temporal variation of macro-, meso- and microplastic abundance on a remote coral Island of the Maldives, Indian Ocean. *Marine Pollution Bulletin*, 116(1–2), 340–347. http://dx.doi.org/10.1016/j.marpolbul.2017.01.010.

Jabeen, K., Li, B., Chen, Q. et al. (2018). Effects of virgin microplastics on goldfish (*Carassius Auratus*). *Chemosphere*, 213, 323–332. https://doi.org/10.1016/j.chemosphere.2018.09.031.

Jasinski, J., Völkl, M., Wilde, M.V. et al. (2024). Influence of the polymer type of a microplastic challenge on the reaction of murine cells. *Journal of Hazardous Materials*, 465. https://doi.org/10.1016/j.jhazmat.2023.133280.

Jayalath, S., Larsen, S.C., & Grassian, V.H. (2018). Surface adsorption of nordic aquatic fulvic acid on amine-functionalized and non-functionalized mesoporous silica nanoparticles. *Environmental Science: Nano,* 5(9), 2162–2172. https://doi.org/10.1039/C8EN00618K.

Jeon, S., Clavadetscher, J., Lee, D.K. et al. (2018). Surface charge-dependent cellular uptake of polystyrene nanoparticles. *Nanomaterials*, 8(12), 1–11. https://doi.org/10.3390/nano8121028.

Jeong, J., & Choi, J. (2020). Development of AOP relevant to microplastics based on toxicity mechanisms of chemical additives using ToxCast™ and deep learning models combined approach. *Environment International*, 137, 105557. https://doi.org/10.1016/j.envint.2020.105557.

Jia, J., Liu, Q., Zhao, E. et al. (2024). Biofilm formation on microplastics and interactions with antibiotics, antibiotic resistance genes and pathogens in aquatic environment. *Eco-Environment & Health*. 3(1), 516-528. https://doi.org/10.1016/j.eehl.2024.05.003.

Johansson, K.S. (2017). Surface modification of plastics. *In: Applied Plastics Engineering Handbook: Processing, Materials, and Applications* (2nd Edn.) (pp. 443–487). William Andrew Publishing. https://doi.org/10.1016/B978-0-323-39040-8.00020-1.

Junaid, M., & Wang, J. (2021). Interaction of nanoplastics with extracellular polymeric substances (EPS) in the aquatic environment: A special reference to eco-corona formation and associated impacts. *Water Research*, 201. https://doi.org/10.1016/j.watres.2021.117319.

Kahane-Rapport, S.R., Czapanskiy, M.F., Fahlbusch, J.A. et al. (2022). Field measurements reveal exposure risk to microplastic ingestion by filter-feeding megafauna. *Nature Communications*, 13(1), 1–11. https://doi.org/10.1038/s41467-022-33334-5.

Kasmuri, N., Tarmizi, N.A.A., & Mojiri, A. (2022). Occurrence, impact, toxicity, and degradation methods of microplastics in environment: A review. *Environmental Science and Pollution Research*, 29(21), 30820–30836. https://doi.org/10.1007/s11356-021-18268-7.

Kernchen, S., Löder, M.G., Fischer, F. et al. (2022). Airborne microplastic concentrations and deposition across the Weser River catchment. *Science of the Total Environment*, 818. https://doi.org/10.1016/j.scitotenv.2021.151812.

Kernchen, S., Schmalz, H., Löder, M.G.J. et al. (2024). Atmospheric deposition studies of microplastics in central Germany. *Air Quality, Atmosphere, and Health*. https://doi.org/10.1007/s11869-024-01571-w.

Kholodovych, V., & Welsh, W.J. (2007). Densities of amorphous and crystalline polymers. *In:* Mark, J.E. (Ed.), *Physical Properties of Polymers Handbook*. Springer, New York, NY. https://doi.org/10.1007/978-0-387-69002-5_37.

Kibria, G., Nugegoda, D., & Haroon, A.K.Y. (2022). Microplastic (MP) pollution in the context of occurrence, distribution, composition and concentration in surface waters and sediments: A global overview. *In:* Hashmi, M.Z. (Ed.), *Microplastic Pollution. Emerging Contaminants, and Associated Treatment Technologies*. Springer, Cham. https://doi.org/10.1007/978-3-030-89220-3_7.

Kim, S., Hyeon, Y., & Park, C. (2023). Microplastics' shape and morphology analysis in the presence of natural organic matter using flow imaging microscopy. *Molecules*, 28(19). https://doi.org/10.3390/molecules28196913.

Kinjo, A., Mizukawa, K., Takada, H. et al. (2019). Size-dependent elimination of ingested microplastics in the Mediterranean Mussel Mytilus Galloprovincialis. *Marine Pollution Bulletin*, 149, 110512. https://doi.org/10.1016/j.marpolbul.2019.110512.

Kniese, J., Ritschar, S., Bünger, L. et al. (2023). Localisation and identification of polystyrene particles in tissue sections using Raman spectroscopic imaging. *NanoImpact*, 30. https://doi.org/10.1016/j.impact.2023.100465.

Koelmans, A.A., Redondo-Hasselerharm, P.E., Nor, N.H.M. et al. (2022). Risk assessment of microplastic particles. *Nature Reviews Materials*, 7(2), 138–152. https://doi.org/10.1038/s41578-021-00411-y.

Kokalj, A.J., Kalčíková, G., Selonen, S. et al. (2024). Strategy towards producing relevant and reliable data for the hazard assessment of micro- and nanoplastics in agricultural soils. *TrAC – Trends in Analytical Chemistry*, 172. https://doi.org/10.1016/j.trac.2024.117567.

Koval, M., Preiter, K., Adles, C. et al. (1998). Size of IgG-opsonized particles determines macrophage response during internalization. *Experimental Cell Research*, 242(1), 265–273. https://doi.org/10.1006/excr.1998.4110.

Kretzschmar, R., Sticher, H., & Hesterberg, D. (1997). Effects of adsorbed humic acid on surface charge and flocculation of kaolinite. *Soil Science Society of America Journal*, 61(1), 101–108. https://doi.org/10.2136/sssaj1997.03615995006100010016x.

Kwan, C.S., & Takada, H. (2016). Release of additives and monomers from plastic wastes. *In:* Takada, H., & Karapanagioti, H.K. (Eds.), *Hazardous Chemicals Associated with Plastics in the Marine Environment: The Handbook of Environmental Chemistry* (Vol. 78). Springer, Cham. https://doi.org/10.1007/698_2016_122.

Lacerda, A.L.D.F., Rodrigues, L.D.S., van Sebille, E. et al. (2019). Plastics in sea surface waters around the Antarctic Peninsula. *Scientific Reports*, 9(1), 1–12. https://doi.org/10.1038/s41598-019-40311-4.

Laforsch, C., Ramsperger, A.F.R.M., Mondellini, S. et al. (2021). Microplastics: A novel suite of environmental contaminants but present for decades. *In:* Reichl, F.X., & Schwenk, M. (Eds.), *Regulatory Toxicology*. Springer, Berlin, Heidelberg. https://doi.org/10.1007/978-3-642-36206-4_138-1.

Lagarde, F., Olivier, O., Zanella, M. et al. (2016). Microplastic interactions with freshwater microalgae: Hetero-aggregation and changes in plastic density appear strongly dependent on polymer type. *Environmental Pollution*, 215, 331–339. https://doi.org/10.1016/j.envpol.2016.05.006.

Lambert, S., Scherer, C., & Wagner, M. (2017). Ecotoxicity testing of microplastics: Considering the heterogeneity of physicochemical properties. *Integrated Environmental Assessment and Management*, 13(3), 470–475. https://doi.org/10.1002/ieam.1901.

Lara-Topete, G.O., Castanier-Rivas, J.D., Bahena-Osorio, M. et al. (2024). Compounding one problem with another? A look at biodegradable microplastics. *Science of The Total Environment*, 173735. https://doi.org/10.1016/j.scitotenv.2024.173735.

Lehmann, M., Häusl, F.P., & Gekle, S. (2023). Modeling of vertical microplastic transport by rising bubbles. *Microplastics and Nanoplastics*, 3(1), 0–5. https://doi.org/10.1186/s43591-023-00053-7.

Lei, L., Wu, S., Lu, S. et al. (2018). Microplastic particles cause intestinal damage and other adverse effects in zebrafish *Danio Rerio* and *Nematode Caenorhabditis* elegans. *Science of The Total Environment*, 619–620, 1–8. https://doi.org/10.1016/j.scitotenv.2017.11.103.

Lesniak, A., Salvati, A., Santos-Martinez, M.J. et al. (2013). Nanoparticle adhesion to the cell membrane and its effect on nanoparticle uptake efficiency. *Journal of the American Chemical Society*, 135(4), 1438–1444. https://doi.org/10.1021/ja309812z.

Li, Y., Chen, P., Tang, Y. et al. (2024). Microplastics in water: A review of characterization and removal methods. *Sustainability (Switzerland)*, 16(10). https://doi.org/10.3390/su16104033.

Lim, X.Z. (2021). Microplastics are everywhere: But are they harmful? *Nature*, 593(7857). https://www.nature.com/articles/d41586-021-01143-3.

Lin, Q., Pang, L., Ngo, H.H., Guo, W. et al. (2023). Occurrence of microplastics in three types of household cleaning products and their estimated emissions into the aquatic environment. *Science of The Total Environment*, 902, 165903. https://doi.org/10.1016/j.scitotenv.2023.165903.

Lin, W., Wu, Z., Wang, Y. et al. (2024). Size-dependent vector effect of microplastics on the bioaccumulation of polychlorinated biphenyls in Tilapia: A tissue-specific study. *Science of The Total Environment*, 915, 170047. https://doi.org/10.1016/j.scitotenv.2024.170047.

Liu, P., Zhan, X., Wu, X. et al. (2020). Effect of weathering on environmental behavior of microplastics: Properties, sorption, and potential risks. *Chemosphere*, 242, 125193. https://doi.org/10.1016/j.chemosphere.2019.125193.

Liu, X., Fang, L., Yan, X. et al. (2023). Surface functional groups and biofilm formation on microplastics: Environmental implications. *Science of The Total Environment*, 903, 166585. https://doi.org/10.1016/j.scitotenv.2023.166585.

Löder, M.G.J., Kuczera, M., Mintenig, S. et al. (2015). Focal plane array detector-based micro-fourier-transform infrared imaging for the analysis of microplastics in environmental samples. *Environmental Chemistry*, 12(5), 563–581. https://doi.org/10.1071/EN14205.

Lu, Y., Zhang, Y., Deng, Y. et al. (2016). Uptake and accumulation of polystyrene microplastics in Zebrafish (*Danio Rerio*) and toxic effects in liver. *Environmental Science and Technology*, 50(7), 4054–4060. https://doi.org/10.1021/acs.est.6b00183.

Luan, Y., Li, M., Zhou, W. et al. (2023). The fish microbiota: Research progress and potential applications. *Engineering*, 29, 137–146. https://doi.org/10.1016/j.eng.2022.12.011.

Luo, H., Liu, C., He, D. et al. (2022). Environmental behaviors of microplastics in aquatic systems: A systematic review on degradation, adsorption, toxicity, and biofilm under aging conditions. *Journal of Hazardous Materials*, 423(PA), 126915. https://doi.org/10.1016/j.jhazmat.2021.126915.

Maddison, C., Sathish, C.I., Lakshmi, D. et al. (2023). An advanced analytical approach to assess the long-term degradation of microplastics in the marine environment. *npj Materials Degradation*, 7(1), 1–11. https://doi.org/10.1038/s41529-023-00377-y.

Mao, R., Lang, M., Yu, X. et al. (2020). Aging mechanism of microplastics with UV irradiation and its effects on the adsorption of heavy metals. *Journal of Hazardous Materials*, 393, 122515. https://doi.org/10.1016/j.jhazmat.2020.122515.

Maric, B., Schuster, S., & Machnik, P. (2024). Exposure to phthalate plasticizer compromises normal brain function in an adult vertebrate. *Ecotoxicology and Environmental Safety*, 286. https://doi.org/10.1016/j.ecoenv.2024.117187.

Marturano, V., Cerruti, P., & Ambrogi, V. (2019). Polymer additives. *Physical Sciences Reviews*, 2(6), 1–20. https://doi.org/10.1515/psr-2016-0130.

Meides, N., Menzel, T., Poetzschner, B. et al. (2021). Reconstructing the environmental degradation of polystyrene by accelerated weathering. *Environmental Science & Technology*, 55(12), 7930–7938. https://doi.org/10.1021/acs.est.0c07718.

Meng, W., Sun, H., & Su, G. (2023). Plastic packaging-associated chemicals and their hazards: An overview of reviews. *Chemosphere*, 331, 138795. https://doi.org/10.1016/j.chemosphere.2023.138795.

Menzel, T., Meides, N., Mauel, A. et al. (2022). Degradation of low-density polyethylene to nanoplastic particles by accelerated weathering. *Science of The Total Environment*, 826. https://doi.org/10.1016/j.scitotenv.2022.154035.

Möller, J.N., Heisel, I., Satzger, A. et al. (2022). Tackling the challenge of extracting microplastics from soils: A protocol to purify soil samples for spectroscopic analysis. *Environmental Toxicology and Chemistry*, 41(4), 844–857. https://doi.org/10.1002/etc.5024.

Mondellini, S., Schwarzer, M., Völkl, M. et al. (2024). Size dependent uptake and trophic transfer of polystyrene microplastics in unicellular freshwater eukaryotes. *Science of The Total Environment*, 929. https://doi.org/10.1016/j.scitotenv.2024.172470.

Moses, S.R., Roscher, L., Primpke, S. et al. (2023). Comparison of two rapid automated analysis tools for large FTIR microplastic datasets. *Analytical and Bioanalytical Chemistry*, 415(15), 2975–2987. https://doi.org/10.1007/s00216-023-04630-w.

Moura, D.S., Pestana, C.J., Moffat, C.F. et al. (2023). Characterisation of microplastics is key for reliable data interpretation. *Chemosphere*, 331, 138691. https://doi.org/10.1016/j.chemosphere.2023.138691.

Moyal, J., Dave, P.H., Wu, M. et al. (2023). Impacts of biofilm formation on the physicochemical properties and toxicity of microplastics: A concise review. *Reviews of Environmental Contamination and Toxicology*, 261(1), 1–17. https://doi.org/10.1007/s44169-023-00035-z.

Nasser, F., & Lynch, I. (2016). Secreted protein eco-corona mediates uptake and impacts of polystyrene nanoparticles on *Daphnia magna*. *Journal of Proteomics*, 137, 45–51. https://doi.org/10.1016/j.jprot.2015.09.005.

Nava, V., Chandra, S., Aherne, J. et al. (2023). Plastic debris in lakes and reservoirs. *Nature*, 619(7969), 317–322. https://doi.org/10.1038/s41586-023-06168-4.

Nerin, C., Alfaro, P., Aznar, M. et al. (2013). The challenge of identifying non-intentionally added substances from food packaging materials: A review. *Analytica Chimica Acta*, 775, 14–24. https://doi.org/10.1016/j.aca.2013.02.028.

Ni, Z., Tan, L., Wang, J. et al. (2023). Toxic effects of pristine and aged polystyrene and their leachate on marine microalgae *Skeletonema costatum*. *Science of The Total Environment*, 857, 159614. https://doi.org/10.1016/j.scitotenv.2022.159614.

Oehlschlägel, L.M., Schmid, S., Lehmann, M. et al. (2024). Water–air transfer rates of microplastic particles through bubble bursting as a function of particle size. *Microplastics and Nanoplastics*, 4(1), 1–12. https://doi.org/10.1186/s43591-023-00079-x.

Ogonowski, M., Wagner, M., Rogell, B. et al. (2023). Microplastics could be marginally more hazardous than natural suspended solids: A meta-analysis. *Ecotoxicology and Environmental Safety*, 264, 115406. https://doi.org/10.1016/j.ecoenv.2023.115406.

Oßmann, B.E., Sarau, G., Holtmannspötter, H. et al. (2018). Small-sized microplastics and pigmented particles in bottled mineral water. *Water Research*, 141, 307–316. https://doi.org/10.1016/j.watres.2018.05.027.

Otto, D.P., Otto, A., & De Villiers, M.M. (2015). Differences in physicochemical properties to consider in the design, evaluation and choice between microparticles and nanoparticles for drug delivery. *Expert Opinion on Drug Delivery*, 12(5), 763–777. https://doi.org/10.1517/17425247.2015.988135.

Panagiotidis, K., Engelmann, B., Krauss, M. et al. (2023). The impact of amine and carboxyl functionalised microplastics on the physiology of daphnids. *Journal of Hazardous Materials*, 458, 132023. https://doi.org/10.1016/j.jhazmat.2023.132023.

Pannetier, P., Cachot, J., Clérandeau, C. et al. (2019). Toxicity assessment of pollutants sorbed on environmental sample microplastics collected on beaches: Part I – Adverse effects on fish cell line. *Environmental Pollution*, 248. https://doi.org/10.1016/j.envpol.2018.12.091.

Parrish, K., & Fahrenfeld, N.L. (2019). Microplastic biofilm in fresh and wastewater as a function of microparticle type and size class. *Environmental Science: Water Research and Technology*, 5(3), 495–505. https://doi.org/10.1039/C8EW00712H.

Paul, D., Achouri, S., Yoon, Y.Z. et al. (2013). Phagocytosis dynamics depends on target shape. *Biophysical Journal*, 105(5), 1143–1150. http://dx.doi.org/10.1016/j.bpj.2013.07.036.

Pérez-Albaladejo, E., Solé, M., & Porte, C. (2020). Plastics and plastic additives as inducers of oxidative stress. *Current Opinion in Toxicology*, 20–21, 69–76. https://doi.org/10.1016/j.cotox.2020.07.002.

Pilapitiya, P.N.T., & Ratnayake, A.S. (2024). The world of plastic waste: A review. *Cleaner Materials*, 11, 100220. https://doi.org/10.1016/j.clema.2024.100220.

Pinlova, B., & Nowack, B. (2024). From cracks to secondary microplastics: Surface characterization of polyethylene terephthalate (PET) during weathering. *Chemosphere*, 352. https://doi.org/10.1016/j.chemosphere.2024.141305.

PlasticsEurope (2024). *The Circular Economy for Plastics: A European Analysis*. https://plasticseurope.org/knowledge-hub/the-circular-economy-for-plastics-a-european-analysis-2024/.

Platt, N., & Gordon, S. (2001). Is the class A macrophage scavenger receptor (SR-A) multifunctional? The mouse's tale. *Journal of Clinical Investigation*, 108(5), 649–654. https://doi.org/10.1172/JCI13903.

Porter, A., Barber, D., Hobbs, C. et al. (2023). Uptake of microplastics by marine worms depends on feeding mode and particle shape but not exposure time. *Science of The Total Environment*, 857, 159287. https://doi.org/10.1016/j.scitotenv.2022.159287.

Prajapati, A., Narayan Vaidya, A., & Kumar, A.R. (2022). Microplastic properties and their interaction with hydrophobic organic contaminants: A review. *Environ. Sci. Pollut. Res.*, 29, 49490–49512. https://doi.org/10.1007/s11356-022-20723-y.

Prata, J.C. (2018). Airborne microplastics: Consequences to human health? *Environmental Pollution*, 234, 115–126. https://doi.org/10.1016/j.envpol.2017.11.043.

Prokić, M.D., Gavrilović, B.R., Radovanović, T.B. et al. (2021). Studying microplastics: Lessons from evaluated literature on animal model organisms and experimental approaches. *Journal of Hazardous Materials*, 414. https://doi.org/10.1016/j.jhazmat.2021.125476.

Qiao, R., Deng, Y., Zhang, S. et al. (2019). Accumulation of different shapes of microplastics initiates intestinal injury and gut microbiota dysbiosis in the gut of Zebrafish. *Chemosphere*, 236, 124334. https://doi.org/10.1016/j.chemosphere.2019.07.065.

Rahman, A.M.N.A.A., Rusli, A., Abdullah, M.K. et al. (2024). A review of microplastic surface interactions in water and potential capturing methods. *Water Science and Engineering*. 17(4), 361-370. https://doi.org/10.1016/j.wse.2023.11.008.

Ramsperger, A.F.R.M., Narayana, V.K.B., Gross, W. et al. (2020). Environmental exposure enhances the internalization of microplastic particles into cells. *Science Advances*, 6(50), 1–10. https://www.science.org/doi/10.1126/sciadv.abd1211.

Ramsperger, A.F.R.M., Jasinski, J., Völkl, M. et al. (2022). Supposedly identical microplastic particles substantially differ in their material properties influencing particle-cell interactions and cellular responses. *Journal of Hazardous Materials*, 425. https://doi.org/10.1016/j.jhazmat.2021.127961.

Ramsperger, A.F.R.M., Bergamaschi, E., Panizzolo, M. et al. (2023). Nano- and microplastics: A comprehensive review on their exposure routes, translocation, and fate in humans. *NanoImpact*, 29. https://doi.org/10.1016/j.impact.2022.100441.

Reineccius, J., Schönke, M., & Waniek, J.J. (2023). Abiotic long-term simulation of microplastic weathering pathways under different aqueous conditions. *Environmental Science and Technology*, 57(2), 963–975. https://doi.org/10.1021/acs.est.2c05746.

Revel, M., Châtel, A., & Mouneyrac, C. (2018). Micro(nano)plastics: A threat to human health. *Current Opinion in Environmental Science and Health*, 1, 17–23. https://doi.org/10.1016/j.coesh.2017.10.003.

Robertson, G.L. (2016). *The Stability and Shelf Life of Food Packaging and Food and Beverage Shelf Life*. Elsevier Ltd. http://dx.doi.org/10.1016/B978-0-08-100435-7.00003-4.

Ross, P.S., Chastain, S., Vassilenko, E. et al. (2021). Pervasive distribution of polyester fibres in the Arctic Ocean is driven by atlantic inputs. *Nature Communications*, 12(1), 4–12. http://dx.doi.org/10.1038/s41467-020-20347-1.

Rossatto, A., Arlindo, M.Z.F., de Morais, M.S. et al. (2023). Microplastics in aquatic systems: A review of occurrence, monitoring and potential environmental risks. *Environmental Advances*, 13(June), 100396. https://doi.org/10.1016/j.envadv.2023.100396.

Rozman, U., & Kalčíková, G. (2022). Seeking for a perfect (non-spherical) microplastic particle: The most comprehensive review on microplastic laboratory research. *Journal of Hazardous Materials*, 424. https://doi.org/10.1016/j.jhazmat.2021.127529.

Rudolph, J., Völkl, M., Jérôme, V. et al. (2021). Noxic effects of polystyrene microparticles on murine macrophages and epithelial cells. *Scientific Reports*, 11(1), 1–16. https://doi.org/10.1038/s41598-021-95073-9.

Rummel, C.D., Jahnke, A., Gorokhova, E. et al. (2017). Impacts of biofilm formation on the fate and potential effects of microplastic in the aquatic environment. *Environmental Science and Technology Letters*, 4(7), 258–267. https://doi.org/10.1021/acs.estlett.7b00164.

Schmid, O., & Stoeger, T. (2016). Surface area is the biologically most effective dose metric for acute nanoparticle toxicity in the lung. *Journal of Aerosol Science*, 99, 133–143. http://dx.doi.org/10.1016/j.jaerosci.2015.12.006.

Schrank, I., Trotter, B., Dummert, J. et al. (2019). Effects of microplastic particles and leaching additive on the life history and morphology of *Daphnia magna*. *Environmental Pollution*, 255. https://doi.org/10.1016/j.envpol.2019.113233.

Schrank, I., Löder, M.G., Imhof, H.K. et al. (2022). Riverine microplastic contamination in Southwest Germany: A large-scale survey. *Frontiers in Earth Science*, 10, 1–17. https://doi.org/10.3389/feart.2022.794250.

Schwarzer, M., Brehm, J., Vollmer, M. et al. (2022). Shape, size, and polymer dependent effects of microplastics on *Daphnia magna*. *Journal of Hazardous Materials*, 426. https://doi.org/10.1016/j.jhazmat.2021.128136.

Setälä, O., Norkko, J., & Lehtiniemi, M. (2016). Feeding type affects microplastic ingestion in a coastal invertebrate community. *Marine Pollution Bulletin*, 102(1): 95–101. http://dx.doi.org/10.1016/j.marpolbul.2015.11.053.

Sharma, P., Sharma, P., & Abhishek, K. (2024). Sampling, separation, and characterization methodology for quantification of microplastic from the environment. *Journal of Hazardous Materials Advances*, 14, 100416. https://doi.org/10.1016/j.hazadv.2024.100416.

Solhaug, A., Vlegels, S., & Eriksen, G.S. (2024). Atlantic salmon gill epithelial cell line ASG-10, an *in vitro* model for studying effects of microplastics in gills. *Aquatic Toxicology*, 272. https://doi.org/10.1016/j.aquatox.2024.106946.

Sutkar, P.R., Gadewar, R.D., & Dhulap, V.P. (2023). Recent trends in degradation of microplastics in the environment: A state-of-the-art review. *Journal of Hazardous Materials Advances*, 11, 100343. https://doi.org/10.1016/j.hazadv.2023.100343.

Swanson, J., & Watts, C. (1995). Macropinocytosis. *Trends in Cell Biology*, 5, 424–428. https://doi.org/10.1016/S0962-8924(00)89101-1.

Tabata, Y., & Ikada, Y. (1988). Effect of the size and surface charge of polymer microspheres on their phagocytosis by macrophage. *Biomaterials*, 9(4), 356–362. https://doi.org/10.1016/0142-9612(88)90033-6.

Tao, S., Li, T., Li, M. et al. (2024). Research advances on the toxicity of biodegradable plastics derived micro/nanoplastics in the environment: A review. *Science of The Total Environment*, 170299. https://doi.org/10.1016/j.scitotenv.2024.170299.

Teng, J., Zhao, J., Zhu, X. et al. (2021). Toxic effects of exposure to microplastics with environmentally relevant shapes and concentrations: Accumulation, energy metabolism, and tissue damage in oyster *Crassostrea Gigas*. *Environmental Pollution*, 269, 116169. https://doi.org/10.1016/j.envpol.2020.116169.

Thompson, R.C., Courtene-Jones, W., Boucher, J. et al. (2024). Twenty years of microplastics pollution research: What have we learned? *Science*, 2746, 1–18. https://doi.org/10.1126/science.adl2746.

Thornton Hampton, L.M., Brander, S.M., Coffin, S. et al. (2022). Characterizing microplastic hazards: Which concentration metrics and particle characteristics are most informative for understanding toxicity in aquatic organisms? *Microplastics and Nanoplastics*, 2(1). https://doi.org/10.1186/s43591-022-00040-4.

Tuvo, B., Scarpaci, M., Bracaloni, S. et al. (2023). Microplastics and antibiotic resistance: The magnitude of the problem and the emerging role of hospital wastewater. *International Journal of Environmental Research and Public Health*, 20(10). https://doi.org/10.3390/ijerph20105868.

Uribe-Echeverría, T., & Beiras, R. (2022). Acute toxicity of bioplastic leachates to paracentrotus lividus sea urchin larvae. *Marine Environmental Research*, 176, 105605. https://doi.org/10.1016/j.marenvres.2022.105605.

Völkl, M., Jérôme, V., Weig, A. et al. (2022). Pristine and artificially-aged polystyrene microplastic particles differ in regard to cellular response. *Journal of Hazardous Materials*, 435. https://doi.org/10.1016/j.jhazmat.2022.128955.

Vroom, R.J., Koelmans, A.A., Besseling, E. et al. (2017). Aging of microplastics promotes their ingestion by marine zooplankton. *Environmental Pollution*, 231, 987–996. https://doi.org/10.1016/j.envpol.2017.08.088.

Wagner, M., Monclús, L., Arp, H.P.H. et al. (2024). State of the science on plastic chemicals: Identifying and addressing chemicals and polymers of concern. *PlastChem.*, Version v1 (Mar 14, 2024). Zenodo. https://doi.org/10.5281/zenodo.10701706.

Wang, C., Zhao, J., & Xing, B. (2021). Environmental source, fate, and toxicity of microplastics. *Journal of Hazardous Materials*, 407, 124357. https://doi.org/10.1016/j.jhazmat.2020.124357.

Wang, J., Lee, J., Kwon, E.E., & Jeong, S. (2023). Quantitative analysis of polystyrene microplastic and styrene monomer released from plastic food containers. *Heliyon*, 9(5), e15787. https://doi.org/10.1016/j.heliyon.2023.e15787.

Wang, J., Wu, F., Dong, S. et al. (2024). Meta-analysis of the effects of microplastic on fish: Insights into growth, survival, reproduction, oxidative stress, and gut microbiota diversity. *Water Research*, 267, 122493. https://doi.org/10.1016/j.watres.2024.122493.

Wang, Y., Xiang, L., Amelung, W. et al. (2023). Micro- and nanoplastics in soil ecosystems: Analytical methods, fate, and effects. *TrAC – Trends in Analytical Chemistry*, 169. https://doi.org/10.1016/j.trac.2023.117309.

Weig, A.R., Löder, M.G., Ramsperger, A.F.R.M. et al. (2021). *In situ* prokaryotic and eukaryotic communities on microplastic particles in a small headwater stream in Germany. *Frontiers in Microbiology*, 12, 1–14. https://doi.org/10.3389/fmicb.2021.660024.

White, A., & Lockyer, S. (2020). Removing plastic packaging from fresh produce: What's the impact? *Nutrition Bulletin*, 45(1), 35–50. https://doi.org/10.1111/nbu.12420.

Wieland, S., Balmes, A., Bender, J. et al. (2022). From properties to toxicity: Comparing microplastics to other airborne microparticles. *Journal of Hazardous Materials*, 428, 128151. https://doi.org/10.1016/j.jhazmat.2021.128151.

Wieland, S., Ramsperger, A.F.R.M., Gross, W. et al. (2024). Nominally identical microplastic models differ greatly in their particle-cell interactions. *Nature Communications*, 15(1), 1–17. https://doi.org/10.1038/s41467-024-45281-4.

Witzmann, T., Ramsperger, A.F.R.M., Wieland, S. et al. (2022). Repulsive interactions of eco-corona-covered microplastic particles quantitatively follow modeling of polymer brushes. *Langmuir*, 38(29), 8748–8756. https://doi.org/10.1021/acs.langmuir.1c03204.

Wu, B., Wu, X., Liu, S. et al. (2019). Size-dependent effects of polystyrene microplastics on cytotoxicity and efflux pump inhibition in human caco-2 cells. *Chemosphere*, 221, 333–341. https://doi.org/10.1016/j.chemosphere.2019.01.056.

Xia, Y., Niu, S., & Yu, J. (2023). Microplastics as vectors of organic pollutants in aquatic environment: A review on mechanisms, numerical models, and influencing factors. *Science of The Total Environment*, 887, 164008. https://doi.org/10.1016/j.scitotenv.2023.164008.

Xiong, X., Tu, Y., Chen, X. et al. (2019). Ingestion and egestion of polyethylene microplastics by goldfish (*Carassius auratus*): Influence of color and morphological features. *Heliyon*, 5(12), e03063. https://doi.org/10.1016/j.heliyon.2019.e03063.

Xu, S., Ma, J., Ji, R. et al. (2020). Microplastics in aquatic environments: Occurrence, accumulation, and biological effects. *Science of The Total Environment*, 703. https://doi.org/10.1016/j.scitotenv.2019.134699.

Xu, Y., Huang, D., Liu, P. et al. (2022). The characteristics of dissolved organic matter release from UV-aged microplastics and its cytotoxicity on human colonic adenocarcinoma cells. *Science of The Total Environment*, 826. https://doi.org/10.1016/j.scitotenv.2022.154177.

Yao, J., Wen, J., Li, H. et al. (2022). Surface functional groups determine adsorption of pharmaceuticals and personal care products on polypropylene microplastics. *Journal of Hazardous Materials*, 423, 127131. https://doi.org/10.1016/j.jhazmat.2021.127131.

Yao, S., Li, X., Wang, T. et al. (2023). Soil metabolome impacts the formation of the eco-corona and adsorption processes on microplastic surfaces. *Environmental Science and Technology*, 57(21), 8139–8148. https://doi.org/10.1021/acs.est.3c01877.

Yu, F., Qin, Q., Zhang, X. et al. (2024). Characteristics and adsorption behavior of typical microplastics in long-term accelerated weathering simulation. *Environmental Science: Processes and Impacts*, 26(5), 882–890. https://doi.org/10.1039/D4EM00062E.

Yu, H., Yang, B., Waigi, M.G. et al. (2020). The effects of functional groups on the sorption of naphthalene on microplastics. *Chemosphere*, 261, 127592. https://doi.org/10.1016/j.chemosphere.2020.127592.

Yu, Y., Kumar, M., Bolan, S. et al. (2024). Various additive release from microplastics and their toxicity in aquatic environments. *Environmental Pollution*, 343, 123219. https://doi.org/10.1016/j.envpol.2023.123219.

Zhai, X., Zhang, X.H., & Yu, M. (2023). Microbial colonization and degradation of marine microplastics in the plastisphere: A review. *Frontiers in Microbiology*, 14, 1–15. https://doi.org/10.3389/fmicb.2023.1127308.

Zhang, D., Zhang, Z., Liu, H. et al. (2024). Insights into the effect of crystallinity on the sorption of organic pollutants to microplastics. *Environmental Science and Pollution Research*, 31(29), 42202–42211. https://doi.org/10.1007/s11356-024-33929-z.

Zhang, H., Cheng, H., Wang, Y. et al. (2022). Influence of functional group modification on the toxicity of nanoplastics. *Frontiers in Marine Science*, 8, 1–11. https://doi.org/10.3389/fmars.2021.800782.

Zhang, Y., Paul, T., Brehm, J. et al. (2023). Role of residual monomers in the manifestation of (cyto)toxicity by polystyrene microplastic model particles. *Environmental Science and Technology*, 57(27). https://doi.org/10.1021/acs.est.3c01134.

Zhao, M., Huang, L., Arulmani, S.R.B. et al. (2022). Adsorption of different pollutants by using microplastic with different influencing factors and mechanisms in wastewater: A review. *Nanomaterials*, 12(13), 2256. https://doi.org/10.3390/nano12132256.

Zhao, K., Li, C., & Li, F. (2024). Research progress on the origin, fate, impacts, and harm of microplastics and antibiotic resistance genes in wastewater treatment plants. *Scientific Reports*, 14(1), 1–19. https://doi.org/10.1038/s41598-024-60458-z.

Zhong, H., Wu, M., Sonne, C. et al. (2023). The hidden risk of microplastic-associated pathogens in aquatic environments. *Eco-Environment and Health*, 2(3), 142–151. https://doi.org/10.1016/j.eehl.2023.07.004.

Zhou, N., Wang, Z., Yang, L. et al. (2023). Size-dependent toxicological effects of polystyrene microplastics in the shrimp *Litopenaeus Vannamei* using a histomorphology, microbiome, and metabolic approach. *Environmental Pollution*, 316(P2), 120635. https://doi.org/10.1016/j.envpol.2022.120635.

Zhu, K., Jia, H., Sun, Y. et al. (2020). Long-term phototransformation of microplastics under simulated sunlight irradiation in aquatic environments: Roles of reactive oxygen species. *Water Research*, 173, 115564. https://doi.org/10.1016/j.watres.2020.115564.

Zhu, L., Xie, C., Chen, L. et al. (2023). Transport of microplastics in the body and interaction with biological barriers, and controlling of microplastics pollution. *Ecotoxicology and Environmental Safety*, 255, 114818. https://doi.org/10.1016/j.ecoenv.2023.114818.

Zimmermann, L., Göttlich, S., Oehlmann, J. et al. (2020). What are the drivers of microplastic toxicity? Comparing the toxicity of plastic chemicals and particles to *Daphnia magna*. *Environmental Pollution*, 267. https://doi.org/10.1016/j.envpol.2020.115392.

4 | Beyond the Surface: Biofilms and Microplastics in Aquatic Systems

Fernanda Miranda Zoppas*, Nicolás Alejandro Sacco and Fernanda Albana Marchesini

Institute of Catalysis and Petrochemistry Research (INCAPE-CONICET)
Santa Fe, Argentina

Abstract: Microplastics, ubiquitous fragments of plastic pollution, have invaded aquatic ecosystems, raising alarms about their long-term impacts. But they're not alone on this voyage. Invisible communities of microorganisms, known as biofilms, colonize their surfaces, altering their fate and potential risk pathways. This review delves into this intricate and dynamic relationship, exploring how biofilms influence the transport, degradation, and potential toxicity of microplastics in aquatic environments. We examine the unique composition and activities of these "microplastic hitchhikers", revealing their role in enhancing pollutant attachment, facilitating biodegradation, and even promoting microplastic aggregation and sinking. Further, we address the concerning role of biofilms in amplifying antibiotic resistance and harboring pathogenic microbes, raising potential human health implications. By comprehensively unraveling the tangled web of microplastic-biofilm interactions, this chapter aims to pave the way for informed mitigation strategies and environmental protection measures, towards a future where aquatic ecosystems are no longer burdened by these insidious hitchhikers.

1. Introduction

Microplastics (MPs), minuscule plastic fragments dwarfed by a sesame seed, have seamlessly integrated themselves as omnipresent companions within our aquatic ecosystems (Pan et al., 2023). Their extensive prevalence, numbering in the trillions solely within the oceans, poses a threat to biodiversity and potentially jeopardizes human health (Samak et al., 2020). While this distressing pollution

*Corresponding author: fzoppas@fiq.unl.edu.ar

crisis demands our unwavering attention, a fascinating narrative unfolds when we shift our focus to the emergence of biofilms (Rajcoomar et al., 2023) as pivotal actors influencing the destiny and conceivable risks associated with MPs.

Biofilms, imperceptible communities of microorganisms, readily establish colonies on the surfaces of MPs (Rummel et al., 2017), transforming them into microscopic metropolises that alter their trajectory. These resilient companions wield a profound influence on how MPs traverse water, undergo degradation (Golmohammadi et al., 2023), and interact with their surroundings. They possess the capability to augment aggregation and sinking, potentially alleviating surface contamination, yet simultaneously facilitate fragmentation and the release of concealed harmful chemicals. Furthermore, biofilms can serve as sanctuaries for pathogens and antibiotic-resistance genes (Fajardo et al., 2023; Tavşanoğlu et al., 2020), potentially amplifying health hazards.

This chapter delves into the enchanting alliance between biofilms and MPs, exploring the intricate web of interactions between these two entities. We will unveil the governing factors behind biofilm formation on these diminutive voyagers, scrutinize their impact on transport and degradation, and dissect their role in either amplifying or mitigating potential risks. By unraveling this complex dance, we can formulate more effective strategies to combat MP pollution, safeguard aquatic ecosystems, and ultimately protect our own health.

2. Microplastics in Aquatic Systems

MPs encompass a diverse array of particles, including microbeads from personal care products, fragments from larger plastic items, and microfibers shed from various products. They originate from anthropogenic sources such as industrial discharges, urban runoff, and plastic degradation. Their small size, diverse shapes, and buoyancy make them particularly challenging to manage and monitor effectively in aquatic environments (Miranda Zoppas et al., 2023; Sacco et al., 2023).

Plastic transport mechanisms in aquatic environments involve various factors such as human activities (Rose et al., 2023), flood and storm events, hydrodynamics (Schreyers et al., 2023), and combinations thereof. Urban areas, transportation infrastructure, and wastewater treatment plants are consistently identified as sources of plastic pollution in rivers and oceans (Horton et al., 2017). Plastics can be transported over land and through rivers, but most of them do not make it to the open sea and instead accumulate on land, floating in coastal waters, or beach (Kuizenga et al., 2023). MPs can be transported and deposited by gravity-driven flows called turbidity currents, which can move sediment and organic matter over long distances in the deep sea (Jambeck Jenna et al., 2015). Coral reefs can act as a sink for MP pollution through various mechanisms such as physical barriers, turbulence modification, and incorporation within coral tissue and skeletons (Lamb et al., 2018). Waves play a significant role in the transport and mixing of MPs in the nearshore zone.

MPs exhibit complex transport dynamics in aquatic systems. They can be transported horizontally by currents and vertically through water columns, influencing their distribution across various water bodies. Their persistence is notable, as they resist natural degradation processes, resulting in long-term exposure to aquatic ecosystems (Li et al., 2023). Furthermore, their ability to adsorb other pollutants further exacerbates their environmental impact.

Rivers play a crucial role as conduits for plastic waste originating from land sources to the oceans. This transport route is significant, particularly in urban areas, where human activities and the lack of proper waste management contribute to the accumulation of plastics in river systems. The concentration of plastics near urban areas, identified in studies, underscores the need for effective waste management approaches in urban environments to mitigate the influx of plastics into rivers (van Emmerik et al., 2023).

Climate events, such as floods and storms, also play a significant role in plastic transport dynamics. The mobilization of plastics during flood events has been identified as a major source of plastic introduction into watercourses. These extreme events can transport large quantities of accumulated plastics from urban areas to rivers and, ultimately, to the oceans. Additionally, the transport capacity of plastics during climate events varies geographically, with densely populated coastal regions being particularly susceptible (van Emmerik et al., 2023).

One of the mechanisms for the dispersion of plastics in aquatic environments is transportation by ocean currents. Plastics can be carried by maritime currents from warmer regions to polar areas, where they end up accumulating due to climatic and geographical conditions. Additionally, human activities such as fishing (Habib & Thiemann, 2021), aquaculture, maritime transport, and offshore operations also contribute to the introduction of plastics into aquatic environments, further increasing plastic pollution on a global scale (Bergmann et al., 2022). Another phenomenon that contributes to MP formation is weathering, as depicted in Figure 1.

Weathering denotes the gradual deterioration of material integrity attributable to both abiotic and biotic elements, encompassing material degradation. Particularly in the context of plastic waste, weathering predominantly stems from photooxidation as a leading abiotic degradation pathway, notably affecting debris exposed to solar radiation. The process of photooxidation unfolds in three primary phases: initiation, involving the fragmentation of polymeric chains induced by ultraviolet light and the consequent formation of free radicals; propagation, characterized by autoxidation; and termination, culminating in the formation of inert byproducts. This degradation predominantly manifests on the material's surface exposed to UV radiation, inciting alterations in surface morphology, heightened surface roughness, and shifts in surface chemistry. These alterations foster microbial adhesion, fostering the development of biofilms and shaping microbial community dynamics. Moreover, the incremental fragmentation of plastic waste into diminutive particles is pivotal for facilitating subsequent

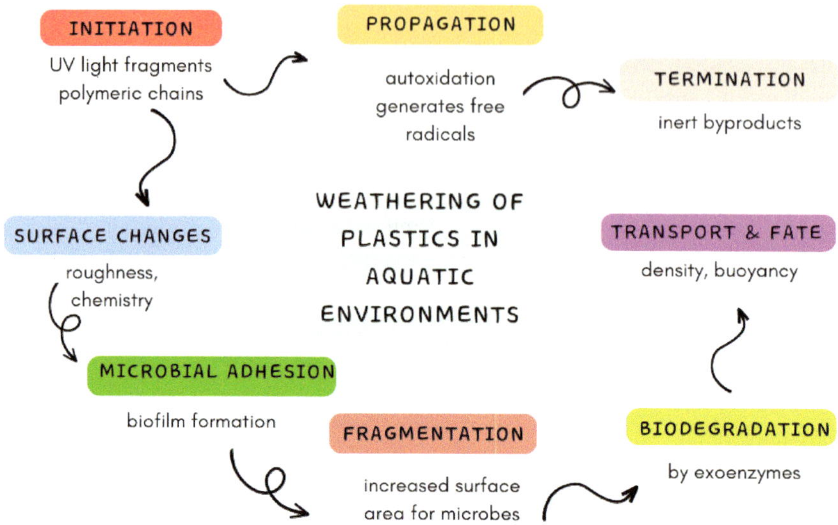

Figure 1: A schematic diagram of the weathering process of plastics in aquatic environments.

biodegradation, augmenting the available surface area for microbial colonization, and dictating the density, buoyancy, and subsidence rate of plastic particles.

Concurrently, the weathering process may be modulated by biofilm formation, potentially shielding plastic from UV exposure or influencing its vertical positioning within the water column. In addition to abiotic weathering, polymers are also susceptible to biological degradation, as exoenzymes discharged by colonizing organisms facilitate polymer disintegration into shorter-chain fragments amenable to mineralization. A comprehensive comprehension of weathering's ramifications on plastic waste is indispensable for elucidating the intricacies of MP transport dynamics and their eventual expulsion from aquatic ecosystems (Rummel et al., 2017).

MPs pose several established risks to aquatic life. Ingestion by marine organisms, ranging from zooplankton to larger species, can lead to physical harm, reduced feeding efficiency, and bioaccumulation of toxic substances within the food chain. The presence of MPs may also alter the behavior and reproductive success of aquatic organisms. Additionally, the potential for these particles to act as vectors for pathogenic microorganisms raises concerns about their role in disease transmission within aquatic ecosystems.

3. Biofilms in Aquatic Ecosystems: Dynamic Colonization of Microplastics

Biofilms, often unseen but omnipresent in our natural surroundings, represent intricate microbial communities that play pivotal roles in various ecosystems

(Tavşanoğlu et al., 2020). They are complex assemblies of microorganisms, including bacteria, fungi, and algae, that adhere to surfaces and secrete a matrix of extracellular polymeric substances (EPS) (Shree et al., 2023). Unlike free-floating microbial cells, these biofilms form cohesive and resilient communities, creating a unique microenvironment with distinct characteristics and behaviors (Sharma et al., 2023).

Biofilms are ubiquitous, thriving in diverse habitats, from natural water bodies to industrial pipelines. In aquatic environments, biofilms adhere to surfaces such as rocks, sediments, and submerged vegetation. Their functions include nutrient cycling, organic matter decomposition, and the provision of a substrate for other organisms (Li et al., 2022). Biofilms also contribute to water purification by removing pollutants and stabilizing sediments (Miao et al., 2020). Figure 2 provides a visual overview of biofilms in aquatic environments, including their structure, composition, ecological functions, and applications.

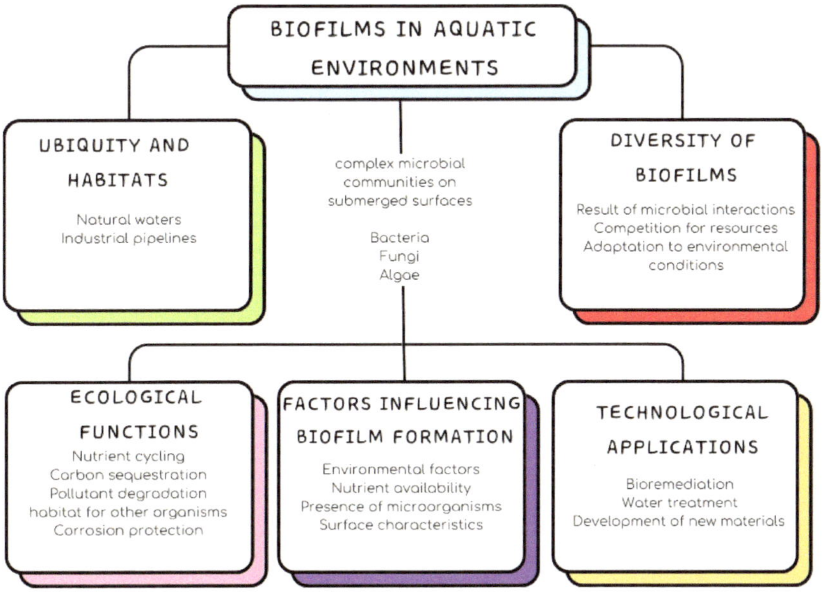

Figure 2: Overview of biofilms in aquatic environments.

Biofilm formation is influenced by a myriad of factors, including environmental conditions, nutrient availability, and the presence of other microorganisms. Surface characteristics and the type of substrate also play crucial roles (Fajardo et al., 2023). The diversity of biofilms is a result of interactions between microbial species, competition for resources, and adaptations to specific environmental conditions. Understanding these factors is essential for unraveling the complexities of biofilm dynamics in aquatic ecosystems.

MPs, prevalent throughout aquatic environments, engage in a captivating biological interaction known as adsorption. During this initial phase, a diverse array of microorganisms, including bacteria, algae, and fungi, establish connections with the microplastic surface (Vishwakarma et al., 2021). Intermolecular forces like Van der Waals and hydrogen bonding govern this interaction, with the surface properties of the microplastic—roughness, hydrophobicity, and surface charge—playing a critical role in the adsorption process (Ramsperger et al., 2020).

As the microorganisms adhere to the MP, the subsequent stage, colonization, unfolds. The adsorbed microorganisms undergo multiplication, forming microcolonies on the MP's surface. To fortify their position, these microorganisms secrete exopolysaccharides (EPS), creating a gelatinous matrix that safeguards cells and facilitates the adhesion of additional microorganisms (Saygin & Baysal, 2020). This culminates in the formation of a complex microbial community within the biofilm.

The maturation phase witnesses the biofilm's structural development, becoming thicker and more intricate, featuring distinct layers and ecological niches. Simultaneously, the diversity of microorganisms within the biofilm increases, giving rise to various ecological functions such as organic matter decomposition and nutrient cycling (Oberbeckmann et al., 2015).

Several factors influence biofilm formation on MPs. The composition of the MP, including different polymers, attracts specific types of microorganisms (Darabi et al., 2021). Water properties, such as temperature, pH, salinity, and nutrient levels, significantly impact microbial colonization and growth (Rummel et al., 2017). The duration of microplastic exposure to water is a crucial factor in biofilm development. Additionally, sunlight exposure stimulates the growth of algae and photosynthetic organisms within the biofilm (Rose et al., 2023).

The impact of biofilm on MPs is multifaceted. Firstly, the biofilm can expedite MP degradation, particularly through microorganisms secreting polymer-degrading enzymes (Tang et al., 2022). This biological interaction has environmental consequences, as the biofilm enhances the adsorption of contaminants on the microplastic (Kong et al., 2023), exacerbating its harmful effects on the environment (Zhou et al., 2023). Moreover, the biofilm's presence makes MPs more appealing to marine organisms, posing adverse health effects through ingestion (Zhan et al., 2016).

The intricate process of adsorption, colonization, and maturation results in the formation of biofilms on MPs, shaping a dynamic microbial community with significant ecological implications. Understanding these processes is crucial for comprehending the broader environmental impact of MPs in aquatic ecosystems.

4. Fateful Journeys: Biofilms and the Transformation of Microplastics

In the intricate dance of aquatic ecosystems, MPs find unexpected partners in biofilms—microbial communities that significantly influence their fate and transport. This exploration delves into the multifaceted interactions between biofilms and MPs, unraveling their impact on aggregation, sinking, biodegradation, and the potential release of harmful chemicals.

Biofilms play a pivotal role in shaping the fate of MPs in aquatic environments. The microbial communities within biofilms readily colonize the surfaces of microplastics, transforming them into miniature ecosystems. This biofilm-encrusted facade alters the transport dynamics, influencing aggregation and sinking. Biofilms act as tenacious hitchhikers, potentially mitigating surface contamination by promoting the settling of MPs to the seafloor.

The prevailing assumption is that biofilms increase the effective mass density of plastics, facilitating their sinking. However, research conducted by Pete et al. challenges this notion, revealing that biofilm formation alone is insufficient to predict the fate of plastics (Pete et al., 2023). A study was conducted on the biofilm formation and vertical transport of polyethylene MPs in the presence of three microorganisms: *Alcanivorax borkumensis*, *Anabaena* sp., and *Synechococcus elongatus*. Interestingly, while all three microorganisms formed biofilms on microplastics, only *Alcanivorax* induced their sinking. This sinking phenomenon was attributed to the production of highly active biosurfactants by *Alcanivorax*, which adsorbed onto MPs, unlike *Anabaena* and *Synechococcus*. These findings underscore the importance of considering not just the formation but also the properties of biofilms in determining the fate of plastics in aquatic environments.

In a study conducted by Magid et al. (2021) the adsorption of polystyrene nanoplastics (PSNPs) onto corncob biochar was investigated under varying conditions. The research examined the effects of pyrolysis temperature and aging, simulating oxidative aging in the environment, on the adsorption process. Results indicated that the adsorption capacity of biochar for PSNPs increased with higher pyrolysis temperatures and after aging. This enhancement was attributed to the increased specific surface area and changes in the surface properties of the biochar. The mechanisms involved in PSNPs adsorption included pore filling, hydrophobic interaction, and hydrogen bonding, with the latter being more prominent in oxidized biochar. Additionally, pH experiments revealed that electrostatic repulsion played a significant role in the adsorption process, with decreased adsorption observed at higher pH levels. Overall, the study suggests that biochar has the potential to immobilize plastic particles in terrestrial ecosystems, and its adsorption capacity can be further enhanced through artificial oxidation or aging processes, offering promising avenues for mitigating plastic pollution in the environment (Abdoul Magid et al., 2021).

In the study of Werner et al. (2016), they investigated the role of marine litter as a vector for the transport of biota, particularly non-indigenous species, across ecological barriers. They found that floating litter, especially plastics, serves as a habitat and transport medium for various organisms, facilitating their movement into new territories. The research highlighted the potential consequences of this phenomenon, including alterations in ecosystem composition and genetic diversity, as well as the spread of diseases. Additionally, the study emphasized the importance of understanding the dynamics of organism colonization on marine litter, as it could help predict and mitigate the ecological impacts of plastic pollution in marine environments.

Guasch et al. (2022) conducted a comprehensive review focusing on the interaction between MPs and biofilms in fluvial ecosystems. They highlighted the significance of biofilms as a sink for MPs and explored the mechanisms governing the retention, degradation, and potential ecological impacts of microplastics within biofilm matrices (Guasch et al., 2022). The study identified key gaps in knowledge regarding the seasonal and hydrological variations in MP storage, the influence of biofilm characteristics on MP retention and degradation, and the exchange of microbial assemblages between MPs and biofilms. Furthermore, the researchers emphasized the need to investigate the effects of MPs on biofilm consumers and the potential implications for food web dynamics.

4.1. Biofilms Facilitating Biodegradation

The partnership between biofilms and MPs extends beyond mere transport dynamics. Certain microbial communities within biofilms exhibit remarkable abilities to facilitate the biodegradation of specific types of microplastics. Through enzymatic activities and metabolic processes, these microbes break down the polymer structures of microplastics, initiating their transformation into smaller fragments (Bacha et al., 2021; Rajcoomar, 2022). This process not only alters the physical characteristics of MPs but also raises questions about the ecological consequences of the resulting breakdown products.

Bacha et al. (2021) discuss the biodegradation of polymers and their associated environmental impacts. The physicochemical properties of polymers play a crucial role in biodegradation, with the degradation rate depending on surface properties, first-order structure, and higher-order structure. Degradation occurs in three main stages: abiotic or biotic breakdown, fragmentation, and mineralization. Microorganisms such as bacteria, fungi, and algae are involved in degradation, facilitating the transformation of polymers into carbon dioxide, water, and biomass. Aerobic conditions are more efficient for polymer degradation, with oxygen acting as the final electron acceptor. Initial abiotic degradation may increase surface area for microbial colonization or reduce polymer molecular weight. Microbial degradation involves the formation of microbial biofilms on polymer surfaces, followed by fragmentation into small fragments. Enzymatic decomposition results in the formation of monomers and oligomers, which can be mineralized. Enzymes such as lipases and esterases play a significant role in polyester hydrolysis. However, the decomposition of polymers without functional groups is more challenging for microorganisms (Bacha et al., 2021). Figure 3 presents a visual representation highlighting the potential of algae, bacteria, and fungi in microplastic biodegradation, underscoring their critical role in environmental remediation processes.

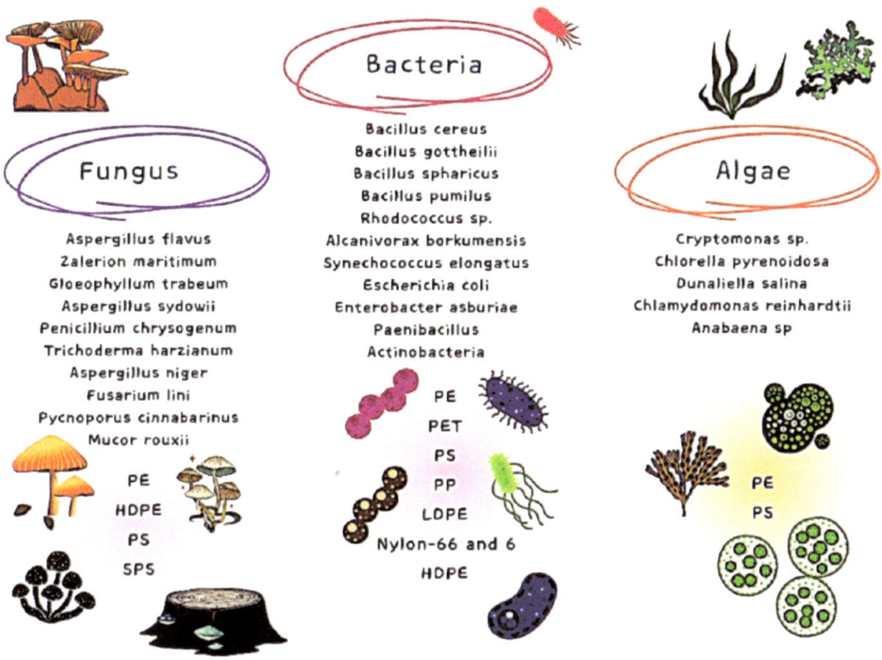

Figure 3: Algae, bacteria, and fungi as potential agents for microplastic biodegradation. Abbreviations: PE – Polyethylene, PET – Polyethylene terephthalate, PS – Polystyrene, PP – Polypropylene, LDPE – Low-density polyethylene, Nylon-66 and 6 – Nylon-66 and Nylon-6, HDPE – High-density polyethylene, SPS – Polystyrene sulfonate.

The following sections delve into the specific mechanisms and capabilities of each group, highlighting their unique approaches to MP degradation and their potential applications in environmental solutions.

Bacteria secrete extracellular hydrolytic enzymes, such as lipase, protease, and others, which play a significant role in the degradation of MPs. Degradation products with carbonyl groups can be metabolized intracellularly through beta-oxidation and the tricarboxylic acid cycle. Some studies highlight the use of bacterial species such as *Bacillus cereus* and *Bacillus gottheilii* (Bule Možar et al., 2023; Chettri et al., 2023) for microplastic degradation, with significant mass reduction observed in laboratory experiments. Other bacterial species, such as *Rhodococcus* (Pathak & Navneet, 2017), have also demonstrated the ability to grow and assimilate MPOs, albeit more gradually. Furthermore, attempts have been made to degrade MPs by bacterial consortia on a laboratory scale. Although these studies show promise, the effectiveness of bacterial degradation of MPs is still partial and requires further investigation for practical applications.

Fungi possess a variety of intracellular and extracellular enzymes that play a significant role in polymer decomposition. Enzymes such as esterases and lipases have the ability to hydrolyze polymers like polycaprolactone and polyethylene adipate. The fungal intracellular enzymatic system, mediated by the cytochrome P450 family, is involved in oxidation and conjugation reactions. Several fungal species have been studied for plastic decomposition, notably *Aspergillus flavus*, *Zalerion maritimum*, and *Gloeophyllum trabeum* (Paço et al., 2017). These species have demonstrated the capacity to degrade polymers such as high-density polyethylene (HDPE), polystyrene, and polystyrene sulfonate. Laccase enzymes and multicopper oxidase-type laccases play a significant role in polyethylene degradation, reaching peak activity after 21 days.

Fungi such as *Aspergillus sydowii*, *Penicillium chrysogenum* (Chettri et al., 2023), *Trichoderma harzianum* (Jasińska et al., 2022), *Aspergillus niger* (Esmaeili et al., 2013), *Aspergillus flavus* (Zhang et al., 2020), *Fusarium lini* (Sankhla et al., 2020), *Pycnoporus cinnabarinus*, and *Mucor rouxii* (Pathak & Navneet, 2017) have shown capabilities in breaking down MPs. These fungi utilize enzymes like depolymerases to break down long-chain polymers into low-molecular-weight oligomers, which can then be assimilated by microbial cells or converted into carbon dioxide (De Jesus & Alkendi, 2022). Additionally, fungi produce hydrophobins that aid in their adhesion to hydrophobic surfaces, facilitating the penetration of hyphae into plastic substrates for degradation (Lahiri et al., 2022). The ability of fungi to interact with MPs and produce enzymes highlights their potential to contribute to the biological treatment of plastic waste.

The literature discusses the degradation of polymers by algae and other microorganisms, as well as the effects of this degradation on the environment. Algae, particularly microalgae like *Dunaliella salina*, have shown potential in degrading MPs. Research indicates that microalgae plays a crucial role in the biodegradation process of various types of plastics, including oxidized oxium and HDPE plastics (Hadiyanto et al., 2022). Furthermore, studies emphasize the role of microalgae in promoting the breakdown of MPs (Chia et al., 2020), offering insights into eco-friendly techniques for plastic biodegradation. Unlike some bacterial systems, microalgae do not produce endotoxins and can utilize polymers like polyethylene (PE) as a carbon source. They produce ligninolytic and exopolysaccharide enzymes that facilitate plastic breakdown and have been found to colonize plastic surfaces without being hazardous. Furthermore, genetic modification of microalgae allows them to produce plastic-degrading enzymes like PETase (Di Rocco et al., 2023), enhancing their biodegradation capabilities. Algae also show potential as a sustainable biomass source for manufacturing bioplastics, offering a solution that does not compete with food production and can sequester carbon dioxide during production.

In continuation of the work mentioned above, Rocco et al. (2023) expanded on the application of genetic modification in microalgae for sustainable plastic degradation. They specifically exploited the microalgal chloroplast to synthesize PETase enzyme, a polyethylene terephthalate hydrolase capable of enzymatic degradation of PET, one of the most prevalent plastics globally. Using a photosynthetic-restoration strategy, they developed a marker-free transformant line of *Chlamydomonas reinhardtii*, expressing PETase from *Ideonella sakaiensis* in the chloroplast. This approach allowed them to investigate PETase activity against both PET and post-consumer plastics using atomic force microscopy, confirming the degradation of the plastics. This innovative study by Rocco et al. further underscores the potential of genetic modification in microalgae for enhancing plastic biodegradation (Di Rocco et al., 2023), offering a promising avenue for sustainable plastic waste management.

Algae such as *Cryptomonas* sp. (Bacha et al., 2021) and *Chlamydomonas reinhardtii* (Cunningham et al., 2023; Jasińska et al., 2022) have been studied for their ability to degrade PE and PS, respectively. The presence of MPs can affect the growth and photosynthesis of algae (Khoironi et al., 2023; W. Wang et al., 2023; Yang & Ma, 2023), and the interaction between algae and microplastics needs further investigation to find solutions to MP pollution in aquatic environments. Furthermore, studies on the ingestion of MPs by organisms such as *Daphnia magna* indicate the accumulation of MPs in increasing concentrations, but there was no significant impact on survival or reproduction rates (Wu et al., 2019).

The study of Zhao et al. (2023) shows that *Chlorella pyrenoidosa* interacts with MPs, especially polystyrene (PS) and polyvinyl chloride (PVC), affecting their growth and photosynthesis. Although *Chlorella pyrenoidosa* does not directly degrade the MPs in this study, its ability to adapt and resist the toxic effects of MPs, as well as its capacity to produce exopolysaccharides (EPS) that promote microplastic aggregation, suggests the potential to influence the behavior and dispersion of MPs in the environment. This indicates that *Chlorella pyrenoidosa* could play a role in biodegradation or at least in the interaction and accumulation of MPs in aquatic ecosystems.

4.2. Enhanced Leaching of Chemicals

As biofilms establish their microscopic metropolises on the surfaces of MPs, they may inadvertently contribute to the leaching of chemicals and additives from these synthetic particles. The biofilm matrix and microbial activities can accelerate the release of potentially harmful substances, increasing the toxic effects of MPs on the surrounding environment. This phenomenon raises concerns about the potential impact on aquatic organisms and ecosystems, emphasizing the need for a comprehensive understanding of the interplay between biofilms and the chemical components of microplastics.

Research by Bridson et al. (2023) highlighted the variability in additive leaching across different plastics, with release kinetics spanning over 64 days and differing by several orders of magnitude (Bridson et al., 2023). Their study emphasized the need for environmentally relevant leaching methods to assess the chemical impacts of MP pollution accurately. This variability in leaching behavior extends to biodegradable plastics as well, as demonstrated by Carteny et al. (2023) who found that while biodegradable MPs accumulated fewer additives, they released a larger amount of contaminants into seawater compared to conventional plastics (Catarci Carteny et al., 2023).

Priya's work underscores the magnitude of plastic pollution, detailing its adverse effects on terrestrial and aquatic ecosystems (Priya et al., 2022). The unique properties of plastics that make them durable and versatile also make them a formidable environmental hazard. As plastics fragment into MPs, their interactions with microalgae become an area of intense research focus. Priya examined the interactions between microalgae and MPs, questioning how the surface characteristics of algae or the presence of a 'hard' substrate influence these interactions. The potential role of algae in promoting MP breakdown is also explored, shedding light on possible removal strategies.

Chen Tu et al. (2020) study further elucidates the complex mechanisms through which biofilms interact with synthetic polymeric materials (Tu et al., 2020). Biofilms disrupt the structure and function of MPs through processes like fouling, which alters surface properties and releases damaging additives. Enzymatic and microbial degradation further accelerates the leaching of chemicals and monomers from the polymer matrix, leading to a loss of mechanical stability and embrittlement. The degradation pathways for different plastic polymers vary based on their main chain components, with factors such as ultraviolet radiation and oxygen playing crucial roles in the initial stages of MP degradation.

Chettri et al. (2023) delve into the intricate phases of MP breakdown by biofilms, detailing how microbial attachment alters surface characteristics, leading to increased leaching and degradation. The process involves enzyme-mediated attacks on MPs and additives, resulting in brittleness and mechanical instability (Chettri et al., 2023). Ultimately, water and microbial filaments permeate the MPs, facilitating their decomposition and reintroduction into the microbial loop.

The intricate relationship between biofilms and MPs significantly influences the transport, fate, and transformation of these synthetic particles in aquatic ecosystems. While biofilms can aid in the aggregation and sinking of microplastics, they also contribute to their biodegradation and the release of associated chemicals. These interactions underscore the complex role biofilms play as both mitigators and contributors to the challenges posed by plastic pollution. Understanding how biofilms influence aggregation, sinking, biodegradation, and chemical release is crucial for developing effective mitigation strategies. As researchers continue to explore this environmental concern, standardized methods for environmentally relevant studies can enhance our understanding. This knowledge will inform robust risk assessments and regulatory measures aimed at safeguarding the health and resilience of aquatic ecosystems.

5. Double-edged Sword: Biofilms and the Potential Risks of Microplastics

The intricate relationship between biofilms and MPs, while influencing transport and degradation, unveils a darker side—a potential amplification of hazards. This exploration delves into how biofilms, typically seen as defenders of aquatic ecosystems, can paradoxically heighten the risks associated with MPs, concentrating pollutants, harboring pathogens, and contributing to the spread of hazards.

5.1. Amplification of Hazards

Biofilms, in their role as microscopic architects on the surfaces of MPs, act as concentration zones for pollutants. As microplastics traverse aquatic environments, they accumulate various contaminants, ranging from heavy metals to organic pollutants (Wang et al., 2021). The biofilm matrix boosts this accumulation, forming areas with high concentrations of pollutants by altering the properties of MPs, affecting the transfer of contaminants in aquatic environments. As a result, the ecological risks associated with MPs are heightened, endangering the well-being of aquatic organisms and ecosystems.

Scientific research on contaminant bioaccumulation in microbial biofilms is fragmented, with most studies focusing on specific contaminants like metals, pesticides, or pharmaceuticals, or on biofilms growing on specific substrates like cobbles, sediments, or leaf litter (Bonnineau et al., 2020; J. Wang et al., 2021). Organic contaminants tend to adsorb to organic matter in the biofilm (De Jesus & Alkendi, 2022), EPS, and cell membranes due to their chemical properties and high lipid content. This bioaccumulation is driven by their hydrophobic nature (Hadiyanto et al., 2022). Additionally, certain organic contaminants, like halogenated compounds (e.g., hexachlorobenzene, PCBs, per- and polyfluoroalkyl substances), show high bioconcentration factors (BCF), indicating their strong affinity with living cells, with BCF values reaching up to 147,000 L/g. This suggests that high electron affinity also plays a significant role in their accumulation (Bonnineau et al., 2020).

In the Cartenys et al.'s (2023) study, researchers aimed to quantify and compare the accumulation of waterborne pollutants and release of chemical additives from both conventional and biodegradable plastic polymers after exposure to a marine harbor on the Belgian coast for 64 days. The plastics studied included low-density PE, and two biodegradable polymers, polyhydroxyalkanoates (PHA) and polybutylene succinate (PBS). The pollutants targeted were polycyclic aromatic hydrocarbons (PAHs), polychlorinated biphenyls (PCBs), organophosphorus flame retardants (PFRs), phthalates, and alternative plasticizers. The results indicated that all types of plastics accumulated PAHs from the seawater, with PBS showing the highest accumulation, followed closely by PE and PHA. Notably, some PAHs were found exclusively in PHA, possibly due to contamination during its synthesis. PCBs were below detectable levels in all plastics, suggesting a need for longer exposure times to observe potential accumulation. PFRs, on the other hand, were frequently detected in all plastics, with higher levels observed in PBS, likely due to its use in 3D printing and its flammable nature requiring more flame retardants during production. After exposure to seawater, PFR concentrations generally decreased, indicating potential release into the environment. However,

PHA showed an increase in some PFRs, suggesting possible accumulation from seawater (Catarci Carteny et al., 2023).

Heavy metal pollutants (HMs) in water have long been recognized as one of the most significant environmental threats globally (Chen et al., 2023; Gola et al., 2021; Naqash et al., 2020; Pourebrahimi & Pirooz, 2023; Ricardo et al., 2021; Wang et al., 2015). Originating from various industries such as mining, electroplating, and paint production, these pollutants, including Cd, Co, Cu, Ni, Pb, Zn, and Hg, are non-degradable and highly toxic. Their persistence and bioaccumulation in aquatic ecosystems pose serious risks to both aquatic life and human health. MPs can act as carriers for HMs, transporting these toxic pollutants into aquatic organisms, particularly fish. When fish ingest these HM-laden MPs, the pollutants can accumulate in their digestive systems, causing a range of harmful effects. These include reduced activity, developmental issues, reproductive problems, and even mass mortality. Moreover, as these pollutants move up the food chain, they can biomagnify, posing risks to larger predators and, ultimately, humans. As MPs migrate through water bodies, the HMs attached to their surfaces can transfer to sediments and groundwater, leading to long-term contamination. This combined pollution amplifies the toxic effects on aquatic life, particularly fish.

The adsorption of HMs by MPs in water involves various mechanisms, primarily physical and chemical adsorption (Naqash et al., 2020). Physical adsorption includes processes like electrostatic interactions and diffusion, while chemical adsorption involves complexation and cation-π bonds. The adsorption capacity of MPs for HMs is influenced by their properties, including species, surface area, and aging status, as well as environmental factors like pH and ionic strength. For example, Barus et al.'s (2021) study delved into the adsorption and release dynamics of heavy metals by polystyrene (PS) particles of varying sizes and salinities. The heavy metals studied were lead (Pb), cadmium (Cd), copper (Cu), and zinc (Zn). The research found that larger PS particles (250 μm) had the highest adsorption capacity for heavy metals. Specifically, Pb exhibited the highest adsorption, followed by Cu, Zn, and Cd. As the initial heavy metal concentration increased, so did the adsorption capacity, albeit with a decrease in the adsorption rate. Salinity also played a role: higher salinity resulted in slower adsorption and reduced adsorption capacity (Barus et al., 2021).

In this sense, biofilm-covered plastics exhibit higher adsorption capacities for metals like lead, cadmium, and zinc compared to virgin plastics (Liu et al., 2022), with biofilms mainly contributing through complexation of functional groups like oxygen and nitrogen groups (Moyal et al., 2023). The adsorption of heavy metals on biofilms is influenced by factors like temperature and pH, with biofilms showing stronger adsorption at specific conditions (Chi et al., 2022). Additionally, the aging process of MPs, including exposure to UV light and bacterial culture, can significantly impact the adsorption of heavy metals, with bacterial culture and aging processes synergistically increasing the adsorption capacity of MPs for heavy metals (Teng et al., 2023).

Environmental factors alter the properties of MPs, affecting their adsorption capacity for HMs. The study of Seyfi et al. (2021) investigates the adsorption process of heavy metals in sands containing MPs through aging and bacterial culture. Experimental conditions were established, determining 25 g of adsorbent in a 250 ml solution with 50 mg/l of six heavy metals and 750 micrograms of MPs. Bacterial culture lasted two days, and aging occurred over 14 days with UV light exposure. Results indicate that virgin MPs minimally enhance adsorption strength, except for nickel which reduces it. Aging amplifies metal absorption by up to 60%. Bacterial culture alone reduces nickel and cadmium uptake, but when combined with aging, adsorption power increases by 60 to 80% for all metals (Seyfi et al., 2021).

5.2. Harboring Pathogenic Microorganisms

Biofilms, as thriving communities of microorganisms, can inadvertently serve as havens for pathogenic species. The microenvironments created within biofilms offer shelter and resources for pathogens, potentially amplifying health threats in aquatic ecosystems (Kalčíková & Bundschuh, 2022; Rummel et al., 2017). They are prevalent in bacterial infections across various animal body systems, including fish, sport, and companion animals. Biofilms have been found in systems like auditory, cardiovascular, digestive, and respiratory, among others. Biofilms play multiple roles in disease. They can induce inflammation, facilitate intracellular invasion, and hinder healing by acting as a physical barrier. Their long-term protection can lead to chronic subclinical infections. Additionally, existing biofilms can shield other pathogens from the immune system. In biofilms, bacteria communicate via signaling molecules and undergo gene expression changes. Horizontal gene transfer can also occur. The protective matrix of biofilms can hide bacteria from the immune system, allowing chronic infections. However, this can also lead to tissue damage due to inflammation (Nesse et al., 2023).

Pathogenic species detected in MP biofilms include *Mycobacterium arupense* and *Methylobacterium adhaesivum* on LDPE and PP surfaces (Rajcoomar, 2022). Additionally, potentially pathogenic bacteria such as *Enterobacter* spp., *Helicobacter* spp., *Arcobacter* spp., *Clostridium perfringens*, and *Escherichia coli* were found on polyethylene MPs in a freshwater ecosystem study (Pham et al., 2021). This symbiotic relationship between microplastics and pathogenic biofilms introduces an additional layer of complexity to the already intricate web of interactions within aquatic environments.

5.3. Promotion of Antibiotic Resistance

The emergence of antibiotic resistance is a global health concern, and biofilms associated with MPs may contribute to its spread. As microbial communities adapt to the synthetic surfaces of MPs, the potential for genetic exchange and horizontal gene transfer increases (Nath et al., 2023). This phenomenon may facilitate the transmission of antibiotic resistance genes within biofilms, potentially leading to the dissemination of antibiotic resistance in aquatic environments.

Biofilms are resistant to antibiotics, altering their metabolism to tolerate nutrient scarcity and anoxia. This tolerance makes them challenging to treat compared to free-floating bacteria. Infections involving biofilms are typically chronic and persistent, with inconsistent responses to treatment (Nesse et al., 2023).

Recent studies have highlighted the extensive presence of both known and latent antibiotic resistance genes (ARGs) in various environments, emphasizing the importance of understanding the resistome beyond established databases like ARDB and CARD. Metagenomic analyses have revealed that latent ARGs, not cataloged in existing databases, are abundant and diverse, potentially posing emerging threats to human health (Davis et al., 2023). Furthermore, the mobilization of ARGs, often located on mobile genetic elements, is a key factor in the spread of antibiotic resistance across bacterial populations (Inda-Díaz et al., 2023). Advances in next-generation sequencing technologies have enabled the rapid identification and characterization of ARGs in complex environmental microbiomes, addressing challenges such as quantification and mobility potential assessment (Li et al., 2024).

Early plastic biofilm communities are dominated by specific microbial groups with genes associated with adhesion, biofilm formation, and motility, facilitating early colonization and genetic exchange on plastic surfaces (Bos et al., 2023). A comparative analysis of these metagenomes revealed differences in mobile genetic elements (MGEs) and antibiotic resistance genes (ARGs) across environments. For example, UV-aged polystyrene MPs were found to increase horizontal gene transfer (HGT) of ARGs in *E. coli* (Nath et al., 2023). MGE-related sequences were less common in river biofilms compared to WWTPs and river water. WWTPs and river biofilms showed higher proportions of transposons and β-lactam antibiotic resistance genes. Tetracycline resistance genes were also prevalent but less so than β-lactams. Acid mine drainage biofilms exhibited high β-lactamase proportions due to increased transposons in acidic environments (Balcázar et al., 2015).

5.4. Human Health Implications

The potential risks associated with biofilm-mediated hazards linked to MPs extend beyond the aquatic realm, raising concerns about human health. Consumable aquatic organisms may accumulate pollutants and pathogens present in biofilms, creating a pathway for these hazards to enter the human food chain. Additionally, the spread of antibiotic resistance through MP-associated biofilms poses a direct threat to human health by diminishing the effectiveness of antibiotics.

The intertwined relationship between biofilms and MPs, while offering insights into transport and degradation dynamics, presents a double-edged sword.

Biofilms, typically viewed as defenders of aquatic ecosystems, can unintentionally amplify the risks associated with MPs. By concentrating pollutants, harboring pathogenic microorganisms, and contributing to antibiotic resistance, biofilms become unwitting accomplices in the broader challenge of MP pollution. Understanding these complex interactions is crucial for devising strategies to mitigate the potential risks, safeguarding both the health of ecosystems and the well-being of human populations dependent on them.

6. Future Directions: Unveiling the Microplastic-biofilm Mystery

MPs pose a growing threat. A critical knowledge gap lies in their complex interactions with biofilms, communities of microorganisms that colonize surfaces. This analysis of a scientific word cloud (Figure 4) highlights key research areas and underscores the need for innovative approaches to unravel this mystery. In Figure 4, a word cloud is presented, derived from the Scopus database, selecting the last 10 years of research containing the words 'microplastic' and 'biofilm'. Works with a minimum of 20 occurrences of the mentioned words were selected.

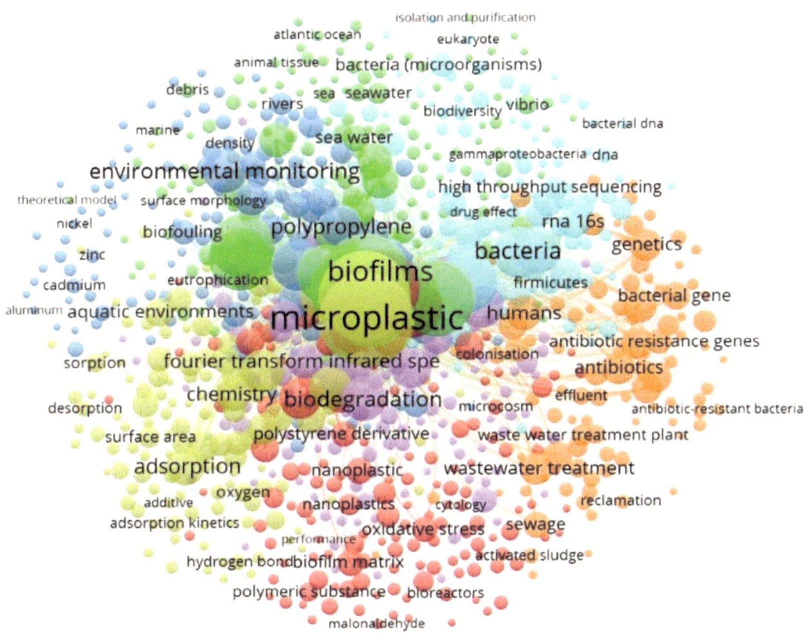

Figure 4: A cluster map of words showing different words used for microplastic and biofilm research. using Vosviewer version 1.6.20.

The word cloud (Figure 4) serves as a map of ongoing investigations. Prominent terms like 'bacteria', 'microplastics', and "antibiotic resistance" reveal a focus on potential health risks. Notably, "biodegradation" suggests a growing interest in understanding how biofilms might break down MPs. The color-coded analysis offers deeper insights. Yellow keywords emphasize fundamental aspects like surface properties and their influence on biofilm formation. These factors can impact the adsorption of pollutants onto MPs, potentially increasing toxicity. Green highlights the intricate relationship between MPs and biofilms. Bacteria in biofilms secrete enzymes that degrade MPs, while biofilm formation can also alter MP properties and influence their transport in water bodies.

Understanding the ecological and health implications is crucial. Dark blue highlights the need to identify MP sources and monitor their distribution to assess environmental risks. MPs can harm aquatic organisms, potentially impacting human health through the food chain. Light blue emphasizes the importance of source reduction and monitoring strategies to protect aquatic ecosystems.

Purple highlights a critical area for future research: the complex interplay between MPs and biofilms. Biofilm formation can alter MP properties, potentially promoting aggregation and hindering degradation. This can exacerbate the ecological risks posed by MPs and potentially introduce contaminants and pathogens, posing human health risks.

Moving forward, innovative technologies are needed to illuminate the "Microplastic-Biofilm Mystery". Advanced microscopy techniques can provide high-resolution images of biofilm-MP interactions. Metagenomic analysis can reveal the composition of microbial communities within biofilms (Davis et al., 2023), offering insights into their role in degradation processes. Additionally, laboratory experiments that simulate real-world conditions are essential to understanding the combined effects of biofilms and environmental factors on MP fate and behavior.

6.1. Knowledge Gaps and Ongoing Research

Despite significant strides, critical knowledge gaps persist in our understanding of biofilm-MP interactions. The specific mechanisms governing the formation and dynamics of these partnerships remain elusive, as does the full extent of their ecological consequences. There is a need to unravel the intricacies of microbial communities within biofilms, the factors influencing their colonization on MP surfaces, and the varied impacts on transport, degradation, and potential risks. Ongoing research efforts are actively addressing these gaps, employing interdisciplinary approaches to combine expertise in microbiology, ecology, chemistry, and materials science.

According to Kalčíková and Bundschuh (2022), a deeper understanding is sought regarding the interaction between MPs and particle-associated biofilms. Current studies aim to elucidate how biofilms develop on MP surfaces and how this interaction affects the ecotoxicity and fate of MPs in aquatic ecosystems. Research endeavors explore the potential of biofilms to serve as sinks for MPs and their contribution to MP degradation in the environment. Regarding substrate-associated biofilms, ongoing research focuses on unraveling the mechanisms and nature of interaction between MPs and biofilms associated with substrates, such as periphyton. Studies investigate how MPs integrate into these biofilms and how this interaction may impact the retention, degradation, and remobilization of MPs in aquatic ecosystems. Identifying the roles of biofilms in MP retention and release is crucial for fully understanding their fate and effects in the environment (Kalčíková & Bundschuh, 2022).

The work by Cunningham et al. (2023) focuses on addressing significant data gaps concerning nanoplastics (Cunningham et al., 2023). It recognizes the existing knowledge in the fields of MPs and nanomaterials that can inform nanoplastic risk assessment yet highlights substantial uncertainties due to data gaps in nanoplastic sources, transport, and toxicological impacts. The necessity to prioritize data collection for plastic waste, specifically accounting for nanoplastic contributions, is emphasized. The paper suggests that conceptual frameworks and preliminary risk assessments can aid in identifying these gaps and highlighting future research needs. It notes the inadequacy of current toxicity studies for nanoplastics in providing exposure-response information suitable for risk modeling, stressing the importance of experimental designs mapping exposure-response curves. Additionally, it underscores the lack of quantification of exposure to MPs and nanoplastics across various endpoints, emphasizing the need for specific transport models for nanoplastics. The paper concludes by emphasizing the importance of conducting a probabilistic, quantitative, and descriptive risk assessment for MPs and nanoplastics, considering complex interactions with other stressors and environmental factors.

The study by Guasch et al. (2022) highlights significant knowledge gaps in the interactions between MPs and benthic biofilms in rivers. Unanswered questions are raised regarding the temporary storage of MPs in biofilms, including how characteristics such as the spatial organization of biofilms and the physicochemical properties of MPs affect their transformation. It is questioned whether denser biofilms retain plastics better than thinner ones and how factors like pH and dissolved oxygen influence MP degradation. Although the release of chemicals from MPs and the adsorption of contaminants have been studied, their impact on biofilms remains unknown. Moreover, the need to understand the exchange of microbial communities between MPs and biofilms is emphasized, along with its effects on key characteristics such as antibiotic resistance and biodiversity. It is expected that MPs also affect biofilm consumers, but research in this area is limited, especially in fluvial environments. Uncertainty arises about whether consumers can select and avoid MPs in their food, and how the shape and physicochemical characteristics of MPs influence their ingestion and effects on consumers (Guasch et al., 2022). Overall, there is an emphasis on the need to investigate the mechanisms of MP absorption and their effects on the food chain to develop efficient management strategies for MP pollution in river ecosystems.

6.2. Technological Advancements and Approaches

The elucidation of the MP-biofilm mystery requires cutting-edge technologies and innovative approaches. Advanced imaging techniques, such as high-resolution microscopy and imaging mass spectrometry, offer glimpses into the microscopic world of biofilm-MP interactions (Wang et al., 2023). Metagenomic and metatranscriptomic analyses unravel the genetic blueprint and functional potential of microbial communities associated with MPs (Davis et al., 2023; Inda-Díaz et al., 2023; Z. Li et al., 2024). The integration of artificial intelligence and machine learning further enhances our ability to process vast datasets and discern patterns in biofilm development and dynamics. These technological advancements hold the key to unraveling the complexities of this ecological puzzle.

The formation of biofilms on MPs contributes to the accumulation of contaminants in aquatic environments. These biofilms can adhere firmly to MPs, providing a conducive surface for the adsorption of chemical contaminants. This accumulation increases the persistence and toxicity of MPs, exacerbating environmental impacts. In this context, antimicrobial peptides (AMPs) can play a significant role. AMPs have the ability to disrupt the formation and stability of biofilms, thereby inhibiting the accumulation of contaminants on MPs. The research on AMPs for biofilm mitigation includes studies on AMP@PDA@AgNPs nanocomposites, which effectively destroy bacterial biofilms by inhibiting biofilm-related genes (Xu et al., 2021). Additionally, AMP-17 derived from *Musca domestica* exhibits significant antifungal activity against *C. albicans* biofilms, inhibiting biofilm formation, yeast-to-hypha transition, and biofilm cell adhesion (Sun et al., 2022). These findings collectively highlight the potential of AMPs, such as AMP@PDA@AgNPs nanocomposites and AMP-17, in combating biofilm formation, including those associated with MPs.

By acting directly on biofilms, AMPs can reduce the MPs' ability to adsorb and transport chemical contaminants, decreasing their bioaccumulation and environmental impact. Moreover, the action of AMPs could make biofilms less resistant to conventional treatments, facilitating the removal of both the biofilms and adsorbed contaminants. In summary, AMPs offer a promising approach to mitigate biofilm formation on MPs, thereby reducing the accumulation of contaminants and the associated environmental risks. This application of AMPs highlights their potential not only in human and animal health but also in protecting aquatic ecosystems from MP contamination and its associated impacts.

6.3. Importance of Future Research for Mitigation Strategies

Advancing research on biofilm-MP interactions holds significance beyond academia; it forms the foundation for informed mitigation strategies and environmental safeguarding measures. Understanding how biofilms influence MP transport, fate, and associated risks is pivotal for devising targeted interventions.

As we unravel the pathways through which biofilms concentrate pollutants, harbor pathogens, and spread antibiotic resistance, we gain crucial insights to develop effective mitigation strategies. These may encompass engineered solutions, policy directives, and public awareness initiatives aimed at addressing the broader issue of MP pollution and its environmental impacts.

The ongoing quest to decipher the intricacies of the MP-biofilm relationship requires collaborative endeavors, technological breakthroughs, and a steadfast commitment to bridging existing knowledge gaps. As researchers delve deeper into this complex dynamic, the knowledge acquired will guide the formulation of mitigation strategies and environmental protection measures. By unlocking the mysteries surrounding biofilm-MP interactions, we pave the way for a more holistic comprehension of the MP predicament, fostering a future where aquatic ecosystems can thrive devoid of the encumbrances of plastic pollution.

7. Conclusion

The interplay between biofilms and MPs in aquatic ecosystems presents both opportunities and challenges. As biofilms colonize MP surfaces, they influence their fate, transport, and degradation dynamics, acting as both defenders and amplifiers of plastic pollution challenges. While biofilms offer potential avenues for mitigation, such as promoting microplastic settling or facilitating biodegradation, they can also exacerbate risks by concentrating pollutants and harboring pathogens. As research delves deeper into biofilm-MP interactions, it unveils pathways for understanding and addressing the complexities of plastic pollution challenges. Technological innovations and interdisciplinary collaborations are pivotal in deciphering this relationship and formulating targeted mitigation strategies. Ultimately, unlocking the secrets of biofilm-MP interactions guides mitigation efforts and fosters a future where aquatic ecosystems thrive free from the impacts of plastic pollution.

References

Abdoul Magid, A.S.I., Islam, M.S., Chen, Y., Weng, L., Li, J., Ma, J., & Li, Y. (2021). Enhanced adsorption of polystyrene nanoplastics (PSNPs) onto oxidized corncob biochar with high pyrolysis temperature. *Science of the Total Environment*, 784. https://doi.org/10.1016/j.scitotenv.2021.147115.

Bacha, A.-U.-R., Nabi, I., & Zhang, L. (2021). Mechanisms and the engineering approaches for the degradation of microplastics. *ACS ES&T Engineering*, 1(11), 1481–1501. https://doi.org/10.1021/acsestengg.1c00216.

Balcázar, J.L., Subirats, J., & Borrego, C.M. (2015). The role of biofilms as environmental reservoirs of antibiotic resistance. *Frontiers in Microbiology*, 6(Oct), 1–9. https://doi.org/10.3389/fmicb.2015.01216.

Barus, B.S., Chen, K., Cai, M., Li, R., Chen, H., Li, C. et al. (2021). Heavy metal adsorption and release on polystyrene particles at various salinities. *Frontiers in Marine Science*, 8(June), 1–14. https://doi.org/10.3389/fmars.2021.671802.

Bergmann, M., Collard, F., Fabres, J., Gabrielsen, G.W., Provencher, J.F., Rochman, C.M. et al. (2022). Plastic pollution in the Arctic. *Nature Reviews Earth and Environment*, 3(5), 323–337. https://doi.org/10.1038/s43017-022-00279-8.

Bonnineau, C., Artigas, J., Chaumet, B., Dabrin, A., Faburé, J., Ferrari, B.J.D. et al. (2020). Role of biofilms in contaminant bioaccumulation and trophic transfer in aquatic ecosystems: Current state of knowledge and future challenges. *In:* P. de Voogt (Ed.), *Reviews of Environmental Contamination and Toxicology* (pp. 115–153). Springer Nature Switzerland. https://doi.org/10.1007/398_2019_39.

Bos, R.P., Kaul, D., Zettler, E.R., Hoffman, J.M., Dupont, C.L., Amaral-Zettler, L.A., & Mincer, T.J. (2023). Plastics select for distinct early colonizing microbial populations with reproducible traits across environmental gradients. *Environmental Microbiology*, 25(12), 2761–2775. https://doi.org/10.1111/1462-2920.16391.

Bridson, J.H., Abbel, R., Smith, D.A., Northcott, G.L., & Gaw, S. (2023). Release of additives and non-intentionally added substances from microplastics under environmentally relevant conditions. *Environmental Advances*, 12(March), 100359. https://doi.org/10.1016/j.envadv.2023.100359.

Bule Možar, K., Miloloža, M., Martinjak, V., Cvetnić, M., Kušić, H., Bolanča et al. (2023). Potential of advanced oxidation as pretreatment for microplastics biodegradation. *Separations*, 10(2). https://doi.org/10.3390/separations10020132.

Catarci Carteny, C., Amato, E.D., Pfeiffer, F., Christia, C., Estoppey, N., Poma, G. et al. (2023). Accumulation and release of organic pollutants by conventional and biodegradable microplastics in the marine environment. *Environmental Science and Pollution Research*, 30(31), 77819–77829. https://doi.org/10.1007/s11356-023-27887-1.

Chen, Q., Zhao, H., Liu, Y., Jin, L., & Peng, R. (2023). Factors affecting the adsorption of heavy metals by microplastics and their toxic effects on fish. *Toxics*, 11(6). https://doi.org/10.3390/toxics11060490.

Chettri, D., Pati, T., & Verma, A.K. (2023). Microbe-mediated biodegradation of microplastics from wastes. *Water and Environment Journal*, 37(4), 671–685. https://doi.org/10.1111/wej.12882.

Chi, Y., Wang, Y., & Ma, X. (2022). Study on the mechanism of influence of water biofilm on heavy metal-polycyclic aromatic hydrocarbon complex pollution. *In: Advances in Energy Materials and Environment Engineering*. CRC Press.

Chia, W.Y., Ying Tang, D.Y., Khoo, K.S., Kay Lup, A.N., & Chew, K.W. (2020). Nature's fight against plastic pollution: Algae for plastic biodegradation and bioplastics production. *Environmental Science and Ecotechnology*, 4, 100065. https://doi.org/10.1016/j.ese.2020.100065.

Cunningham, B.E., Sharpe, E.E., Brander, S.M., Landis, W.G., & Harper, S.L. (2023). Critical gaps in nanoplastics research and their connection to risk assessment. *Frontiers in Toxicology*, 5(April), 1–16. https://doi.org/10.3389/ftox.2023.1154538.

Darabi, M., Majeed, H., Diehl, A., Norton, J., & Zhang, Y. (2021). A review of microplastics in aquatic sediments: Occurrence, fate, transport, and ecological impact. *Current Pollution Reports*, 7(1), 40–53. https://doi.org/10.1007/s40726-020-00171-3.

Davis, B.C., Brown, C., Gupta, S., Calarco, J., Liguori, K., Milligan, E. et al. (2023). Recommendations for the use of metagenomics for routine monitoring of antibiotic resistance in wastewater and impacted aquatic environments. *Critical Reviews in*

Environmental Science and Technology, 53(19), 1731–1756. https://doi.org/10.1080/10643389.2023.2181620.

De Jesus, R., & Alkendi, R. (2022). A mini review on the bioremediative potential of microbial enzymes as a solution to emerging microplastic pollution. *Frontiers in Microbiology*, 13(March), 1–9. https://doi.org/10.3389/fmicb.2022.1066133.

Di Rocco, G., Taunt, H.N., Berto, M., Jackson, H.O., Piccinini, D., Carletti, A. et al. (2023). A PETase enzyme synthesized in the chloroplast of the microalga *Chlamydomonas reinhardtii* is active against post-consumer plastics. *Scientific Reports*, 13(1), 1–10. https://doi.org/10.1038/s41598-023-37227-5.

Esmaeili, A., Pourbabaee, A.A., Alikhani, H.A., Shabani, F., & Esmaeili, E. (2013). Biodegradation of low-density polyethylene (LDPE) by mixed culture of *Lysinibacillus xylanilyticus* and *Aspergillus niger* in soil. *PLoS ONE*, 8(9). https://doi.org/10.1371/journal.pone.0071720.

Fajardo, C., Sánchez-Fortún, S., Videira-Quintela, D., Martin, C., Nande, M., D´ors, A. et al. (2023). Biofilm formation on polyethylene microplastics and their role as transfer vector of emerging organic pollutants. *Environmental Science and Pollution Research*, 30(35), 84462–84473. https://doi.org/10.1007/s11356-023-28278-2.

Gola, D., Kumar Tyagi, P., Arya, A., Chauhan, N., Agarwal, M., Singh, S.K., & Gola, S. (2021). The impact of microplastics on the marine environment: A review. *In: Environmental Nanotechnology, Monitoring and Management* (Vol. 16). Elsevier B.V. https://doi.org/10.1016/j.enmm.2021.100552.

Golmohammadi, M., Fatemeh Musavi, S., Habibi, M., Maleki, R., Golgoli, M., Zargar, M. et al. (2023). Molecular mechanisms of microplastics degradation: A review. *In: Separation and Purification Technology* (Vol. 309). Elsevier B.V. https://doi.org/10.1016/j.seppur.2022.122906.

Guasch, H., Bernal, S., Bruno, D., & Almroth, B.C. (2022). *Interactions between Microplastics and Benthic Biofilms in Fluvial Ecosystems: Knowledge Gaps and Future Trends*, 41(3). The University of Chicago Press.

Habib, R.Z., & Thiemann, T. (2021). Microplastic in commercial fish in the Mediterranean Sea, the Red Sea and the Arabian Gulf. Part 1: The Mediterranean Sea. *Journal of Water Resource and Protection*, 13(08), 563–587. https://doi.org/10.4236/jwarp.2021.138031.

Hadiyanto, H., Khoironi, A., Dianratri, I., Huda, K., Suherman, S., & Muhammad, F. (2022). Biodegradation of oxidized high-density polyethylene and oxo-degradable plastic using microalgae Dunaliella salina. *Environmental Pollutants and Bioavailability*, 34(1), 469–481. https://doi.org/10.1080/26395940.2022.2128884.

Horton, A.A., Walton, A., Spurgeon, D.J., Lahive, E., & Svendsen, C. (2017). Microplastics in freshwater and terrestrial environments: Evaluating the current understanding to identify the knowledge gaps and future research priorities. *Science of The Total Environment*, 586, 127–141. Elsevier B.V. https://doi.org/10.1016/j.scitotenv.2017.01.190.

Inda-Díaz, J.S., Lund, D., Parras-Moltó, M., Johnning, A., Bengtsson-Palme, J., & Kristiansson, E. (2023). Latent antibiotic resistance genes are abundant, diverse, and mobile in human, animal, and environmental microbiomes. *Microbiome*, 11(1), 1–16. https://doi.org/10.1186/s40168-023-01479-0.

Jambeck Jenna, R., Geyer Roland, Wilcox Chris, Siegler Theodore, R., Perryman Miriam, Andrady Anthony et al. (2015). Plastic waste inputs from land into the ocean. *Science*, 347(6223), 768–770. http://www.scopus.com/inward/record.url?eid=2-s2.0-84954204572&partnerID=40&md5=28a97ef4a4fdee6db9ef2fe507a1a02a.

Jasińska, A., Różalska, S., Rusetskaya, V., Słaba, M., & Bernat, P. (2022). Microplastic-induced oxidative stress in metolachlor-degrading filamentous fungus *Trichoderma harzianum*. *International Journal of Molecular Sciences*, 23(21). https://doi.org/10.3390/ijms232112978.

Kalčíková, G., & Bundschuh, M. (2022). Aquatic biofilms—Sink or source of microplastics? A critical reflection on current knowledge. *Environmental Toxicology and Chemistry*, 41(4), 838–843. https://doi.org/10.1002/etc.5195.

Khoironi, A., Hadiyanto, H., Hartini, E., Dianratri, I., Joelyna, F.A., & Pratiwi, W.Z. (2023). Impact of disposable mask microplastics pollution on the aquatic environment and microalgae growth. *Environmental Science and Pollution Research*, 30(31), 77453–77468. https://doi.org/10.1007/s11356-023-27651-5.

Kong, X., Zhang, J., Ji, Q., Li, C., Chen, X., Cao, X. et al. (2023). Insights into adsorption mechanisms of nitro polycyclic aromatic hydrocarbons on common microplastic particles: Experimental studies and modeling. *Chemosphere*, 320. https://doi.org/10.1016/j.chemosphere.2023.138050.

Kuizenga, B., Tasseron, P.F., Wendt-Potthoff, K., & van Emmerik, T.H.M. (2023). From source to sea: Floating macroplastic transport along the Rhine river. *Frontiers in Environmental Science*, 11(May), 1–9. https://doi.org/10.3389/fenvs.2023.1180872.

Lahiri, D., Nag, M., Ghosh, S., & Ray, R.R. (2022). No title biodegradation of plastic by fungi, sustainable utilization of fungi in agriculture and industry mycology. *In: Current and Future Developments* (p. 384). Bentham Science Publisher. https://doi.org/https://doi.org/10.2174/97898150403401220400024.

Lamb, J.B., Willis, B.L., Fiorenza, E.A., Couch, C.S., Howard, R., Rader, D.N. et al. (2018). Plastic waste associated with disease on coral reefs. *Science*, 359(6374), 460–462. https://doi.org/10.1126/science.aar3320.

Li, Q., Hu, Y., Kou, D., Yang, W., Tang, W., Chen, Q. et al. (2022). Factors impacting microplastic biofilm community and biological risks posed by microplastics in drinking water sources. *Water, Air, and Soil Pollution*, 233(6). https://doi.org/10.1007/s11270-022-05649-y.

Li, W., Zhao, W., Zhu, H., Li, Z.J., & Wang, W. (2023). State of the art in the photochemical degradation of (micro)plastics: From fundamental principles to catalysts and applications. *Journal of Materials Chemistry A*, 11(6), 2503–2527. https://doi.org/10.1039/d2ta09523h.

Li, Z., Yuan, D., Kou, Y., Li, X., & Du, C. (2024). Metagenome sequencing to unveil the occurrence and distribution of antibiotic resistome and in a wastewater treatment plant. *Environmental Technology*, 45(10), 1933–1942. https://doi.org/10.1080/09593330.2022.2158758.

Liu, Z., Adyel, T.M., Wang, Z., Wu, J., Liu, J., Miao, L., & Hou, J. (2022). Effects of biofilms on trace metal adsorption on plastics in freshwater systems. *International Journal of Environmental Research and Public Health*, 19(21). https://doi.org/10.3390/ijerph192113752.

Miao, L., Wang, C., Adyel, T.M., Wu, J., Liu, Z., You, G. et al. (2020). Microbial carbon metabolic functions of biofilms on plastic debris influenced by the substrate types and environmental factors. *Environment International*, 143(July), 106007. https://doi.org/10.1016/j.envint.2020.106007.

Miranda Zoppas, F., Sacco, N., Soffietti, J., Devard, A., Akhter, F., & Marchesini, F.A. (2023). Catalytic approaches for the removal of microplastics from water: Recent advances and future opportunities. *Chemical Engineering Journal Advances*, 16, 100529. https://doi.org/10.1016/J.CEJA.2023.100529.

Moyal, J., Dave, P.H., Wu, M., Karimpour, S., Brar, S.K., Zhong, H., & Kwong, R.W.M. (2023). Impacts of biofilm formation on the physicochemical properties and toxicity of microplastics: A concise review. *Reviews of Environmental Contamination and Toxicology*, 261(1), 1–17. https://doi.org/10.1007/s44169-023-00035-z.

Naqash, N., Prakash, S., Kapoor, D., & Singh, R. (2020). Interaction of freshwater microplastics with biota and heavy metals: A review. *Environmental Chemistry Letters*, 18(6), 1813–1824. Springer Science and Business Media Deutschland GmbH. https://doi.org/10.1007/s10311-020-01044-3.

Nath, J., De, J., Sur, S., & Banerjee, P. (2023). Interaction of microbes with microplastics and nanoplastics in the agroecosystems: Impact on antimicrobial resistance. *Pathogens*, 12(7), 1–18. https://doi.org/10.3390/pathogens12070888.

Nesse, L.L., Osland, A.M., & Vestby, L.K. (2023). The role of biofilms in the pathogenesis of animal bacterial infections. *Microorganisms*, 11(3). https://doi.org/10.3390/microorganisms11030608.

Oberbeckmann, S., Löder, M.G.J., & Labrenz, M. (2015). Marine microplastic-associated biofilms: A review. *Environmental Chemistry*, 12(5), 551–562. https://doi.org/10.1071/EN15069.

Paço, A., Duarte, K., da Costa, J.P., Santos, P.S.M., Pereira, R., Pereira, M.E. et al. (2017). Biodegradation of polyethylene microplastics by the marine fungus *Zalerion maritimum*. *Science of The Total Environment*, 586, 10–15. https://doi.org/10.1016/j.scitotenv.2017.02.017.

Pan, Y., Gao, S.H., Ge, C., Gao, Q., Huang, S., Kang, Y. et al. (2023). Removing microplastics from aquatic environments: A critical review. *Environmental Science and Ecotechnology*, 13. Editorial Board, Research of Environmental Sciences. https://doi.org/10.1016/j.ese.2022.100222.

Pathak, V.M., & Navneet. (2017). Review on the current status of polymer degradation: A microbial approach. *Bioresources and Bioprocessing*, 4(1). https://doi.org/10.1186/s40643-017-0145-9.

Pete, A.J., Brahana, P.J., Bello, M., Benton, M.G., & Bharti, B. (2023). Biofilm formation influences the wettability and settling of microplastics. *Environmental Science & Technology Letters*, 10(2), 159–164. https://doi.org/10.1021/acs.estlett.2c00728.

Pham, D.N., Clark, L., & Li, M. (2021). Microplastics as hubs enriching antibiotic-resistant bacteria and pathogens in municipal activated sludge. *Journal of Hazardous Materials Letters*, 2(December 2020), 100014. https://doi.org/10.1016/j.hazl.2021.100014.

Pourebrahimi, S., & Pirooz, M. (2023). Microplastic pollution in the marine environment: A review. *Journal of Hazardous Materials Advances*, 10(May). https://doi.org/10.1016/j.hazadv.2023.100327.

Priya, A.K., Jalil, A.A., Dutta, K., Rajendran, S., Vasseghian, Y., Karimi-Maleh, H., & Soto-Moscoso, M. (2022). Algal degradation of microplastic from the environment: Mechanism, challenges, and future prospects. *Algal Research*, 67(March), 102848. https://doi.org/10.1016/j.algal.2022.102848.

Rajcoomar, S. (2022). *The Factors Affecting Bacterial Colonisation on Microplastics and the Impact of Tertiary Treatment of Wastewater on the Attached Bacteria and Microplastics*. Durban University of Technology.

Rajcoomar, S., Amoah, I.D., Abunama, T., Mohlomi, N., Bux, F., & Kumari, S. (2023). Biofilm formation on microplastics in wastewater: Insights into factors, diversity and inactivation strategies. *International Journal of Environmental Science and Technology*, 0123456789. https://doi.org/10.1007/s13762-023-05266-0.

Ramsperger, A.F.R.M., Stellwag, A.C., Caspari, A., Fery, A., Lueders, T., Kress, H. et al. (2020). Structural diversity in early-stage biofilm formation on microplastics depends on environmental medium and polymer properties. *Water* (Switzerland), 12(11), 1–21. https://doi.org/10.3390/w12113216.

Ricardo, I.A., Alberto, E.A., Silva Júnior, A.H., Macuvele, D.L.P., Padoin, N., Soares, C. et al. (2021). A critical review on microplastics, interaction with organic and inorganic pollutants, impacts, and effectiveness of advanced oxidation processes applied for their removal from aqueous matrices. *Chemical Engineering Journal*, 424. Elsevier B.V. https://doi.org/10.1016/j.cej.2021.130282.

Rose, P.K., Jain, M., Kataria, N., Sahoo, P.K., Garg, V.K., & Yadav, A. (2023). Microplastics in multimedia environment: A systematic review on its fate, transport, quantification, health risk, and remedial measures. *Groundwater for Sustainable Development*, 20. https://doi.org/10.1016/j.gsd.2022.100889.

Rummel, C.D., Jahnke, A., Gorokhova, E., Kühnel, D., & Schmitt-Jansen, M. (2017). Impacts of biofilm formation on the fate and potential effects of microplastic in the aquatic environment. *Environmental Science and Technology Letters*, 4(7), 258–267. https://doi.org/10.1021/acs.estlett.7b00164.

Sacco, N.A., Zoppas, F.M., Devard, A., Gonzalez Muñoz, P., García, G., & Marchesini, F.A. (2023). Recent advances in microplastics removal from water with special attention given to photocatalytic degradation: Review of scientific research. *Microplastics*, 2, 278–303.

Samak, N.A., Jia, Y., Sharshar, M.M., Mu, T., Yang, M., Peh, S., & Xing, J. (2020). Recent advances in biocatalysts engineering for polyethylene terephthalate plastic waste green recycling. *Environment International*, 145, 106144. https://doi.org/10.1016/j.envint.2020.106144.

Sankhla, I.S., Sharma, G., & Tak, A. (2020). Fungal degradation of bioplastics: An overview. *New and Future Developments in Microbial Biotechnology and Bioengineering: Recent Advances in Application of Fungi and Fungal Metabolites: Environmental and Industrial Aspects*, 35–47. https://doi.org/10.1016/B978-0-12-821007-9.00004-8.

Saygin, H., & Baysal, A. (2020). Biofilm formation of clinically important bacteria on bio-based and conventional micro/submicron-sized plastics. *Bulletin of Environmental Contamination and Toxicology*, 105(1), 18–25. https://doi.org/10.1007/s00128-020-02876-z.

Schreyers, L., van Emmerik, T., Bui, K., Van Le Thi, K., Vermeulen, B., Nguyen, H.-Q., & van der Ploeg, M. (2023). *Tidal Plastic Recycling: Net River Plastic Transport Limited by Tidal Dynamics*.

Seyfi, S., Katibeh, H., & Heshami, M. (2021). Investigation of the process of adsorption of heavy metals in coastal sands containing microplastics, with special attention to the effect of aging process and bacterial spread in microplastics. *Archives of Environmental Protection*, 47(3), 50–59. https://doi.org/10.24425/aep.2021.138463.

Sharma, S., Mohler, J., Mahajan, S.D., Schwartz, S.A., Bruggemann, L., & Aalinkeel, R. (2023). Microbial biofilm: A review on formation, infection, antibiotic resistance, control measures, and innovative treatment. *Microorganisms*, 11(6). https://doi.org/10.3390/microorganisms11061614.

Shree, P., Singh, C.K., Sodhi, K.K., Surya, J.N., & Singh, D.K. (2023). Biofilms: Understanding the structure and contribution towards bacterial resistance in antibiotics. *Medicine in Microecology*, 16, 100084. https://doi.org/10.1016/J.MEDMIC.2023.100084.

Sun, C., Zhao, X., Jiao, Z., Peng, J., Zhou, L., Yang, L. et al. (2022). The antimicrobial peptide AMP-17 derived from *Musca domestica* inhibits biofilm formation and eradicates mature biofilm in *Candida albicans*. *Antibiotics*, 11(11). https://doi.org/10.3390/antibiotics11111474.

Tang, K.H.D., Lock, S.S.M., Yap, P.S., Cheah, K.W., Chan, Y.H., Yiin, C.L. et al. (2022). Immobilized enzyme/microorganism complexes for degradation of microplastics: A review of recent advances, feasibility, and future prospects. *Science of The Total Environment*, 832. Elsevier B.V. https://doi.org/10.1016/j.scitotenv.2022.154868.

Tavşanoğlu, Ü.N., Başaran Kankılıç, G., Akca, G., Çırak, T., & Erdoğan, Ş. (2020). Microplastics in a dam lake in Turkey: Type, mesh size effect, and bacterial biofilm communities. *Environmental Science and Pollution Research*, 27(36), 45688–45698. https://doi.org/10.1007/s11356-020-10424-9.

Teng, W., Zhang, X., Yang, Y., Zeng, J., Lian, J., & Qin, X. (2023). Study on adsorption of heavy metals Cu and Zn by microplastics under different aged factors. *E3S Web of Conferences*, 393. https://doi.org/10.1051/e3sconf/202339303034.

Tu, C., Zhou, Q., Zhang, C., Liu, Y., & Luo, Y. (2020). Biofilms of microplastics. *Microplastics in Terrestrial Environments* (Issue: March), 299–317. Springer. https://doi.org/10.1007/698.

van Emmerik, T.H.M., González-Fernández, D., Laufkötter, C., Blettler, M., Lusher, A., Hurley, R., & Ryan, P.G. (2023). Focus on plastics from land to aquatic ecosystems. *Environmental Research Letters*, 18(4). https://doi.org/10.1088/1748-9326/acc086.

Vishwakarma, A., Dang, F., Ferrell, A., Barton, H.A., & Joy, A. (2021). Peptidomimetic polyurethanes inhibit bacterial biofilm formation and disrupt surface established biofilms. *Journal of the American Chemical Society*, 143(25), 9440–9449. https://doi.org/10.1021/jacs.1c02324.

Wang, F., Shih, K.M., & Li, X.Y. (2015). The partition behavior of perfluorooctanesulfonate (PFOS) and perfluorooctanesulfonamide (FOSA) on microplastics. *Chemosphere*, 119, 841–847. https://doi.org/10.1016/j.chemosphere.2014.08.047.

Wang, J., Guo, X., & Xue, J. (2021). Biofilm-developed microplastics as vectors of pollutants in aquatic environments. *Environmental Science & Technology*, 55(19), 12780–12790. https://doi.org/10.1021/acs.est.1c04466.

Wang, M.F., Joignant, A.N., Sohn, A.L., Garrard, K.P., & Muddiman, D.C. (2023). Time of acquisition and high spatial resolution mass spectrometry imaging. *Journal of Mass Spectrometry*, 58(3), 1–6. https://doi.org/10.1002/jms.4911.

Wang, W., Liu, H., Liu, H., Chen, J., Xu, X., Xia, J., & Zhang, P. (2023). Effects of polystyrene microparticles on growth and physiological metabolism of microalgae *Scendesmus obliquus*. *Sustainability* (Switzerland), 15(14). https://doi.org/10.3390/su151411223.

Werner, S., Budziak, A., Van Franeker, J., Galgani, F., Hanke, G., Maes, T. et al. (2016). Harm caused by marine litter. *In: JRC Technical Report* (Issue March 2017). https://doi.org/10.2788/19937.

Wu, J., Jiang, R., Lin, W., & Ouyang, G. (2019). Effect of salinity and humic acid on the aggregation and toxicity of polystyrene nanoplastics with different functional groups and charges. *Environmental Pollution*, 245, 836–843. https://doi.org/10.1016/j.envpol.2018.11.055.

Xu, J., Li, Y., Wang, H., Zhu, M., Feng, W., & Liang, G. (2021). Enhanced antibacterial and anti-biofilm activities of antimicrobial peptides modified silver nanoparticles. *International Journal of Nanomedicine*, 16, 4831–4846. https://doi.org/10.2147/IJN.S315839.

Yang, L., & Ma, C. (2023). Toward a better understanding of microalgal photosynthesis in medium polluted with microplastics: A study of the radiative properties of microplastic particles. *Frontiers in Bioengineering and Biotechnology*, 11(May), 1–8. https://doi.org/10.3389/fbioe.2023.1193033.

Zhan, Z., Wang, J., Peng, J., Xie, Q., Huang, Y., & Gao, Y. (2016). Sorption of 3,3′,4,4′-tetrachlorobiphenyl by microplastics: A case study of polypropylene. *Marine Pollution Bulletin*, 110(1), 559–563. https://doi.org/10.1016/j.marpolbul.2016.05.036.

Zhang, J., Gao, D., Li, Q., Zhao, Y., Li, L., Lin, H. et al. (2020). Biodegradation of polyethylene microplastic particles by the fungus *Aspergillus flavus* from the guts of wax moth *Galleria mellonella*. *Science of The Total Environment*, 704, 1–8. https://doi.org/10.1016/j.scitotenv.2019.135931.

Zhao, S., Qian, J., Wang, P., Tang, S., Lu, B., He, Y., & Xu, K. (2023). The effects of microplastics on growth and photosynthetic activity of *Chlorella pyrenoidosa*: The role of types and sizes. *Water, Air, and Soil Pollution*, 234(10), 1–14. https://doi.org/10.1007/s11270-023-06642-9.

Zhou, Y., Ashokkumar, V., Amobonye, A., Bhattacharjee, G., Sirohi, R., Singh, V. et al. (2023). Current research trends on cosmetic microplastic pollution and its impacts on the ecosystem: A review. *Environmental Pollution*, 320. Elsevier Ltd. https://doi.org/10.1016/j.envpol.2023.121106.

5 | Evidence of Microplastics in the Mediterranean (Adriatic) Sea and the Pacific Ocean

Antonietta Morena Gatti* **and Stefano Montanari**
Nanodiagnostics Foundation, Modena, Italy

Abstract: The chapter concerns the problem of microplastic fragments dispersed in the sea, starting with how that material reaches the sea and how it is reduced to very small, in many cases micro- and nano-sized, dimensions. The damage that those materials cause to marine life is also illustrated. Analyses on plastic particles collected both in Italy and in Australia were performed using a Field Emission Gun Environmental Scanning Electron Microscope coupled with the X-ray microsensor of an Energy Dispersive System to assess their elemental chemical composition. The analyses show how those particles are re-inhabited by organisms such as diatoms and algae. In conclusion, the projects proposed for cleaning the seas are briefly discussed along with what the most obvious problems involving those proposals are.

1. Introduction

The surface of planet Earth is 71% occupied by the sea and, for a relatively small share, by the fresh waters of lakes, ponds, and various waterways. Traditionally, without real borders existing, marine surfaces are indicated with different names, from the various oceans to their fractions such as, for example, the Mediterranean and, in an even more detailed way, in their smaller sections such as the Adriatic Sea, the Tyrrhenian Sea, the Ionian Sea, etc. In fact, the sea surface has neither interruptions nor natural barriers, so that the same water, moved mainly by currents as well as by the normal mixing of all liquids, touches all the coasts of the emerged lands. Leaving aside the unlikely bottles containing romantically mysterious messages and used for their stories by quite

*Corresponding author: gatti@nanodiagnostics.it

a few writers, the sea has always carried at least part of the materials capable of floating ashore. Driftwood, i.e., wood that is floating on the sea and sooner or later brought onto the beach by the sea, is among these. Figure 1 shows an attempt to construct a hut with wood transported by the sea currents.

Figure 1: An attempt to construct a hut on a beach with wood transported by the sea current.

And it is precisely with the logs that arrived with the waves and recovered on the beaches that some of the Icelandic and Greenlandic houses, the so-called torfbæir, were built in the Middle Ages. And with the sea also arrived a good part of the amber, the fossil terpene resin of *Pinus succinifera*, a now-extinct tree. Thanks to its specific weight lower than that of water and, therefore, capable of floating, amber arrived from the Baltic forests of origin to the coasts of the Baltic Sea and the North Sea. In short, the sea surrounds and unites all the emerged lands, and tends to homogenize its contents.

For the ancient Greeks, the sea was 'incorruptible', however, perhaps, they had not considered the destructive power of the only being present on the Planet and actually capable of, or in any case, potentially capable of, corrupting any aspect of Nature: the *Homo sapiens*. More and fairly often, *Homo sapiens* got rid of its waste by disposing of it everywhere and, when possible, throwing it into waterways, so as not to come into contact with it. With the growth of the population and, above all, with the sophistication of technologies, waste has become not only more and more present in the environment but progressively distant from Nature, that is, more and more difficult to transform naturally into something tolerable for the environment, until transformation is impossible. For decades, huge quantities of nuclear waste have ended up in the sea, and it is impossible to rule out that the practice persists even now. Accidents involving oil tankers and oil wells located at sea are far from rare news. Added to this is the illegal washing of oil transport tanks carried out with seawater. Furthermore, among the many polluting products that sooner or later end up in the sea are many variations of waste plastic products.

In 1967, the film *The Graduate* was released, based on the novel of the same title by Charles Webb published four years earlier. One of the minor characters, speaking to the boy who had just got his graduation, says that the future for those who want to enter the profitable world of business is to deal with plastic. In fact, the idea, widely shared at that time, was that plastic would somehow save damage to the environment, replacing many materials obtained from Nature, trees first and foremost. The events that followed sadly demonstrated that things went differently. As is often the case, Man does not seem to remember that the accelerator pedal exists in cars, a pedal that must not and cannot always be pressed to the maximum. And he doesn't even remember that next to that pedal there is another one that is used for braking. This should also apply to technological progress. What happens, however, is that we always run at maximum speed, even in competition between the groups taking part in the race. The most serious problem is linked to the fact that a product or technology is studied from its origin only up to the moment and duration of practical use. What happens afterward, i.e., the end of their life, seems to be of no interest and is not taken into consideration.

Not surprisingly, plastic materials and products made from them have suffered the same fate. Thus, we found ourselves with immense quantities of artifacts abandoned in the environment after having been used very little, often only once, all facilitated and justified by extremely low prices which do not encourage saving or reuse. And so, many of those objects, perhaps after various passages, almost none of which are truly without a negative impact on the environment, end up in the sea, forgotten by those who used them, but certainly not by Nature, incapable of reintroducing them into a virtuous circle as happens with everything that is part of natural cycles. So, the circular ocean currents channel the plastic they come across along their path into various marine areas, forming real islands of floating waste called Plastic Vortexes, because they slowly rotate on themselves. These garbage islands do not have precisely defined physical boundaries and can vary in both size and location. The largest of these formations is the so-called Great Pacific Garbage Patch that lies in the middle of the North Pacific Subtropical Gyre, located about 30–40 degrees north latitude and 135–145 degrees west longitude. It has existed since the early 1960s, and is located in the Pacific Ocean, between California and the Hawaiian Archipelago, and, accepting a certain approximation on dimensions we have said to be variable and difficult to define precisely, it would extend for approximately 10 million km^2.

Ignoring smaller islands, there are at least six other large oceanic formations: Sargasso Garbage Patch, in the Atlantic Ocean off Florida, between the Antilles and the Azores; Arctic Garbage Patch, in the Barents Sea near the Arctic Circle; Indian Ocean Garbage Patch, already reported in the Indian Ocean at the end of the 1980s; South Atlantic Garbage Patch, between South America and southern Africa; North Atlantic Garbage Patch, the second largest island by extension area, located in the North Atlantic Ocean region; South Pacific Garbage Patch, off Chile and Peru. It should be noted that all these formations are constantly increasing in size and that the material that composes them remains only partly floating, descending towards the depths of the sea to the deepest seabed, so much so that a study by Newcastle University (UK) detected them in the Mariana Trench at almost 11 km depth (Jamieson, 2017).

The greater aggressiveness of this material, at least as far as biological life is concerned, is because the objects that have become waste are chopped up over time, often reducing themselves to particulates of various sizes, which can even be less than one micron, on which other waste also sticks such as light metals and decomposing organic residues.

Various factors contribute to the breaking of plastic objects, the first of which is probably aging. Wave motion, UV exposure, oxidation, temperature variations (for example, the heating and freezing of the water, thus changing physical state), and chemical reactions with what is present in the water, both for natural reasons and what Man pours into it, contribute to the process.

A variety of natural pollution that cannot help but affect somehow the transformations of plastics is caused by underwater volcanoes, estimated to be approximately 75% of those existing on the Planet. To date, around 5,000 of them are known, but at least one million, likely, exist. Like volcanoes on land, underwater ones also form mainly along plate margins, that is, in those areas where two plates collide, move apart, or slide past each other. Magmas and gases emerge from these eruptive vents, with chemical differences depending on the area in which they are located, thus triggering different chemical transformations.

To evaluate chemical reactions, it must be considered that there are numerous types of plastic molecules, each with different chemical and physical characteristics. Then we should know how much is occasionally or chronically present in various sea areas due to natural, industrial, or urban pollution, which is extremely difficult to consider. Organic products, chromium, cadmium, and lead are increasingly present, and pollution phenomena that have had very serious consequences on the human population should not be forgotten, such as the famous ecological disaster that occurred in Minamata (Japan) (Harada, 1995), due to the release of methylmercury in the wastewater of the Chisso Corporation chemical industry, a release that lasted from 1932–1968. The toxic problem started to be understood when babies born in that area showed various symptoms, among which is dementia. The Japanese scientists found that the fish eaten by the parents of those children, fish polluted by the mercury present in the water and that the fish were forced to take in with their food, was responsible.

It is impossible to exclude the fact that the various pollutants and their combinations which, in fact, we do not have sufficient knowledge of, influence the changes undergone by the many plastics used. What we have excellent knowledge of is the salt content of seawater, whose average salinity is 35 g/l, 27 of which are sodium chloride. The rest consists mainly of magnesium, calcium, and potassium salts. Salinity is minimum at the poles and maximum in the seas between the two tropics, due to strong evaporation. Organic substances derived from the many activities of living beings are also dissolved in water, with a concentration of a few mg/l. Also important are dissolved gases (nitrogen, oxygen, carbon dioxide, hydrogen sulfide, methane, etc.), coming both from the atmosphere and from the activity of living beings. All these substances react in a more or less important way on various plastic materials. As is obvious, all the boundary conditions and all the substances present in the water react differently to different types of plastic. Acrylic or Polymethyl Methacrylate (PMMA), Polycarbonate (PC), Polyethylene (PE), Polypropylene (PP), Polyethylene terephthalate (PETE or PET), Polyvinyl chloride (PVC), Acrylonitrile-Butadiene-Styrene (ABS), Polysulfones (PS) Polystyrene, and Silicone plastics (a hybrid between a synthetic rubber and a synthetic plastic polymer) differ mutually for both chemical and physical behavior. The aging chemical mechanisms induce plastic cracking, depolymerization, and chemical oxidation, but in a marine environment, there is also abrasion, light irradiation, and biodegradation (Liu et al., 2020). In some circumstances, when the bacteria present are capable of doing so, a phenomenon of microbial degradation of some of the plastic materials can also occur. Furthermore, the different types of plastics show significant differences between each other, depending, for example, on the length of the molecular chain, the type of polymerization, and any substituents and additives.

As with several organic substances, plastics can be synthesized to make them degradable (Aminabhavi et al., 1990). BioPal was one of the first degradable plastic materials obtained by bacteria. Mater-Bi by Novamont (Italy) was a polymer obtained by the polymerization of corn. Other degradable polymers have the polylactic acid (PLA) as the base. We do not yet know what the actual impact on the marine environment in their degraded form is. The degradation process, comparatively less slow in polymers containing hetero atoms like O, S, N, is extremely long, sometimes lasting longer than a hundred years. The speed at which new plastic waste reaches the sea is far greater than its eventual degradation. So, inevitably, its accumulation constantly increases. Be that as it may, in any case, the particulate matter to which these plastics are reduced represents an enormous quantity of foreign bodies incompatible with life.

It must be considered that the survival of aquatic life largely depends on the concentration of oxygen. The oxygen present in seawater is produced both by the photosynthetic activity of algae in the surface layers and by what dissolves from the atmosphere according to Henry's Law and the temperature: in cold seas, the quantity of dissolved oxygen is greater than that which dissolves in warm seas. It is clear that, if the aquatic surface is occupied by a layer of plastic, the passage will be prevented and the oxygen will not be able to dissolve, with serious damage to marine life (Mazzoli & Gorbi, 2021; Vaseashta, 2022; Vaseashta et al., 2022).

Another important damage caused by plastic in the sea is its entry into the food chain. Plastic, in fact, is ingested by fish both by chance and because, in some circumstances, is mistaken for food. The very small crustaceans called krill, particularly present and abundant in cold seas, can reduce relatively coarse plastic fragments to dimensions even below one micron. Fish and marine mammals usually feed on these tiny invertebrates, thereby acquiring the very small particles of plastic material that enter their tissues without any possibility of being eliminated. Bivalve mollusks such as, for example, mussels, filter very large quantities of water through their organism, water inevitably containing everything dissolved or suspended in it, and the corpuscular fraction made up of plastic particles is only very partially expelled or not at all. In this way, plastic is found in the tissue of the fish and shellfish we eat.

2. Results, Interpretation, and Discussion

Our direct contact with marine plastic waste occurred in the Italian Adriatic Sea and the Pacific Ocean on the coast of Eastern Australia. In the upper Adriatic Sea, in front of the city of Ancona, there is an area where agglomerated waste gathers. GreenPeace dedicated a report on the state of plastic litter (Araujo et al., 2018). The particulate analysis is performed on what remains on the Millipore filter after filtering the seawater sample we had previously taken. Afterward, the residue is inserted into the environmental scanning electron microscope with which our laboratory is equipped (Field Emission Gun Environmental Scanning Electron Microscope, FEGESEM, Quanta 650, Thermofisher), is observed and analyzed using the X-ray microsensor of an Energy Dispersive System (EDS by Thermo Fisher USA). The sample is further analyzed with a Raman spectroscope (Araujo et al., 2018; Nava et al., 2021), since the EDS system only shows the elemental chemical composition of what is analyzed and cannot distinguish between the different plastics, composed largely of the same elements. Detecting chlorine may indicate that it is PVC, detecting fluorine indicates that it is polytetrafluoroethylene, and detecting silicon indicates that it is silicone, but not much more.

Raman spectroscopy shows the ratio and chemical links between carbon and oxygen, so determining which plastic it is much easier. Figure 2 is an Italian sample showing diatoms that were grown on the plastic surface transforming it into a sort of new home. The growth of a marine alga on a plastic substrate means that there has been an adaptation of that vegetable, but there is a limit beyond which no adaptation can be achieved.

Mollusks such as mussels, for example, filter enormous quantities of water every day, so that they can trap particles of plastic materials in their tissues. This is also how plastics can enter the food chain. Something similar happens with fish that ingest plastic particles that get trapped in their tissues become food for

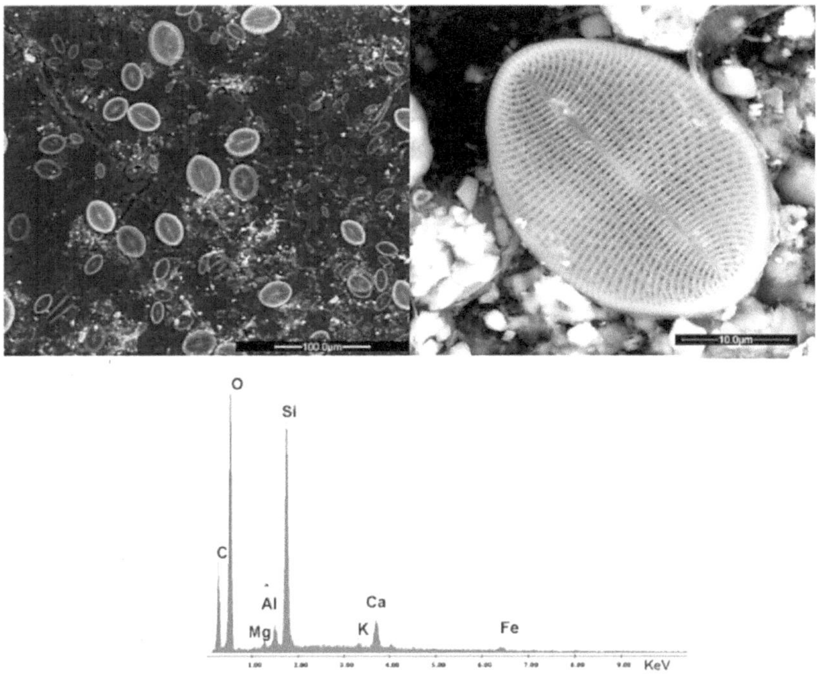

Figure 2: Images of diatoms that were grown on plastic debris. The EDS spectrum shows the chemical composition of the alga (unicellular, eukaryotic entities), which is composed mostly of silica or silicates.

animals and man. Other microplastic particles were found on two Australian beaches: Balmoral Beach North of Sydney City (NSW), and Putty Beach, in the Killcare region (NSW), far from industrial facilities.

Figure 3 shows plastic debris (left image) collected in just a 20-minute search on Balmoral Beach (Sydney, Australia), only spotting them by eye and only picking up what appeared colorful to the eye. In the case of Putty Beach (Killcare, Australia) far from industrial facilities, no similar debris could be spotted, but only a few white, spherical debris, recognized as polystyrene spherules (white arrow). The observation under the FEGESEM verified the dirtiness of the plastic debris from NaCl crystals or alga or other contaminants.

Figure 4 shows a diatom grown on plastic, dirty debris. That demonstrates that there is a certain re-habitation of the plastics by algae. We think that the dirtiness creates a certain surface roughness that helps the anchorage of the algae. Figures 5, 6, and 7 show small algae deposited on plastic debris.

The image of Figure 8 shows the internal part of plastic debris. It appears to be composed of nanoparticles probably of iron, chromium, and lead. Once the debris is physically-chemically degrading these nanoparticles are released into the marine environment, becoming fish food.

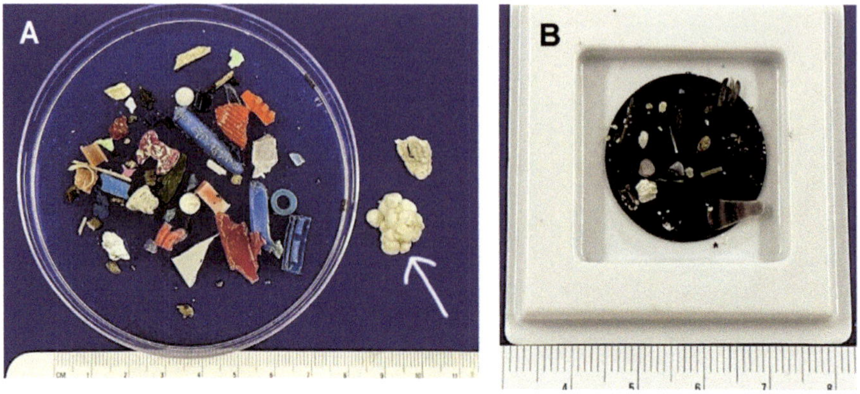

Figure 3: Microplastic debris collected in (A) collected on Balmoral Beach (Sydney). The white arrow indicates two pieces of polystyrene debris found on another beach north of Sydney (Killcare). The right image (B) shows some of the smallest debris deposited on a carbon disk for the FEGESEM analyses.

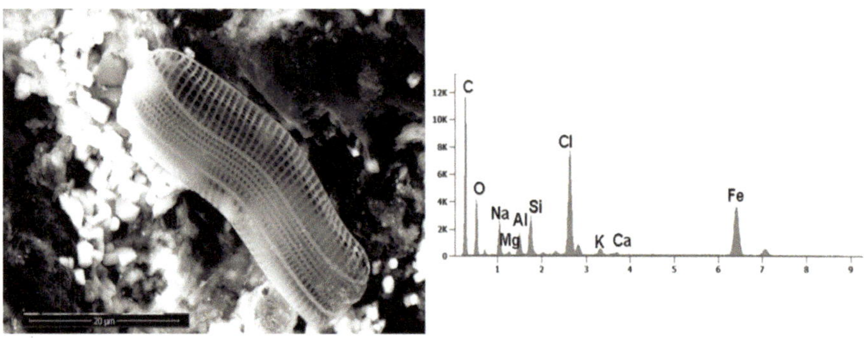

Figure 4: Image of diatom that re-habited the dirty surface of the plastic debris. The surface is composed of NaCl crystals and debris containing aluminum, silicon, iron, magnesium, potassium, and calcium.

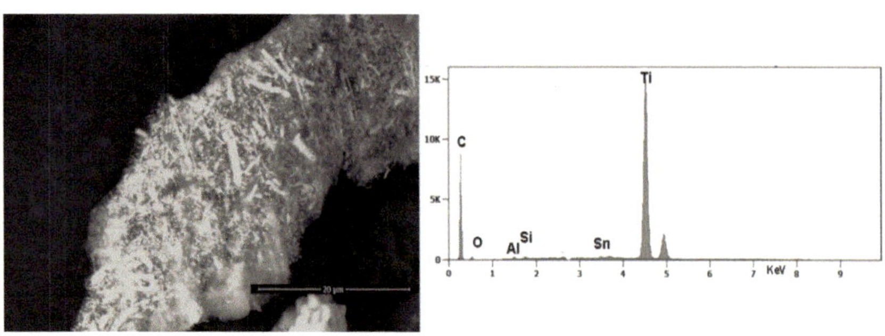

Figure 5: Image of plastic debris contaminated by titanium-tin submicrometric pollution.

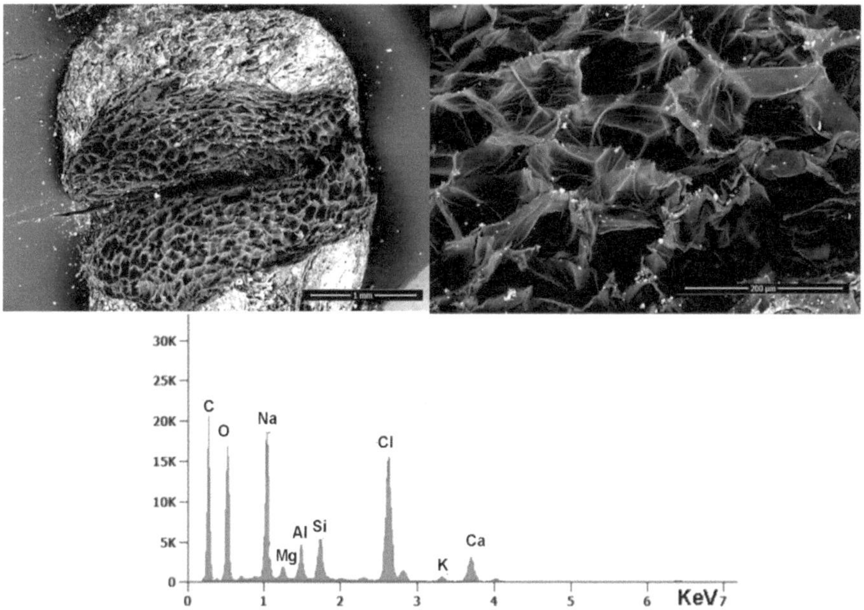

Figure 6: Image of a polystyrene spherule (left) and its internal morphology and contamination (right) in the porous structure. Internal pollution is composed of carbon, oxygen, sodium, chloride, aluminum, silicon, potassium, and calcium.

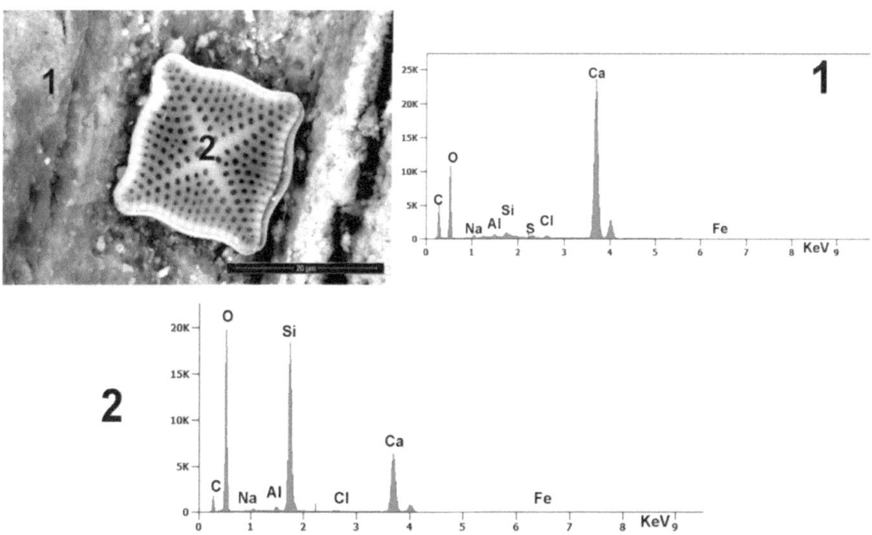

Figure 7: Image shows a diatom living in the ocean, growing on plastic debris. The two EDS spectra are referred to as the polymeric, dirty surface, and to diatom content.

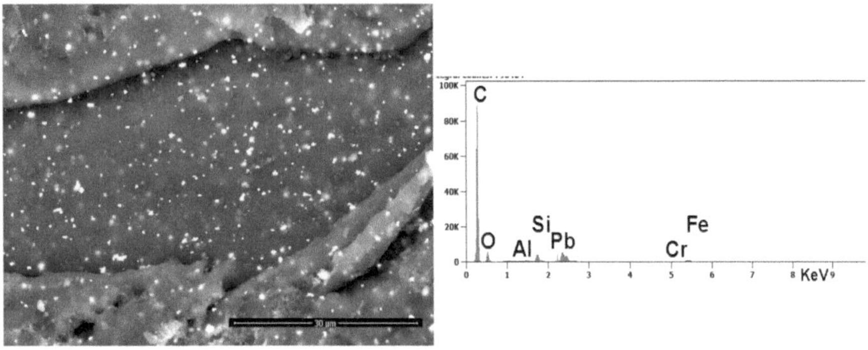

Figure 8: Image of an internal part of plastic debris found on a beach. The plastic particulate matter appears to be a composite. The filler is composed of carbon, oxygen, aluminum, silicon, lead, chromium, and iron.

3. Conclusions

It is obvious that an ever-increasing presence of plastic waste in the marine environment represents a danger to marine life and, consequently, to the planet. We discussed the impact of submicronic particles on marine life and verified that the nanosized particulate matter can induce malformations in the offspring and death or diseases in marine species (Gambardella et al., 2013, 2014, 2016). The question that we posit is: how can we clean up the oceans? Regardless of the obvious fact that plastic, as well as any pollutant, should no longer be allowed into the sea, what is already present in the water, considering the time it takes to be transformed naturally, should be removed. This pollutant is now present in enormous quantities and does not spare any body of water. Furthermore, its presence is demonstrated at any depth. For this reason, the proposed solutions can only concern small extensions. One proposal is that of the non-profit association, such as "The Ocean Cleanup" – a very ambitious project that aims to clean 90% of the oceans. Leaving aside the actual feasibility of the undertaking, everything just below the surface of the water is outside the scope (Cho, 2022; Mountford, 2022; NOAA, 2024). We can very well state that even if an efficient cleaning system were identified, we would still find ourselves with millions of tons of plastic to get rid of. Excluding incineration, if only for the toxicity that would derive from the fumes produced, one could hypothesize forms of recycling, as a solution, however, we must take into account not only the quantities involved but also the fact that the plastic itself is inevitably 'polluted' by animals and plants. Hence, briefly put, plastic provides very useful and extremely cheap uses, but, despite the illusions through the past few decades, also due to the fact that throughout its entire cycle, viz., from production to the moment it is no longer needed, it has never been given serious consideration, and it often ends its lifecycle embarrassingly!

References

Aminabhavi, T.M., Balundgi, R.H., & Cassidy, P.E. (1990). A review on biodegradable plastics. *Polymer-Plastics Technology and Engineering*, 29(3), 235–262. https://doi.org/10.1080/03602559008049843.

Araujo, C.F., Nolasco, M.M., Ribeiro, A.M.P., & Ribeiro-Claro, P.J.A. (2018). Identification of microplastics using Raman spectroscopy: Latest developments and future prospects. *Water Research*, 142, 426–440. https://doi.org/10.1016/j.watres.2018.05.060.

Cho, R. (2022, October 13). *How Do We Clean Up All That Ocean Plastic?* https://news.climate.columbia.edu/2022/10/13/how-do-we-clean-up-all-that-ocean-plastic/.

Gambardella, C., Aluigi, M.G., Ferrando, S., Gallus, L., Ramoino, P., Gatti, A.M. et al. (2013). Developmental abnormalities and changes in cholinesterase activity in sea urchin embryos and larvae from sperm exposed to engineered nanoparticles. *Aquatic Toxicology*, 130–131, 77–85. https://doi.org/10.1016/j.aquatox.2012.12.025.

Gambardella, C., Gallus, L., Gatti, A.M., Faimali, M., Carbone, S., Antisari, L.V. et al. (2014). Toxicity and transfer of metal oxide nanoparticles from microalgae to sea urchin larvae. *Chemistry and Ecology*, 30(4), 308–316. https://doi.org/10.1080/02757540.2013.873031.

Gambardella, C., Ferrando, S., Gatti, A.M., Cataldi, E., Ramoino, P., Aluigi, M.G. et al. (2016). Review: Morphofunctional and biochemical markers of stress in sea urchin life stages exposed to engineered nanoparticles. *Environmental Toxicology*, 31(11), 1552–1562. https://doi.org/10.1002/tox.22159.

Harada, M. (1995). Minamata disease: Methylmercury poisoning in Japan caused by environmental pollution. *Critical Reviews in Toxicology*, 25(1), 1–24. https://doi.org/10.3109/10408449509089885.

Jamieson, A. (2017). *How We Discovered Pollution-poisoned Crustaceans in the Mariana Trench* [The Conversation]. Newcastle University. https://theconversation.com/how-we-discovered-pollution-poisoned-crustaceans-in-the-mariana-trench-72900.

Liu, P., Zhan, X., Wu, X., Li, J., Wang, H., & Gao, S. (2020). Effect of weathering on environmental behavior of microplastics: Properties, sorption and potential risks. *Chemosphere*, 242, 125193. https://doi.org/10.1016/j.chemosphere.2019.125193.

Mazzoli, C., & Gorbi, S. (2021). *Plastica in Adriatico, a che punto siamo?* https://www.greenpeace.org/static/planet4-italy-stateless/2021/06/3bb98793-plastica-in-adriatico-a-che-punto-siamo_.docx.pdf.

Mountford, A.S. (2022). *Plastic Dispersion in the Global Ocean*. Newcastle University. http://theses.ncl.ac.uk/jspui/handle/10443/5831.

Nava, V., Frezzotti, M.L., & Leoni, B. (2021). Raman spectroscopy for the analysis of microplastics in aquatic systems. *Applied Spectroscopy*, 75(11), 1341–1357. https://doi.org/10.1177/00037028211043119.

NOAA. (2024, June 16). *What are Microplastics?* https://oceanservice.noaa.gov/facts/microplastics.html.

Vaseashta, A. (2022). Future of water: Challenges and potential solution pathways using a nexus of exponential technologies and transdisciplinarity. *In:* A. Vaseashta, G. Duca, & S. Travin (Eds.), *Advances in Environmental Engineering and Green Technologies* (pp. 37–63). IGI Global. https://doi.org/10.4018/978-1-7998-7356-3.ch002.

Vaseashta, A., Duca, G., & Travin, S.O. (Eds.). (2022). *Handbook of Research on Water Sciences and Society*. Engineering Science Reference.

6 | Sources and Impacts of Microplastics in Aquatic Environment and Remediation Strategies

Gor Khachatryan[1], Ashok Vaseashta[2,3,4(0000-0002-5649-0067)] and Gor Gevorgyan[1*]

[1] Scientific Center of Zoology and Hydroecology, National Academy of Sciences of Republic of Armenia, Yerevan, Armenia

[2] International Clean Water Institute, Office of Strategic Research, Manassas, VA, USA

[3] Ghitu Institute of Electronic Engineering and Nanotechnology, Technical University of Moldova, and Academy of Sciences of Moldova, Chisinau, Moldova

[4] Institutul de Cercetare al Universităţii din Bucureşti, Strada Atomiştilor, Magurele, Romania

Abstract: Microplastics, originating from various sources such as degraded larger plastics, industrial processes, personal care products, and textile fibers, have become a pervasive pollutant in aquatic environments. These tiny plastic particles pose significant ecological and health risks by entering food chains, disrupting aquatic ecosystems, and releasing toxic chemicals. Their presence affects aquatic organisms, humans, and water quality. Effective remediation strategies include filtration techniques, bioremediation using microorganisms, and advanced oxidation processes. Additionally, public awareness campaigns, regulatory measures, and the promotion of sustainable alternatives to plastics are essential to mitigate microplastic pollution and safeguard aquatic ecosystems. This chapter describes several sources of microplastics and their potential impact not only on human health but also on ecology. An overview of the remediation of microplastics using some conventional and advanced technologies is discussed.

*Corresponding author: gor.gevorgyan@sczhe.sci.am

1. Introduction

Plastic is now considered an integral part of human daily life, aided by its versatility, durability, cost-effectiveness, lightness, hygiene, and safety. Plastic packaging helps preserve the freshness and safety of food and other perishable items, contributing to food storage innovation and convenience. Plastics have largely replaced wood and various types of metals due to their affordability and versatility (Dallyn & Shorten, 1998). Although plastics appear to reduce human impact on the environment by lessening the need to cut down trees, extract metals, and process alloys, which all cause significant damage to the planet, they actually have produced the opposite effect. This will be discussed in more detail in the relevant sections. According to available literature, the world's first synthetic plastic was produced by Leo Baekeland in 1907 (Kiran et al., 2022).

As per environmental problems, the convenience and versatility of plastic have led to its penetration into almost all areas of modern life, significantly contributing to the pollution of oceans, other bodies of water, forests, rivers, and even the air we breathe. As plastic waste accumulates in ecosystems, it poses a threat to wildlife and its delicate habitats (Azevedo-Santos et al., 2021). Microscopic plastic particles, otherwise known as microplastics (MPs), have even penetrated the most remote corners of our planet, reaching as far as the oceans (Yeo et al., 2023). It is not only the physical presence of plastics that is of concern but also the fact that they are composed of a complex mixture of chemicals, many of which are toxic and harmful. As plastics break down over the years, these chemicals bound to the plastic enter the environment, contaminating soil, water sources, and the entire food chain. The effects of this pollution range widely, impacting ecosystems and human health (Li et al., 2023).

Today, the following four main types of plastic dimensions are accepted in the International System of Measurements (SI): nanoplastic, MP, mesoplastic, and macroplastic (Ramkumar et al., 2022). Microplastics are small plastic particles less than 5 millimeters in size, often microscopic in nature (Vaseashta et al., 2021). This work focuses primarily on MPs, which can be classified based on their different properties, including both physical and chemical characteristics. Physically, MPs vary in size, shape, and color, ranging from microscopic particles to larger pieces visible to the naked eye. Chemically, MPs consist of various polymers, such as polyethylene, polypropylene, and polystyrene (Figure 1). Each of these polymers exhibits different properties and behaviors in the environment, resulting in varied effects under different environmental conditions (Sarkar et al., 2023).

Understanding these characteristics allows for the assessment and identification of potential impacts on ecosystems, wildlife, and human health, especially concerning transmission through the food chain (animal-human interaction). The classification of MPs aids in developing and implementing effective monitoring strategies, which are crucial for mitigating their negative effects on the environment and preventing further environmental degradation (Chen et al., 2023).

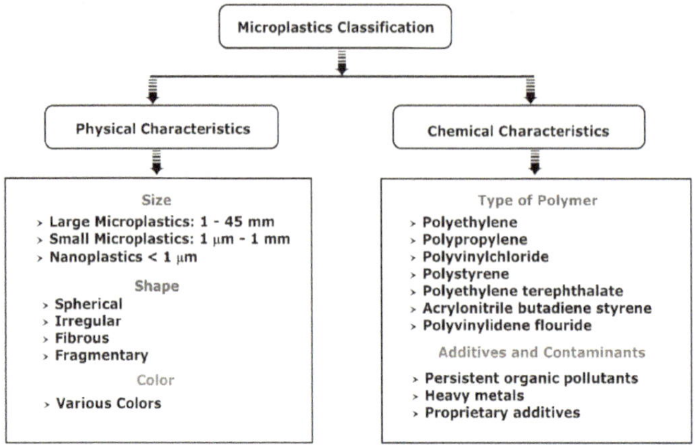

Figure 1: MP classification: General scheme.

As shown in Figure 1, MPs are classified according to color; they can be black, white, red, light blue, etc. Color grading of MPs is often done under a microscope or by visual inspection with thc unaided eye, depending on the size of the MPs. Many researchers perform color grading using standardized color charts or scales that help accurately describe the color of MPs (Hernández-Fernández et al., 2023). Although, at first glance, it may seem that color classification is simple and not particularly valuable, this is not the case. Color classification can provide useful information for understanding the sources, behavior, and impact of MPs on the environment (Yin et al., 2019). In addition to the aforementioned classification, there are two other main categories of MPs (Figure 2): primary and secondary (Cole et al., 2011).

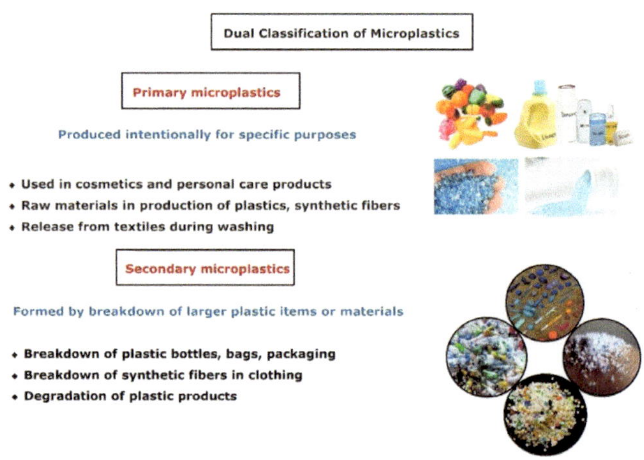

Figure 2: Dual classification of MPs.

Primary MPs are produced as small plastic particles for specific purposes in various industries. Notably, the production of cosmetics and personal care products uses synthetic fibers as raw materials, which enter the environment when released from textiles during washing (Ziani et al., 2023). Another category is secondary MPs, formed by the breakdown of larger plastic items or materials. Over time, and under certain environmental conditions such as sunlight, wind, and water, MPs like plastic bottles, bags, and other packaging materials break down into smaller fragments (Wright et al., 2013). Secondary MPs also result from the degradation of synthetic fibers in clothing, tire wear, and the breakdown of other plastic products (Agbekpornu & Kevudo, 2023). The concentration (quantity) of MPs is a critical factor in each classification (Shanmugam et al., 2020). Various methods are used for the separation, identification, and quantification of MPs in aquatic environments (Browne et al., 2011).

One common method is sieving, which involves passing the sample through a series of mesh sieves with different pore sizes. Larger particles such as macroplastics are retained on the sieve, while smaller particles pass through. This method is often used for the initial separation of MPs from sediment or water samples (Bergmann et al., 2019).

Filtering is another method, which involves passing the sample through a filter with a defined pore size to capture MPs while allowing other particles to pass through. Filters with pore sizes ranging from micrometers to nanometers are used depending on the targeted MP size and the research objectives (Dris et al., 2016). Visual sorting, based on the manual identification of MPs by their size, shape, and color, is another method. Although labor-intensive, it can provide valuable information about the types, amounts, and composition of MPs present in the sample (Horton et al., 2017). Spectroscopic methods such as Fourier transform infrared spectroscopy and Raman spectroscopy are used to identify MPs based on their chemical composition, which provides information about the polymer type and facilitates their characterization (Semmouri et al., 2022).

Microscopic techniques like optical microscopy, scanning electron microscopy, and transmission electron microscopy enable the sorting and characterization of MPs based on their morphological and surface characteristics (Huang et al., 2023). Once separated and identified, MPs can be quantified using various methods such as gravimetric analysis, image analysis software, or particle counting techniques (Löder et al., 2017). Scientific studies have shown that MPs are present in every part of the environment, including the polar regions. Human activities have caused MPs to reach even the ice areas in the North and South poles. Sea ice helps MPs floating on the water surface reach these polar regions (Vivekanand et al., 2021). MPs mainly composed of polyvinyl chloride, nylon threads, and polyethylene terephthalate tend to sink in aquatic environments, while those made of polyethylene, polypropylene, and polystyrene are more likely to float due to their lower density and lightness (Auta et al., 2017).

2. Sources of MPs in the Aquatic Environment

Freshwater ecosystems serve as major accumulation and transport routes for plastic pollutants into the global ocean. MPs easily reach and accumulate in the global ocean via rivers. Poorly managed landfills, untreated sewage, inadequate industrial controls, recreational use of coastal areas, tourism, and other activities contribute to MP contamination of freshwater (Mintenig et al., 2017). Land-based sources, particularly rivers, are the primary contributors to sea pollution, accounting for approximately 80% of the total (Torres et al., 2019).

MPs are produced through the direct emission of plastic particles and the degradation of larger objects by environmental conditions (Andrady, 2011). Sources of MPs include packaging materials (Mattsson et al., 2018), garbage, wood, paint preservatives, synthetic textiles (fabrics made of synthetic fibers, as opposed to natural fibers such as cotton or wool) (Zambrano et al., 2019), car tires, personal care products (e.g., face and body scrubs, toothpaste, face wash, shower gels, shampoo, hand soap, hair conditioners, sunscreens, deodorants, cosmetic products), abrasive machining (a mass abrasive machining process carried out using high-velocity compressed air or by centrifugal force), and cleaning agents (mainly for removing stains, grease, and dirt from hard surfaces) (Andrady, 2017). Additionally, some industries, particularly the oil and gas industry and the production of plastic pellets (also known as slags or resin balls), medical devices (including single-use items such as syringes, tubes, and catheters, as well as durable items like implants and prostheses), pharmaceuticals, medical waste, medical personal care products such as lotions, creams, and disinfectants, toys, household items, and food contribute to MP pollution. Figure 3 shows the main sources of MP pollution. Sewage and industrial effluent are also major sources of MP contamination in aquatic environments (Mehra et al., 2021).

Materials used for packaging, vital in maintaining food freshness and ensuring product safety, account for approximately 40% of plastic demand in Europe, with 50% being single-use products (Ramkumar et al., 2022). They are widely used due to their low production cost, lightweight, ease of use, durability, easy transport, and storage. These materials are primarily composed of thermoplastic resins, particularly polyethylene terephthalate, high-density polyethylene, low-density polyethylene, polypropylene, polyvinyl chloride, and polystyrene (Ekta et al., 2021).

Many activities carried out in our daily lives also contribute to environmental pollution with MPs. People are often unaware of how their normal activities contribute to this pollution. Here are some concrete examples to illustrate this. When we open a plastic bag, such as the one used for snacks like chocolate, small particles of plastic can be released into the air and the environment. These particles can then return to the water environment through atmospheric precipitation. Another example is when cutting or tearing adhesive tape, such as the tape used to seal boxes, which can release MPs. The mechanical act of cutting or tearing breaks the plastic into smaller particles, contributing to MP pollution. These

Figure 3: Origins of MP contamination
(images are adapted from http://citymoversandpackers.in/services/quality-packaging-material and https://wisepoint.org/2176).

particles can stick to our hands, or the box, or disperse in the air. Surprisingly, even opening a plastic bottle cap, which involves twisting, can produce MPs. Contact between the cap and bottle threads can cause small plastic fragments to break off and be released into the environment.

Additionally, it's evident that waste accumulation significantly contributes to the spread of MPs in aquatic ecosystems. This claim is supported by comprehensive studies based on Jambeck et al. (2015). Solid waste is one of the main sources of MP pollution in the environment. According to the Resource Conservation and Recovery Act (RCRA), solid waste includes garbage or waste, sewage sludge, industrial or commercial waste, construction and demolition debris, electronic waste, hazardous waste, green waste, bulk waste, and other materials from commercial activities (USEPA, 2018). Landfills are the main sites of accumulation and the spread of MPs (Kabir et al., 2023). Meteorological factors can play a significant role in the transport of solid waste from landfills to water bodies. Among these factors, wind, rain, storm waves, floods, and seasonal variability play significant roles, which will be discussed in more detail below.

Strong winds can transport light solid waste, such as plastic bags, paper, and light garbage scraps, over long distances. Wind direction and speed are considered the main factors determining the direction and extent of waste spreading from landfills to nearby water bodies (Zainuddin et al., 2022). Besides wind, heavy rains are another phenomenon contributing to the spread of plastic waste. These rains can cause runoff from landfills, transporting solid plastic waste into nearby water bodies (streams, rivers, and other water bodies) (Lance et al., 2014). In coastal areas, storm surges associated with tropical storms or severe weather events can inundate landfills near coastal areas, leading to the transport of solid waste into adjacent water bodies as well. Floods can also transport large amounts of debris and pollutants into the aquatic environment. Regardless of their occurrence circumstances (heavy rainfall, snowmelt, or other factors), floods can overwhelm landfill infrastructure and lead to the direct transport of solid waste into water bodies (Rochman et al., 2015). Surprisingly, seasonal weather changes such as dry seasons followed by wet seasons, particularly heavy rainfall or spring snowmelt, also contribute to the movement of MPs from landfills to water bodies. For example, snow accumulated in landfills during the winter can melt quickly in warm weather, causing a large flow of water, which facilitates the transport of MPs to nearby water bodies (Uogintė et al., 2022). These meteorological factors also interact with the topographic features of a given region, along with landfill design and waste management practices, affecting solid waste transport to water bodies (Galloway et al., 2017). It's also intriguing that wood and paint preservatives are recognized as significant sources of MP pollution. Not only the EU but also several countries have begun to closely monitor the potential environmental contamination from paint (Manzoor et al., 2020). On average, 37% of conventional paints comprise plastic polymers (such as acrylic or latex paints), usually added as binders to enhance paint performance and durability. Paints are utilized not only in liquid form but also in solid or gaseous states (aerosols) (Osman et al., 2023). Aerosol paints, commonly found in spray cans, can emit tiny droplets or particles into the air during application. These aerosolized particles can be carried by wind currents and deposited onto the surfaces of water bodies through precipitation (Figure 4).

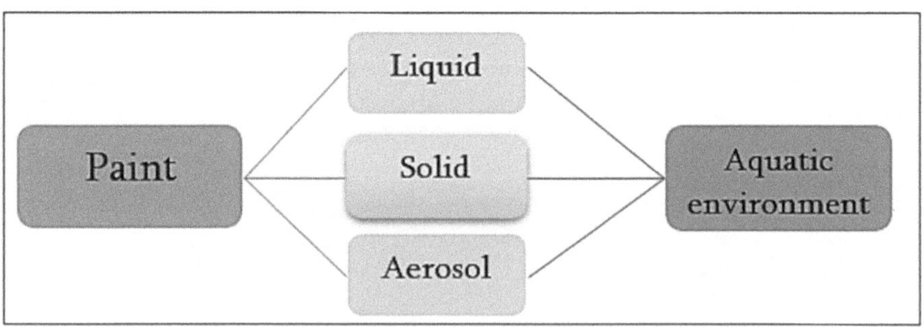

Figure 4: Pathways of MP transfer from paints to aquatic environments (Brännström et al., 2023).

Approximately 8–30% of waste is generated as a result of paint spills and other paint operations. Research suggests that 1.0–5.3 tons per year of MPs are released from paints, with an estimated number of primary MPs entering the aquatic environment from paints being around 0.1–0.5 tons per year (Sukanya et al., 2020). MP pollution resulting from washing synthetic textiles is considered one of the primary sources of primary MP pollution. During the washing of synthetic fabrics, MPs are detached from clothes due to chemical and mechanical processes. Owing to their size, they can pass through water treatment plants and reach the world's oceans. However, it's worth mentioning that very few countries worldwide have the capability to install such treatment plants, and their number is quite limited and largely ineffective in addressing the problem. Countries with developing and transitioning economies, often considered major contributors to global ocean pollution, lack the financial resources to install such treatment plants. While some facilities focus on recovering large plastics, the retrieval of small-sized MPs remains a global environmental challenge (Browne et al., 2011).

Recent research indicates that synthetic clothing accounts for about 35% of primary MPs transported to the global ocean, making it the primary source of MPs (De Falco et al., 2019). Another study suggests that around 8% of European MPs entering the oceans originate from synthetic textiles. Estimates suggest that between 200,000–500,000 tons of textile MPs enter the marine environment annually worldwide (EEA, 2022). Despite legislative efforts by states to eliminate primary MPs from products, synthetic fibers remain a significant concern due to their prevalence in the apparel industry and the challenges associated with removing them from water treatment plants compared to natural fibers. The main fibers found in MPs released during clothes washing are polyester, acrylic, and polyamide. The impact of MPs on the aquatic environment remains a major challenge, as the production of industrial textile fibers continues to increase due to the growing human population and its increasing demand (Acharya et al., 2021).

A review of the literature reveals that tires and road construction material particles are considered to be one of the largest sources of MPs that end up in water bodies and eventually into the world's oceans (Unice et al., 2019). Tire wear particles result from the degradation of automobile tires, creating particles that consist of rubber mixed with asphalt and mineral materials (Tamis et al., 2021). Research shows that more than 200,000 tons of MPs are blown into the oceans from roads every year. These wind-borne particles are a greater source of MP pollution than rivers. They also appear in ice-covered areas (Bergmann et al., 2019). MPs on the surface of ice, which have a dark coloration, absorb sunlight better and increase heat absorption, leading to a reduction in albedo, which accelerates the melting of ice-covered areas. This phenomenon may contribute to the loss of ice mass in polar regions and glaciers, exacerbating the effects of climate change and contributing to sea level rise (Kole et al., 2017). In fact, the impact of MPs is significant in the context of climate change, necessitating consideration of this factor when designing climate risk mitigation, management, and other plans.

Polyethylene (PE), polypropylene (PP), polystyrene, polyethylene terephthalate (PET), polymethyl methacrylate (PMMA), nylon, and polyurethane (PU) are mainly used in personal care products (Bashir et al., 2021). PE microbeads are commonly used in exfoliating scrubs, cleansers, and toothpastes because of their smooth texture and abrasiveness (Eriksen et al., 2013). However, in recent years there has been a decrease in their use due to environmental concerns. PE microbeads have similar properties to PE and are used in similar personal care products. Both PE and PP are seeing a decrease in their use (Guerranti et al., 2019). PET MPs are used in cleaning products. These are less common than the two listed above (Sundt et al., 2014). PMMAs are often used as substitutes for PE and PP because they are considered biodegradable. Although biodegradable, it should be noted that they can persist in the environment and contribute to MP pollution (Shengpei et al., 2004). Nylon (polyamide) microfibers are used in some exfoliating products and toothpaste for their abrasive properties. Unlike PMMA, nylon is not biodegradable but can break down into smaller MPs over time (Auta et al., 2017). PU is less commonly used in personal care products but can be found in some cleaning products. They are biodegradable and cannot accumulate in the aquatic environment (Duis & Coors, 2016). Figure 5 illustrates the dominant types of plastic found in personal care products, highlighting their prevalence in consumer goods.

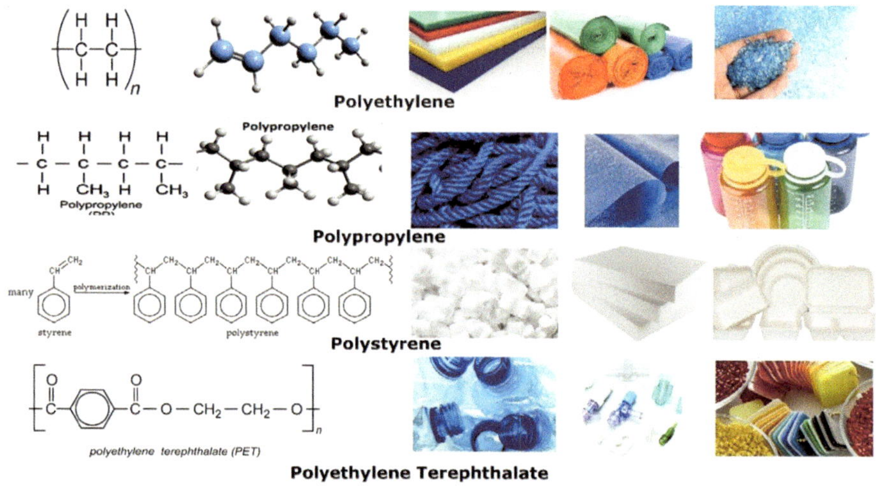

Figure 5: Dominant plastic types in personal care products.

Personal care products such as soap, shampoo, shower gels, many face and hand cleansers, deodorants, and toothpaste often contain particles <300 μm and may contain additives such as dyes, as previously mentioned (Storck et al., 2015). As reiterated multiple times, rivers serve as the primary conduit for transporting MPs. Conducted studies and research documents that MPs can easily accumulate in places such as logging sites and dams (Schmidt et al., 2017). Additionally, factors such as the slope of the river, the shape of the river's relief, and morphological features (e.g., deep or shallow sections) are also important. Unlike rivers, lakes can retain MP fragments for quite a long time, both in bottom accumulation and surface layers (Alimi et al., 2018).

Factors such as large-scale fishing, recreational use of the sea, and demographic shifts to coastal areas further increase the amount of plastic waste entering the oceans (Portman & Brennan, 2017). Recent studies also document that MPs have reached Antarctica, posing serious challenges in the face of global warming and the climate crisis (Andrady, 2011). MPs are widely present in the ocean, resulting from the breakdown of larger plastic items, and have been found not only in the pelagic zone but also accumulated in the bottom layers (Thompson et al., 2004). Cigarette residues, plastic bottles, and bottle caps are the most frequently dumped into the marine environment, followed by plastics used in food packaging, plastic grocery bags, and food containers, among other solid waste residues (Kumar et al., 2021). These are considered the main types of plastic that pollute and continue to pollute the aquatic environment.

As already mentioned, MPs are scattered throughout the world's oceans and can often be found both in the coastal zones of water bodies and in bottom sediments, beaches, and sewage. Some conducted studies prove that MPs are also present in frozen ice (Bergmann et al., 2019). MPs have reached the Arctic and Antarctic due to hydrometeorological phenomena such as ocean currents and winds, and their existence in these places has already been scientifically proven (Nor et al., 2020).

A study of scientific materials carried out shows that plastic pollution of oceans and seas has increased during the COVID-19 pandemic, when millions of single-use plastic products were produced, including medical face masks, protective medical gowns, disposable gloves, medical test kits, hand disinfection bottles, and plastic bags (Wang et al., 2021). These are easily transported from terrestrial environments to seas and oceans by winds or rivers or directly discharged into the marine environment, thus exacerbating plastic pollution problems. Studies have shown that MPs are widespread throughout the world, particularly in rivers and lakes (Silva et al., 2021). Estimates show that 75–90% of plastic in the marine environment originates from terrestrial sources, with the remaining 10–25% from oceanic sources (Jambeck et al., 2015).

3. Effects of MPs in the Aquatic Environment

The impact of MPs on aquatic organisms has become a major concern due to their widespread presence in marine and freshwater environments. When MPs appear in the aquatic environment, they are easily absorbed by aquatic organisms, including crustaceans, mollusks, fish, plankton, and other aquatic animals. These organisms can be affected by MPs in several ways: ingestion, physical damage, toxicity, bioaccumulation, behavioral changes, ecological effects, long-term effects, or a combination of these (Rakib et al., 2023).

One of the main ways MPs affect aquatic organisms is through ingestion. Organisms such as zooplankton often mistake MPs for food particles and ingest them, leading to reduced feeding efficiency (in the form of physical blockages in the digestive tract), nutrient absorption, and energy allocation, ultimately affecting growth, reproduction, and survival (Richon et al., 2022). Small pieces of MPs are ingested by aquatic organisms, passing through the gastrointestinal tract and releasing any toxic substances they carry, which can disrupt the normal biological functions of the organism. Studies have shown that carnivorous species have higher levels of MPs in their bodies compared to detrital species, which feed on dead organic matter (Núñez et al., 2021). Some species of zooplankton are particularly vulnerable. In larger organisms, MPs can become lodged in the digestive tract, leading to blockages, internal injuries, and secondary infections (Lee et al., 2023). The sharp edges or irregular shapes of MPs can damage delicate tissues, leading to infections or organ dysfunction. Additionally, large-sized MPs can settle on the bottom of water bodies, altering the physical properties of sediments and affecting the behavior, habitats, and survival of benthic organisms (Haque & Fan, 2023).

As mentioned, MPs can also absorb and accumulate other toxic pollutants from the environment, such as heavy metals, trace elements, polycyclic aromatic hydrocarbons, and persistent organic pollutants. These chemicals can be released into the tissues of aquatic organisms after ingestion, leading to toxic effects, including reproductive disorders, impaired immune function, and developmental issues (Chatterjee & Sharma, 2019). For example, exposure to polychlorinated biphenyls can disrupt hormone regulation, reproductive function, and immune responses in fish and other aquatic organisms, while heavy metals like mercury and lead can accumulate in animal tissues, causing neurological damage and metabolic disturbances. The combined effects of MPs and other pollutant components can pose a serious threat to the health and survival of aquatic populations. MPs can bioaccumulate and biomagnify through the food web, meaning that organisms at higher trophic levels can accumulate higher concentrations of MPs and associated pollutants than those lower down the food chain (Walkinshaw et al., 2020). This can lead to greater exposure of MPs to predators and top predators, including marine mammals and seabirds. Predatory organisms, as mentioned, occupy a significant position in the animal hierarchy (Botterell et al., 2019).

As predators consume contaminated prey over time, the concentration of MPs and associated contaminants can increase, a process known as biomagnification. Marine mammals and seabirds are particularly vulnerable to the effects of biomagnification because they can accumulate high levels of MPs and associated toxins in their tissues. Studies have shown that MPs can bioaccumulate in aquatic organisms such as plankton, invertebrates, and vertebrates (Xu et al., 2020). MPs can easily bioaccumulate in fish organisms, which are an important food source for humans. Fish are rich in omega-3 fatty acids and vitamins such as D and B2 (riboflavin), as well as calcium, phosphorus, iron, zinc, and iodine, along with magnesium and potassium. Additionally, fish play a crucial role in the marine ecosystem, contributing to material circulation, energy flow, and information transfer. The health of fish populations can directly impact the stability of marine ecosystem structure and function (Li et al., 2021).

Behavioral changes in aquatic organisms refer to alterations in their normal patterns of behavior due to exposure to MPs. For example, fish may exhibit changes in foraging behavior, reduced predator avoidance, or impaired reproductive behavior. These alterations in behavior can have cascading effects on population dynamics and community structure within aquatic ecosystems. For instance, increased feeding rates among fish populations can lead to declines in prey populations, which in turn can affect the abundance and distribution of other organisms in the food web (Candolin & Rahman, 2023). The ecological impact of MPs extends beyond individual organisms to entire ecosystems. Changes in the abundance, distribution, and composition of aquatic populations can disrupt ecological interactions, trophic dynamics, and ecosystem functioning.

Ecosystem processes such as nutrient cycling, energy flow, and habitat connectivity can be disrupted, ultimately affecting overall ecosystem health and sustainability (Yang et al., 2018). The long-term effects of MPs on aquatic organisms and ecosystems are not yet fully understood. Continued studies are needed to assess the cumulative and synergistic effects of MPs alongside other environmental stressors like climate change, habitat degradation, and overfishing (Kalčíková & Rozman, 2022). Long-term monitoring studies are crucial for assessing the persistence and impact of MPs on aquatic ecosystems over time. These studies can provide valuable insights into the chronic effects of MP pollution on aquatic populations, communities, and ecosystems. Additionally, MPs can interact with other chemicals and toxins in water, potentially producing new toxins that can harm aquatic organisms (Sandil & Zaray, 2024).

The relationship between MPs and marine organisms, particularly eukaryotic microbes, is complex and reciprocal. MPs can disrupt marine food webs, affect prey-predator dynamics, and alter the structure and function of microbial communities (Kumar et al., 2024). The risks associated with MPs in aquatic ecosystems can be direct, such as when MPs are ingested by organisms and accumulate in their tissues, or indirect, when MPs release toxic micropollutants into the environment (Chaukura et al., 2021). Figure 6 depicts the impact of MP pollution on the marine ecosystem, as analyzed by Issac and Kandasubramanian (2021), and provides a comprehensive overview of the multiple dimensions of MP pollution, highlighting its detrimental effects on marine life and ecosystems, as well as its potential implications for environmental health.

It illustrates the intricate pathways through which MPs infiltrate marine environments, the adverse impacts on biodiversity and food webs, and the possible risks posed to human health through the consumption of contaminated seafood. MPs exert significant impacts on living organisms through both physical and chemical mechanisms. Physically, marine animals can become entangled in larger plastic debris or ingest smaller particles, leading to blockages in their digestive systems, reduced feeding efficiency, and physical harm (Browne et al., 2015; Rochman et al., 2016). This ingestion of plastic can cause various health issues and even death in marine organisms (Wright et al., 2013).

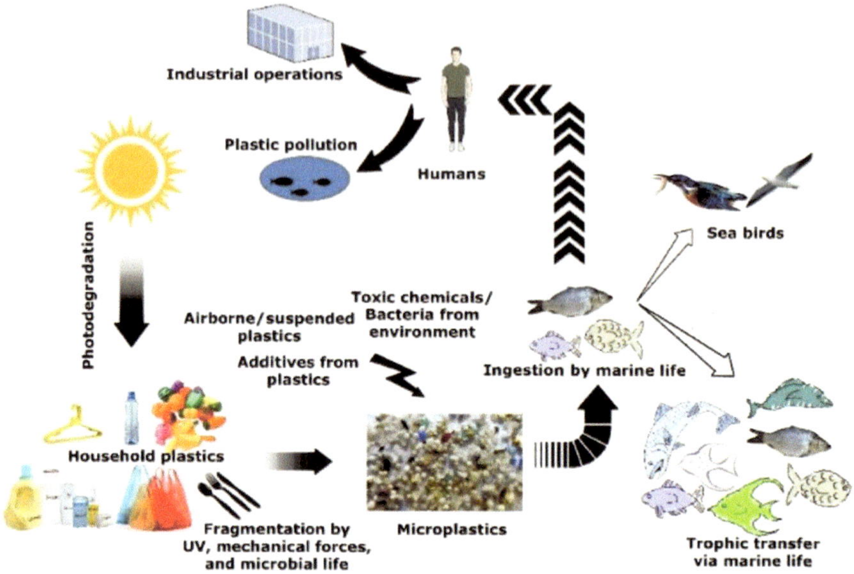

Figure 6: The impact of MP pollution on the marine ecosystem.

Figure 7: Effects of physical and chemical factors on aquatic organisms (images are adopted from https://blog.abamahotelresort.com/tenerife-nature-a-landscape-wins-the-world-press-photo-nature-prize and https://pressug.com/microplastics-found-in-nile-rivers-tilapia-fish-new-study).

Chemically, MPs can leach and adsorb harmful substances from the surrounding environment. These include persistent organic pollutants such as polychlorinated biphenyls, polycyclic aromatic hydrocarbons , and heavy metals, which are known for their carcinogenic, teratogenic, and mutagenic properties (GESAMP, 2016; Andrady, 2017). When ingested, these toxic chemicals can bioaccumulate in the tissues of marine organisms, potentially affecting their health and reproductive success (Teuten et al., 2009; Galloway et al., 2017).

Figure 7 provides a visual representation of these dual impacts, illustrating how physical entanglement and ingestion of MPs, as well as chemical exposure to associated pollutants, affect aquatic organisms. This figure synthesizes current research on the diverse effects of environmental variables such as plastic pollution on marine life and ecosystems, highlighting the urgent need for mitigation strategies and further scientific investigation.

MPs from the aquatic environment can enter the human body through the food chain, representing a potential pathway for human exposure. As MPs pollute water bodies, they are ingested by aquatic organisms at the base of the food chain, such as plankton and small fish. As larger predators consume these contaminated organisms, MPs accumulate and biomagnify, leading to increased concentrations in species higher up the food chain, including fish commonly consumed by humans (Borriello et al., 2023). The potential impact of MPs on human health is a growing concern, although ongoing studies are still in their early stages. These tiny plastic particles, less than 5 mm in size, can enter the human body through a variety of routes, including inhalation, ingestion (through the food chain), and through the skin. As a number of authors note in their works, after entering the

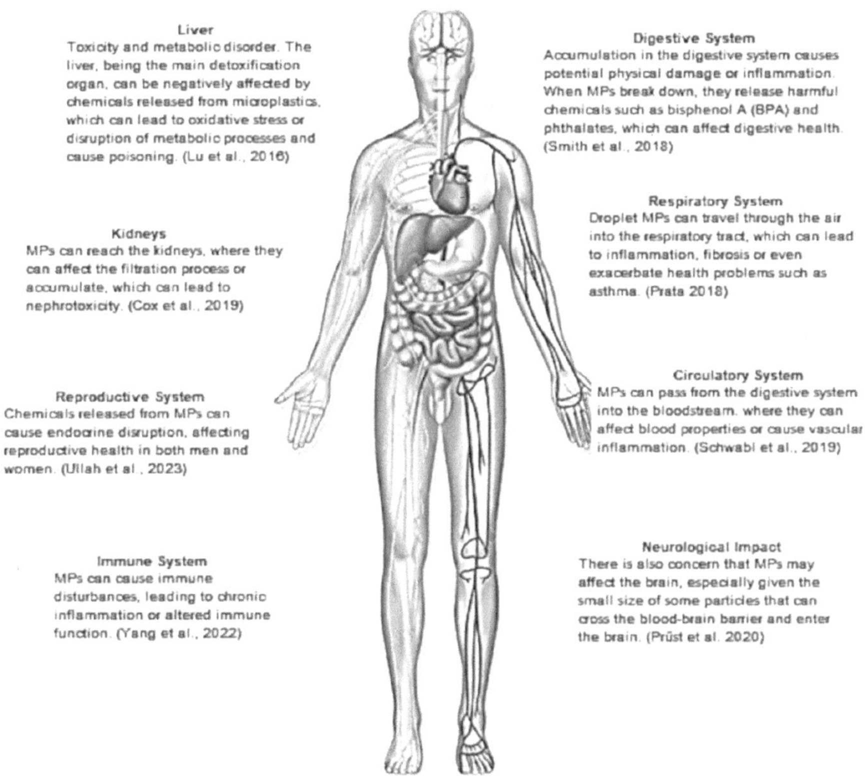

Figure 8: The impact of MPs on the human body
(image is adapted from https://alamyimages.fr/anatomie-image69117744.html).

body, MPs can affect various organs and systems. These include the digestive, respiratory, circulatory, reproductive, immune systems, liver, and kidneys (Fig. 8). Scientific studies have shown the presence of MPs in various seafood such as shellfish, crustaceans and fish, highlighting the potential for human exposure through the consumption of contaminated seafood (Unuofin & Igwaran, 2023). In addition, MPs have been found in drinking water, salt, and other foodstuffs, further expanding human exposure pathways. In general, the presence of MPs in the aquatic environment and their accumulation in the food chain raise concerns about possible effects on human health. Continuous studies are needed to better understand the exposure pathways, bioavailability, and potential health effects associated with MP ingestion. Additionally, efforts to mitigate MP pollution at its source are critical for reducing human exposure and protecting public health (Yu et al., 2023).

4. MP Remediation from the Aquatic Environment

MPs possess properties that make them persist in the environment for extended periods, breaking down into smaller particles and interacting with and absorbing other pollutants, rendering their removal challenging compared to larger plastic items like bottles and containers (Vivekanand et al., 2021). Therefore, there is a pressing need to develop, create, and implement new methods and equipment for MP removal. As previously mentioned, dyes also contain MP particles. Currently, there is no specific legislation addressing the presence of MPs in paints, posing a significant obstacle to their regulation. EU countries are working to develop legislation aimed at limiting the use of MPs in paints. However, developing and transitional economy states face additional challenges due to their limited financial resources (Gaylarde et al., 2021). Despite this, EU member states have established legislation governing the composition, application, and disposal of dyes and materials, which indirectly influences the release of MPs from paints, contributing to the degradation of aquatic environments (Faber et al., 2021).

Directive 2004/42/EC, also known as the Volatile Organic Compounds Directive, is a European Union directive targeting the reduction of emissions from certain paints, varnishes, car refinishing products, and decorative coatings. This directive applies to both professional and non-professional use of decorative paints, varnishes, car refinishing products, and other coatings. EU member states are responsible for transposing this directive into their national legislation and ensuring compliance (DIRECTIVE 2004/42/CE). Adoption and ratification of similar directives and legislation by states may encourage the reduction or complete elimination of MPs in paints.

It is considered a necessary and important step to create closed systems in industrial enterprises, which will make it possible to regulate the pollution of the environment with MPs (Sundt et al., 2014). Studies on MP pollution have attracted considerable attention since the 2000s. Pollution of oceans and seas with MPs has become a global problem and requires urgent steps to solve it. First of all, it is necessary to understand not only the sources and means of transporting MPs to water bodies but also the possible solutions to the problem of pollution and degradation (Jambeck et al., 2015). Although waste stored in landfills in developed countries is usually covered with soil or synthetic materials or surrounded by a fence to prevent the wind from dispersing plastic waste, in developing and transitional economies, waste is simply dumped into the environment without proper management, as is done in developing countries (Jambeck et al., 2015). Figure 9 illustrates MP removal methods, which are classified into four main groups: physical methods, chemical methods, biological removal, and combined methods. Each group contains specific techniques or approaches for MP removal from the aquatic environment, related to the content of the given problem.

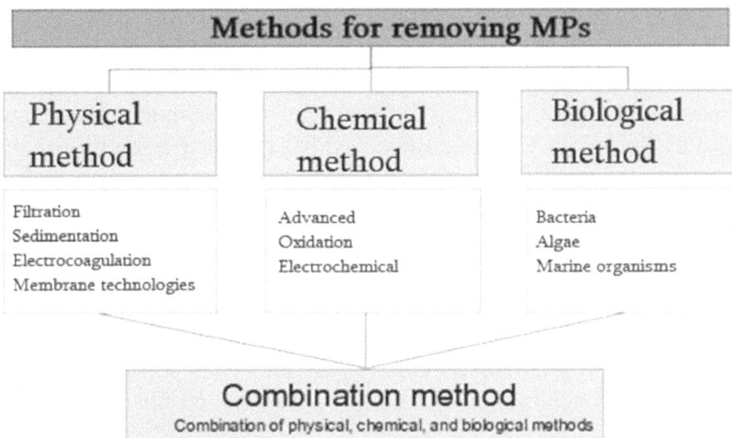

Figure 9: Categorization of techniques for MP removal.

By physical methods, we understand when the physical removal or extraction of MPs from different environments takes place. Filtering is one of the physical methods of removing MPs. Filtration is the passage of water through a filter, which is carried out with different types of filters, for example, membrane, mesh, activated carbon filters, sand filters, ultrafiltration, and nanofiltration membranes, etc. Filtration is considered an effective and less expensive method. Although filtering has advantages, it also has disadvantages. Minute and small MPs may not be captured by some filters, and then it will be necessary to use other filters as well, which requires additional financial resources. It is most effective when filtration is combined with other MP removal methods (Demir et al., 2021; Pizzichetti et al., 2021). Over time, constant use of the filters causes clogging of the filters and the need to replace the filters from time to time. Filtration is generally effective when applied to the filtration of large amounts of MPs, for example, it is not effective in the filtration of nanoplastics because the filters are unable to collect the nanoplastic particles. However, currently developing and transitional economies do not have sufficient financial resources to carry out MP filtration in water (Kau et al., 2024). Another physical method is sedimentation. The essence of this method lies in the fact that MP settles on the bottom of water bodies or, in laboratory conditions, on the bottom of containers or vessels. Coagulants or flocculants are often used to make sedimentation more effective. Coagulants and flocculants are chemicals used in water treatment processes. Common coagulants include aluminum sulfate (alumina), ferric chloride, and polyaluminum chloride. After being added to water, coagulants form fine insoluble (Rani et al., 2023). Flocculants are substances that contribute to the accumulation of small insoluble particles formed by coagulants, thereby increasing their size and facilitating their removal from water. Flocculants usually have a high molecular weight, which enables the settling or filtration of MPs (Jachimowicz & Cydzik-Kwiatkowska, 2022). Another effective method is the method of electrocoagulation. The studies conducted by some researchers prove that the mentioned method provides more than 80% efficiency. The essence of this method lies in the fact that an electric current passes through the water, which leads to the formation of coagulation and coagulability of suspended particles, which can then be separated from the water. However, electrocoagulation also has some limitations, such as high energy consumption and sludge generation (Mao et al., 2023).

Membrane technologies are also considered as one of the methods for removing MPs from the water environment (Agata Egea-Corbacho et al., 2023). Ultrafiltration and nanofiltration are considered advanced membrane filtration methods to remove MPs. Membrane technologies are particularly effective in removing MPs from wastewater. Electrocoagulation-electroflotation processes made it possible to achieve 100% efficiency. Membrane filtration also showed 100% efficiency in removing MPs. In fact, membrane technologies generally show excellent results in removing MPs from wastewater. Sewage treatment will greatly contribute to reducing the percentage of MP contamination of water bodies, as sewage is considered one of the main sources that contaminate water bodies with MPs (Carr et al., 2016). The advantages of the mentioned methods are that they have high removal efficiency, versatility, continuous operation, etc. As with any method of removing MPs from the aquatic environment, this method also has its drawbacks. Membranes can become fouled over time, which reduces their efficiency, other disadvantages include energy requirements, their change over time, and their financial cost, as their initial installation and operating costs can be relatively high (Gao et al., 2022).

Another method of removing MPs from the water environment is biological removal. Some microorganisms are capable of degrading or immediately assimilating MPs. Biological methods are based on the use of various living organisms that are capable of breaking down or absorbing MPs. Such organisms include bacteria, algae, fungi, enzymes, zooplankton, marine clams, corals, and marine microorganisms including archaea and eukaryotes (Nikolaivits et al., 2020). The advantage of this method is that it is considered environmentally friendly because in this method no chemicals are used, and the process of MPs degradation is carried out only through their decomposition by organisms. In addition to these advantages, it is also considered that these organisms can completely decompose MPs, that is, these organisms can assimilate it completely. After decomposition, they synthesize non-harmful substances, such as biomass, water, and carbon dioxide (Shaji et al., 2024). In addition, the mentioned method does not require large financial investments, unlike other methods, which require electrical energy or other infrastructure for cleaning, establishment, operation, and maintenance of the subsequent working regime. The disadvantage of this method is that there are some types of MPs that are degraded by organisms after a long time and may require additional conditions, for example, in the case of polyethylene and polypropylene degradation. If in the case of physical or chemical methods, the degradation or removal of MPs takes place in a short period of time, in the case of the biological method the process can be quite long (Viel et al., 2023). The size and type of MPs, as well as environmental conditions, are considered to be the main determinants that play a decisive role in the biological process of MPs degradation and removal. In addition, it is necessary to carry out the necessary monitoring and research in this water body to identify and address the complex and interconnected relationships between microorganisms, environmental conditions, and types of MPs. Another method of removing MPs from the aquatic environment is absorption. The essence of the absorption method is as follows: materials such as activated carbon or specialized resins, materials with a high affinity for MPs, capable of adsorbing MPs, thus removing MPs from the aquatic environment. There are two types of absorption: physical and chemical (Reza et al., 2024). Chemical absorption proceeds on the basis of chemical bonds, particularly covalent, ionic, and hydrogen bonds. Physical absorption occurs through the interaction of intermolecular forces. Physical absorption is usually performed at low temperatures. The advantages of this method are its high removal efficiency and versatility. It can be used to absorb different types of MPs from various water sources. Additionally, a variety of materials can be employed depending on the specific cleaning purpose. Moreover, this method is effective not only for absorbing MPs in water but also for removing other pollutants, including trace elements, organic compounds, and other chemical substances. After use, absorbent materials can be reused, which makes it possible to avoid unnecessary financial costs, in addition to reducing environmental pollution with absorbent materials (Verma et al., 2024). One of the disadvantages of the method is the limitation of the absorption capacity of the absorbent material since there is a maximum value, after which the absorbent material can no longer absorb MPs and it will be necessary to replace it with a new one, thus increasing the financial costs. Each absorbent absorbs different types of MPs, so it is more appropriate to use a combination of several absorbents. In addition, in order to prevent environmental pollution with absorbent materials, it is necessary to apply proper methods.

Advanced oxidation processes (Duca et al., 2023) are considered one of the methods for removing MPs from the aquatic environment. The basis of this method is the use of powerful oxidizing substances or rays, which are aimed at breaking down MPs. Advanced oxidation with hydrogen peroxide, ozonation, and photocatalysis are examples of these methods. The type, size, and water quality parameters of MPs are the main features that ensure the effectiveness of the mentioned method (Marcharla et al., 2024). (not in ref.)Advanced oxidative processes have a number of advantages. This method is highly effective, especially in removing MPs. In addition, this method is versatile, it can be used to remove MPs from various aquatic environments, both fresh water and sewage. Using this method, it is possible to remove not only primary MPs but also secondary MPs from the water environment. Another advantage is that in addition to removing MPs, persistent organic pollutants and other chemicals associated with MPs can also be broken down and removed. In addition, this method is considered to be somewhat environmentally friendly, because this method can be carried out without the addition of other chemical additives, or to reduce the level of these additives to a minimum (Wang et al., 2023). Combining this method with other methods, such as physical removal methods (particularly filtration and sedimentation), gives higher MP removal results. Among the disadvantages of this method, we can mention the energy demand and the financial need. Photocatalysis is a process that uses a photocatalyst, usually titanium dioxide (TiO_2), that is activated by ultraviolet light. When exposed to light, the photocatalyst generates reactive oxygen species that can break down MP particles. The effectiveness of this method has been proven based on laboratory research and experiments (Esther et al., 2023). Electrochemical oxidation is carried out by electric current, during which chemical reactions occur, which lead to the breakdown of MPs. Currently, electrochemical methods for the identification and removal of MPs are in the early stages of development but have great prospects for development, and with the potential and necessary funding, these methods can be widely used and highly effective (Wilsey et al., 2023). Ozonation is also considered one of the methods. Ozone (O_3) is a powerful oxidizing agent that can be used to break down MPs. The introduction of ozone gas into the water causes it to interact with MPs, leading to their degradation and eventual removal (Wu et al., 2022).

5. Conclusions and Path Forward

Due to the exponential increase of MPs in our immediate environment, reducing and remediating MPs in aquatic environments is a critical environmental challenge. For source reduction, the future directions include improved manufacturing practices that include bio-based plastics, upcycling, and efficient designs to produce reduced shedding of MPs, such as in paints, tires, and fabrics. Other steps include the implementation of stricter bans or restrictions on products using legislative and regulatory controls. Consumer Awareness through education and campaigns to reduce the use of plastics and encourage sustainable alternatives is always a viable step. To further reduce MPs, filtration steps are mentioned in the chapter, however, processes such as nanofiltration that can capture smaller particles are viable alternatives. Other processes include infrastructure improvements to prevent MPs from urban runoff from entering water bodies. We have provided several remediation steps to capture or mitigate MPs, however, several processes are still under investigation and show a great potential to remediate MPs from aquatic environments. The processes include advanced nanofiltration membranes, surface skimmers for water cleanup, and magnetic separation. Chemical processes such as photocatalytic and enzyme-based delegation show great potential. Advanced Oxidation Processes employ chemical methods such as ozone or hydrogen peroxide to degrade MPs and are being scaled up to process large volumes of water. While several innovative processes are still under development, the primary challenge is that each solution must be scalable and cost-effective, without adding any unintended ecological consequences. While it is essential to balance economic, social, and environmental priorities in policy making and implementation, it is equally important that we conjoin innovative technologies, stricter regulations, and community-driven actions toward reducing and remediating MPs from aquatic environments.

Acknowledgments

The work was partly supported by the Higher Education and Science Committee of MESCS RA in the frame of research project no. 23AA-1F022.

References

Aashlesha Chekkala Vivekanand, Sanjeeb Mohapatra, & Vinay Kumar Tyagi. (2021). Microplastics in aquatic environment: Challenges and perspectives. *Chemosphere*, 282, 131151. https://doi.org/10.1016/j.chemosphere.2021.131151.

Acharya, S., Rumi, S.S., Hu, Y., & Abidi, N. (2021). Microfibers from synthetic textiles as a major source of microplastics in the environment: A review. *Textile Research Journal*, 91(17–18), 2136–2156. https://doi.org/10.1177/0040517521991244.

Agata Egea-Corbacho, Ana Pilar Martín-García, Ana Amelia Franco, José María Quiroga, Rune Røjgaard Andreasen, Mads Koustrup Jørgensen, & Morten Lykkegaard Christensen. (2023). Occurrence, identification, and removal of microplastics in a wastewater treatment plant compared to an advanced MBR technology: Full-scale pilot plant. *Journal of Environmental Chemical Engineering*, 11(3), 109644. https://doi.org/10.1016/j.jece.2023.109644.

Agbekpornu, P., & Kevudo, I. (2023). The risks of microplastic pollution in the aquatic ecosystem. In *IntechOpen*. https://doi.org/10.5772/intechopen.108717.

Ahmad, M., Li, J.-L., Wang, P.-D., Hozzein, W., & Li, W.-J. (2020). Environmental perspectives of microplastic pollution in the aquatic environment: A review. *Marine Life Science & Technology*, 2. https://doi.org/10.1007/s42995-020-00056-w.

Akshay Verma, Gaurav Sharma, Amit Kumar, Pooja Dhiman, Genene Tessema Mola, Ali Shan, & Chuanling Si. (2024). Microplastic pollutants in water: A comprehensive review on their remediation by adsorption using various adsorbents. *Chemosphere*, 352, 141365. https://doi.org/10.1016/j.chemosphere.2024.141365.

Alan Shaji, Kamalesh, R., Yuvaraj Dinakarkumar, Saravanan, A., Selvaraj Arokiyaraj, Hari Palaniappan Mani et al. (2024). Microbial degradation of marine plastic debris: A comprehensive review on the environmental effects, disposal, and biodegradation. *Biochemical Engineering Journal*, 201, 109133. https://doi.org/10.1016/j.bej.2023.109133.

Alfaro-Núñez, A., Astorga, D., Cáceres-Farías, L., Bastidas, L., Soto Villegas, C., Macay, K., & Christensen, J.H. (2021). Microplastic pollution in seawater and marine organisms across the Tropical Eastern Pacific and Galápagos. *Scientific Reports*, 11(1), 6424. https://doi.org/10.1038/s41598-021-85957-5.

Andrady, A.L. (2011). Microplastics in the marine environment. *Marine Pollution Bulletin*, 62(8), 1596–1605. https://doi.org/10.1016/j.marpolbul.2011.05.030.

Andrady, A.L. (2017). The plastic in microplastics: A review. *Marine Pollution Bulletin*, 119(1), 12–22. https://doi.org/10.1016/j.marpolbul.2017.01.082.

Auta, H.S., Emenike, C.U., & Fauziah, S.H. (2017). Distribution and importance of microplastics in the marine environment: A review of the sources, fate, effects, and potential solutions. *Environment International*, 102, 165–176. https://doi.org/10.1016/j.envint.2017.02.013.

Azevedo-Santos, V.M., Brito, M.F.G., Manoel, P.S., Perroca, J.F., Rodrigues-Filho, J.L., Paschoal, L.R.P. et al. (2021). Plastic pollution: A focus on freshwater biodiversity. *Ambio*, 50(7), 1313–1324. https://doi.org/10.1007/s13280-020-01496-5.

Bashir, S.M., Kimiko, S., Mak, C.W., Fang, J.K.H., & Gonçalves, D. (2021). Personal care and cosmetic products as a potential source of environmental contamination by microplastics in a densely populated Asian city. *Frontiers in Marine Science*, 8, 683482. https://doi.org/10.3389/fmars.2021.683482.

Bergmann, M., Mützel, S., Primpke, S., Tekman, M.B., Trachsel, J., & Gerdts, G. (2019). White and wonderful? Microplastics prevail in snow from the Alps to the Arctic. *Science Advances*, 5(8), aax1157. https://doi.org/10.1126/sciadv.aax1157.

Borriello, L., Scivicco, M., Cacciola, N.A., Esposito, F., Severino, L., & Cirillo, T. (2023). Microplastics, a global issue: Human exposure through environmental and dietary sources. *Foods*, 12(12), 3396. https://doi.org/10.3390/foods12183396.

Botterell, Z.L.R., Beaumont, N., Dorrington, T., Steinke, M., Thompson, R.C., & Lindeque, P.K. (2019). Bioavailability and effects of microplastics on marine zooplankton: A review. *Environmental Pollution*, 245, 98–110. https://doi.org/10.1016/j.envpol.2018.10.065.

Brännström, S., Rosengren, H., Wrange, A.-L., & Olshammar, M. (2023). Microplastic emissions from paint. *IVL Swedish Environmental Research Institute*, SMED Report No. 7 2023.

Browne, M.A., Crump, P., Niven, S.J., Teuten, E., Tonkin, A., Galloway, T., & Thompson, R. (2011). Accumulation of microplastic on shorelines worldwide: Sources and sinks. *Environmental Science & Technology*, 45(21), 9175–9179. https://doi.org/10.1021/es201811s.

Candolin, U., & Rahman, T. (2023). Behavioral responses of fishes to anthropogenic disturbances: Adaptive value and ecological consequences. *Journal of Fish Biology*, 103(4), 773–783. https://doi.org/10.1111/jfb.15322.

Carr, S.A., Liu, J., & Tesoro, A.G. (2016). Transport and fate of microplastic particles in wastewater treatment plants. *Water Research*, 91, 174–182. https://doi.org/10.1016/j.watres.2016.01.002.

Chatterjee, S., & Sharma, S. (2019). Microplastics in our oceans and marine health. *Field Actions Science Reports* (Special Issue), 19, 54–61.

Chaukura, N., Kefeni, K.K., Chikurunhe, I., Nyambiya, I., Gwenzi, W., Moyo, W. et al. (2021). Microplastics in the aquatic environment: The occurrence, sources, ecological impacts, fate, and remediation challenges. *Pollutants*, 1, 95–118. https://doi.org/10.3390/pollutants1020009.

Chen, Y.-N., Rani, A., Chiang, C.-Y., Kim, H., & Pan, S.-Y. (2023). Monitoring, control and assessment of microplastics in bioenvironmental systems. *Environmental Technology & Innovation*, 32, 103250. https://doi.org/10.1016/j.eti.2023.103250.

Cole, M., Lindeque, P., Halsband, C., & Galloway, T.S. (2011). Microplastics as contaminants in the marine environment: A review. *Marine Pollution Bulletin*, 62(12), 2588–2597. https://doi.org/10.1016/j.marpolbul.2011.09.025.

Dallyn, H., & Shorten, D. (1988). Hygiene aspects of packaging in the food industry. *International Biodeterioration*, 24(4–5), 387–392. https://doi.org/10.1016/0265-3036(88)90025-5

De Falco, F., Di Pace, E., Cocca, M., Avella, M., & Thompson, R.C. (2019). The contribution of washing processes of synthetic clothes to microplastic pollution. *Scientific Reports*, 9, 6633. https://doi.org/10.1038/s41598-019-43023-x.

Demir, D., Vaseashta, A., & Bölgen, N. (2021). Macroporous cryogels for water purification. *In:* Vaseashta, A., Maftei, C. (Eds), *Water Safety, Security, and Sustainability. Advanced Sciences and Technologies for Security Applications*. Springer, Cham. https://doi.org/10.1007/978-3-030-76008-3_12.

Dris, R., Gasperi, J., Saad, M., Mirande, C., & Tassin, B. (2016). Synthetic fibers in atmospheric fallout: A source of microplastics in the environment? *Marine Pollution Bulletin,* 104(1–2), 290–293. https://doi.org/10.1016/j.marpolbul.2016.01.006.

Duca, G., & Vaseashta, A. (Eds.). (2023). *Environmental and Technological Aspects of Redox Processes*. IGI Global. https://doi.org/10.4018/979-8-3693-0512-6.

Duis, K., & Coors, A. (2016). Microplastics in the aquatic and terrestrial environment: Sources (with a specific focus on personal care products), fate, and effects. *Environmental Sciences Europe*, 28(1), 2. https://doi.org/10.1186/s12302-015-0069-y.

Ekta, J.B., Sankhla, M.S., Bhat, R.A., & Bhagat, D.S. (2021). Microplastics from food packaging: An overview of human consumption, health threats, and alternative solutions. *Environmental Nanotechnology, Monitoring & Management,* 16, 100608. https://doi.org/10.1016/j.enmm.2021.100608

Enyoh, C.E., Devi, A., Kadono, H., Wang, Q., & Rabin, M.H. (2023). The plastic within: Microplastics invading human organs and bodily fluids systems. *Environments*, 10(1), 194. https://doi.org/10.3390/environments10110194.

Eriksen, M., Mason, S., Wilson, S., Box, C., Zellers, A., Edwards, W., & Farley, H. (2013). Microplastic pollution in the surface waters of the Laurentian Great Lakes. *Marine Pollution Bulletin*, 77(1–2), 177–182. https://doi.org/10.1016/j.marpolbul.2013.10.007.

European Environment Agency. (2022). Microplastics from textiles: Towards a circular economy for textiles in Europe. Retrieved from https://www.eea.europa.eu/publications/microplastics-from-textiles-towards-a#:~:text=About%208%25%20of%20European%20microplastics,global%20marine%20environment%20each%20year.

European Parliament, & Council of the European Union. (2004). Directive 2004/42/Ce of the European Parliament and of the Council of 21 April 2004 on the limitation of emissions of volatile organic compounds due to the use of organic solvents in certain paints and varnishes and vehicle refinishing products and amending Directive 1999/13/EC. Retrieved from http://data.europa.eu/eli/dir/2004/42/oj

Faber, M., Marinković, M., de Valk, E., & Waaijers-van der Loop, S.L. (2021). *Paints and Microplastics: Exploring Recent Developments to Minimize the Use and Release of Microplastics in the Dutch Paint Value Chain.* RIVM Report 2021-0037. DOI 10.21945/RIVM-2021-0037.

Galloway, T.S., Cole, M., & Lewis, C. (2017). Interactions of microplastic debris throughout the marine ecosystem. *Nature Ecology & Evolution*, 1, 0116. https://doi.org/10.1038/s41559-017-0116.

Gao, W., Zhang, Y., Mo, A., Jiang, J., Liang, Y., Cao, X., & He, D. (2022). Removal of microplastics in water: Technology progress and green strategies. *Green Analytical Chemistry*, 3, 100042. https://doi.org/10.1016/j.greeac.2022.100042.

Gaylarde, C.C., Baptista Neto, J.A., & Monteiro da Fonseca, E. (2021). Paint fragments as polluting microplastics: A brief review. *Marine Pollution Bulletin*, 162, 111847. https://doi.org/10.1016/j.marpolbul.2020.111847.

GESAMP (Joint Group of Experts on the Scientific Aspects of Marine Environmental Protection). (2016). *Sources, Fate, and Effects of Microplastics in the Marine Environment: Part Two of a Global Assessment.* IMO/FAO/UNESCO-IOC/UNIDO/WMO/IAEA/UN/UNEP/UNDP.

Guerranti, C., Martellini, T., Perra, G., Scopetani, C., & Cincinelli, A. (2019). Microplastics in cosmetics: Environmental issues and needs for global bans. *Environmental Toxicology and Pharmacology*, 68, 75–79. https://doi.org/10.1016/j.etap.2019.03.007.

Haque, F., & Fan, C. (2023). Fate and impacts of microplastics in the environment: Hydrosphere, pedosphere, and atmosphere. *Environments*, 10(5), 70. https://doi.org/10.3390/environments10050070.

Hernández-Fernández, J., Puello-Polo, E., & Castro-Suarez, J.R. (2023). Characterization of the morphological and chemical profile of different families of microplastics in samples of breathable air. *Molecules*, 28(3), 1042. https://doi.org/10.3390/molecules28031042.

Horton, Alice A., Alexander Walton, David J. Spurgeon, Elma Lahive, & Claus Svendsen. (2017). Microplastics in freshwater and terrestrial environments: Evaluating the current understanding to identify the knowledge gaps and future research priorities. *Science of The Total Environment*, 586, 127–141. https://doi.org/10.1016/j.scitotenv.2017.01.190.

Huang, Z., Hu, B., & Wang, H. (2023). Analytical methods for microplastics in the environment: A review. *Environmental Chemistry Letters*, 21(1), 383–401. https://doi.org/10.1007/s10311-022-01525-7.

Issac, M.N., & Kandasubramanian, B. (2021). Effect of microplastics in water and aquatic systems. *Environmental Science and Pollution Research International*, 28(16), 19544–19562. https://doi.org/10.1007/s11356-021-13184-2.

Jachimowicz, P., & Cydzik-Kwiatkowska, A. (2022). Coagulation and flocculation before primary clarification as efficient solutions for low-density microplastic removal from wastewater. *International Journal of Environmental Research and Public Health*, 19(20), 13013. https://doi.org/10.3390/ijerph192013013.

Jadhav, E.B., Sankhla, M.S., Bhat, R.A., & Bhagat, D.S. (2021). Microplastics from food packaging: An overview of human consumption, health threats, and alternative solutions. *Environmental Nanotechnology, Monitoring, & Management*, 16, 100608. https://doi.org/10.1016/j.enmm.2021.100608.

Jambeck, J.R., Andrady, A., Geyer, R., Narayan, R., Perryman, M., Siegler, T. et al. (2015). Plastic waste inputs from land into the ocean. *Science*, 347, 768–771. https://doi.org/10.1126/science.1260352.

Kabir, M.S., Wang, H., Luster-Teasley, S., Zhang, L., & Zhao, R. (2023). Microplastics in landfill leachate: Sources, detection, occurrence, and removal. *Environmental Science and Ecotechnology*, 16, 100256. https://doi.org/10.1016/j.ese.2023.100256.

Kalčíková, G., & Rozman, U. (2022). Effects of microplastics on aquatic organisms in receiving waters. In: *Advances in Marine Biology* (pp. 169–203). https://doi.org/10.1002/9781119759379.ch10.

Kau, D., Materić, D., Holzinger, R., Baumann-Stanzer, K., Schauer, G., & Kasper-Giebl, A. (2024). Fine micro- and nanoplastics concentrations in particulate matter samples from the high alpine site Sonnblick, Austria. *Chemosphere,* 352, 141410. https://doi.org/10.1016/j.chemosphere.2024.141410.

Kiran, B.R., Kopperi, H., & Venkata Mohan, S. (2022). Micro/nano-plastics occurrence, identification, risk analysis and mitigation: Challenges and perspectives. *Reviews in Environmental Science and Biotechnology*, 21, 169–203. https://doi.org/10.1007/s11157-021-09609-6.

Kole, P.J., Löhr, A.J., Van Belleghem, F.G.A.J., & Ragas, A.M.J. (2017). Wear and tear of tyres: A stealthy source of microplastics in the environment. *International Journal of Environmental Research and Public Health*, 14(12), 1265. https://doi.org/10.3390/ijerph14101265.

Kumar, R., Verma, A., Shome, A., Sinha, R., Sinha, S., Jha, P. K. et al. (2021). Impacts of plastic pollution on ecosystem services, sustainable development goals, and need to focus on circular economy and policy interventions. *Sustainability*, 13, 9963. https://doi.org/10.3390/su13179963.

Kumar, R., Dhanker, R., Américo-Pinheiro, J.H.P., Kumar, D., & Hwang, J.S. (2023). Effects of microplastics on ecosystem functioning of eukaryotic marine microbes. *Frontiers in Ecology and Evolution*, 12, Article 1390158. https://doi.org/10.3389/fevo.2022.1390158.

Lance, Y.T., Friedel, E.A., Perez-Reyes, A.C., Ghosal, S., & Arthur, C.D. (2014). Microplastics in four estuarine rivers in the Chesapeake Bay, U.S.A. *Environmental Science & Technology,* 48(24), 14195–14202. https://doi.org/10.1021/es5036317

Lee, Y., Cho, J., Sohn, J., & Kim, C. (2023). Health effects of microplastic exposures: Current issues and perspectives in South Korea. *Yonsei Medical Journal*, 64(5), 301–308. https://doi.org/10.3349/ymj.2023.0048.

Li, Y., Sun, Y., Li, J., Tang, R., Miu, Y., & Ma, X. (2021). Research on the influence of microplastics on marine life. *IOP Conference Series: Earth and Environmental Science*, 631, Article 012006. https://doi.org/10.1088/1755-1315/631/1/012006.

Löder, M.G.J., Imhof, H.K., Ladehoff, M., Löschel, L.A., Lorenz, C., Mintenig, S. et al. (2017). Enzymatic purification of microplastics in environmental samples. *Environmental Science & Technology*, 51(24), 14283–14292. https://doi.org/10.1021/acs.est.7b03055.

Manzoor, J., Sharma, M., Sofi, I.R., & Dar, A.A. (2020). Plastic waste: Environmental and human health impacts. In: *Handbook of research on environmental and human health impacts of plastic pollution,* (pp. 29–37). IGI Global. https://doi.org/10.4018/978-1-5225-9452-9.ch002.

Mao, Y., Zhao, Y., & Cotterill, S. (2023). Examining current and future applications of electrocoagulation in wastewater treatment. *Water*, 15, 1455. https://doi.org/10.3390/w15081455.

Marcharla, E., Vinayagam, S., Gnanasekaran, L., Soto-Moscoso, M., Chen, W.-H., Thanigaivel, S. et al. (2024). Microplastics in marine ecosystems: A comprehensive review of biological and ecological implications and its mitigation approach using nanotechnology for the sustainable environment. *Environmental Research,* 256, 119181. https://doi.org/10.1016/j.envres.2024.119181.

Mattsson, K., Hansson, L.A., & Cedervall, T. (2018). Nano-plastics in the aquatic environment. *Environmental Science: Processes & Impacts*, 20(3), 383–394. https://doi.org/10.1039/c7em00455b.

Mehra, S., Sharma, K., Sharma, G., Singh, M., & Chadha, P. (2021). Sources, fate, and impact of microplastics in aquatic environment. *Emerging Contaminants*. IntechOpen. Available at: http://dx.doi.org/10.5772/intechopen.93805.

Milojević, D., Mitrović, A., Smiljanić, D., Jovanović, G., & Sokić, M. (2023). Nanoplastics in aquatic environments: Origin, separation, and characterization: Review. *Tehnika,* 78, 103–108. https://doi.org/10.5937/tehnika2301103M.

Mintenig, S.M., Int-Veen, I., Löder, M.G.J., Primpke, S., & Gerdts, G. (2017). Identification of microplastic in effluents of wastewater treatment plants using focal plane array-based micro-Fourier-transform infrared imaging. *Water Research,* 108, 365–372. https://doi.org/10.1016/j.watres.2016.11.015

Nikolaivits, E., Dimopoulou, P., Maslak, V., Nikodinovic-Runic, J., & Topakas, E. (2020). Discovery and biochemical characterization of a novel polyesterase for the degradation of synthetic plastics. *Chemical Processes*, 2, 33. https://doi.org/10.3390/ECCS2020-07572.

Nor, N.H.M., Obbard, J.P., Cheng, Y., Azid, A., & Mokhtar, M.B. (2020). Occurrence of microplastics in snow from the Arctic to Antarctic regions: Sources, concentrations, and implications. *Water, Air, & Soil Pollution*, 231(6), 308. https://doi.org/10.1007/s11270-020-04755-0.

Osman, A.I., Hosny, M., Eltaweil, A.S., Omar, S., Elgarahy, A.M., Farghali, M. et al. (2023). Microplastic sources, formation, toxicity, and remediation: A review. *Environmental Chemistry Letters*. Advance online publication. https://doi.org/10.1007/s10311-023-01593-3.

Pizzichetti, A.R.P., Pablos, C., Álvarez-Fernández, C., Reynolds, K., Stanley, S., & Marugán, J.C. (2021). Evaluation of membranes performance for microplastic removal in a simple and low-cost filtration system. *Case Studies in Chemical and Environmental Engineering*, 3, 100075. https://doi.org/10.1016/j.cscee.2020.100075.

Portman, M.E., & Brennan, R.E. (2017). Marine litter from beach-based sources: A case study of an Eastern Mediterranean coastal town. *Waste Management*, 69, 535–544. https://doi.org/10.1016/j.wasman.2017.07.040.

Rakib, M., Sarker, A., Ram, K., Uddin, M., Walker, T., Chowdhury, T. et al. (2023). Microplastic toxicity in aquatic organisms and aquatic ecosystems: A review. *Water, Air, and Soil Pollution*. Advance online publication. https://doi.org/10.1007/s11270-023-06062-9.

Ramkumar, M., Balasubramani, K., Santosh, M., & Nagarajan, R. (2022). The plastisphere: A morphometric genetic classification of plastic pollutants in the natural environment. *Gondwana Research*, 108, 4–12. https://doi.org/10.1016/j.gr.2021.07.004.

Reza, T., Mohamad Riza, Z.H., Sheikh Abdullah, S.R., Abu Hasan, H., Ismail, N.I., & Othman, A.R. (2024). Microplastic removal in wastewater treatment plants (WWTPs) by natural coagulation: A literature review. *Toxics*, 12(1), 12. https://doi.org/10.3390/toxics12010012.

Richon, C., Gorgues, T., Paul-Pont, I., & Maes, C. (2022). Zooplankton exposure to microplastics at global scale: Influence of vertical distribution and seasonality. *Frontiers in Marine Science*, 9, Article 947309. https://doi.org/10.3389/fmars.2022.947309.

Rochman, C., Browne, M.A., Halpern, B.S., Hentschel, B.T., Hoh, E., Karapanagioti, H.K. et al. (2013). Classify plastic waste as hazardous. *Nature*, 494(7436), 169–171.

Rochman, C.M., Andrady, A., Dudas, S., Fabres, J., Galgani, F., Lead, D. et al. (2016). *Sources, Fate, and Effects of Microplastics in the Marine Environment: Part 2 of a Global Assessment*. IMO/FAO/UNESCO-IOC/UNIDO/WMO/IAEA/UN/UNEP/UNDP Joint Group of Experts on the Scientific Aspects of Marine Environmental Protection (GESAMP).

Rochman, C., Tahir, A., Williams, S. et al. (2015). Anthropogenic debris in seafood: Plastic debris and fibers from textiles in fish and bivalves sold for human consumption. *Sci Re.*, 5, 14340. https://doi.org/10.1038/srep14340.

Sandil, S., & Zaray, G. (2024). Impact of microplastics and nanoplastics in the aquatic environment. *In:* S.A. Bhat, V. Kumar, F. Li, & S. Kumar (Eds.), *Management of Micro- and Nano-plastics in Soil and Biosolids* (pp. 11–22). Springer. https://doi.org/10.1007/978-3-031-51967-3_2.

Sarkar, S., Diab, H., & Thompson, J. (2023). Microplastic pollution: Chemical characterization and impact on wildlife. *International Journal of Environmental Research and Public Health*, 20(3), 1745. https://doi.org/10.3390/ijerph20031745.

Schmidt, C., Krauth, T., & Wagner, S. (2017). Export of plastic debris by rivers into the sea. *Environmental Science & Technology*, 51(21), 12246–12253. https://doi.org/10.1021/acs.est.7b02368

Semmouri, I., Vercauteren, M., Van Acker, E., et al. (2022). Presence of microplastics in drinking water from different freshwater sources in Flanders (Belgium), an urbanized region in Europe. *Food Contamination*, 9(6). https://doi.org/10.1186/s40550-022-00091-8.

Shanmugam, V., Das, O., Esmaeely Neisiany, R., Babu, N.B., Singh, S., Hedenqvist, M. et al. (2020). Polymer recycling in additive manufacturing: An opportunity for the circular economy. *Materials Circular Economy*, 2, 11. https://doi.org/10.1007/s42824-020-00012-0.

Silva, A.L.P., Prata, J.C., Walker, T.R., Duarte, A.C., Ouyang, W., Barcelò, D., & Rocha-Santos, T. (2021). Increased plastic pollution due to COVID-19 pandemic: Challenges and recommendations. *Chemical Engineering Journal*, 405, 126683. https://doi.org/10.1016/j.cej.2020.126683.

Storck, F.R., Kools, S.A.E., & Rinck-Pfeiffer, S. (2015). *GWRC Science Brief : Microplastics in Fresh Water Resources*. TZW Karlsruhe; KWR Watercycle Research Institute. Retrieved from https://www.gwrc.eu/wp-content/uploads/2015/09/1508-GWRC-Microplastics-in-Fresh-Water-Resources.pdf.

Sundt, P., Schulze, P.-E., & Syversen, F. (2014). *Sources of Microplastic Pollution to the Marine Environment*. Report no M-321/2015. Asker: MepexConsult.

Tamis, J.E., Koelmans, A.A., Dröge, R. et al. (2021). Environmental risks of car tire microplastic particles and other road runoff pollutants. *Microplastics and Nanoplastics*, 1, 10. https://doi.org/10.1186/s43591-021-00008-w.

Teuten, E.L., Saquing, J.M., Knappe, D.R.U., Barlaz, M.A., Jonsson, S., Björn, A. et al. (2009). Transport and release of chemicals from plastics to the environment and to wildlife. *Philosophical Transactions of the Royal Society B: Biological Sciences*, 364(1526), 2027–2045. https://doi.org/10.1098/rstb.2008.0284.

Thompson, R.C., Olsen, Y., Mitchell, R.P., Davis, A., Rowland, S.J., John, A.W.G. et al. (2004). Lost at sea: Where is all the plastic? *Science*, 304(5672), 838.

Torres, P., Augusto, M., & Matos, M. (2019). Antecedents and outcomes of digital influencer endorsement: An exploratory study. *Psychology & Marketing*, 36, 10.1002/mar.21274.

Unice, K.M., Weeber, M.P., Abramson, M.M., Reid, R.C.D., van Gils, J.A.G., Markus, A.A., & Vethaak, A.D. (2019). Characterizing export of land-based microplastics to the estuary – Part II: Sensitivity analysis of an integrated geospatial microplastic transport modeling assessment of tire and road wear particles. *Science of The Total Environment*, 646, 1650–1659. https://doi.org/10.1016/j.scitotenv.2018.08.301.

United States Environmental Protection Agency. (2018). *Resource Conservation and Recovery Act (RCRA) Orientation Manual*. EPA Document Number: 530-R-18-003.

Unuofin, J.O., & Igwaran, A. (2023). Microplastics in seafood: Implications for food security, safety, and human health. *Journal of Sea Research*, 194, 102410. https://doi.org/10.1016/j.seares.2023.102410.

Uogintė, I., Pleskytė, S., Pauraitė, J., & Lujanienė, G. (2022). Seasonal variation and complex analysis of microplastic distribution in different WWTP treatment stages in Lithuania. *Environmental Monitoring and Assessment*, 194(11), 829. https://doi.org/10.1007/s10661-022-10478-x.

Vaseashta, A., Ivanov, V., Stabnikov, V., & Marinin, A. (2021). Environmental safety and security investigations of neustonic microplastic aggregates near water-air interphase. *Polish Journal of Environmental Studies*, 30(4), 3457–3469. https://doi.org/10.15244/pjoes/131947.

Verma, A., Sharma, G., Kumar, A., Dhiman, P., Mola, G.T., Shan, A. et al. (2024). Microplastic pollutants in water: A comprehensive review on their remediation by adsorption using various adsorbents. *Chemosphere,* 352, 141365. https://doi.org/10.1016/j.chemosphere.2024.141365

Viel, T., Manfra, L., Zupo, V., Libralato, G., Cocca, M., & Costantini, M. (2023). Biodegradation of plastics induced by marine organisms: Future perspectives for bioremediation approaches. *Polymers*, 15, 2673. https://doi.org/10.3390/polym15122673.

Walkinshaw, C., Lindeque, P.K., Thompson, R., Tolhurst, T., & Cole, M. (2020). Microplastics and seafood: Lower trophic organisms at highest risk of contamination. *Ecotoxicology and Environmental Safety*, 190, 110066. https://doi.org/10.1016/j.ecoenv.2019.110066.

Wang, Y., Liu, G., Wang, Y., Mu, H., Shi, X., Wang, C., & Wu, N. (2021). Microplastics pollution during the COVID-19 pandemic: Evidence, policy strategies, and recommendations. *Journal of Cleaner Production*, 303, 126988. https://doi.org/10.1016/j.jclepro.2021.126988.

Wilsey, M.K., Taseska, T., Meng, Z., Yu, W., & Möller, A.M. (2023). Advanced electrocatalytic redox processes for environmental remediation of halogenated organic water pollutants. *Chemical Communications*, 59(80), 11895–11922.

Wright, S.L., Rowe, D., Thompson, R.C., & Galloway, T.S. (2013). Microplastic ingestion decreases energy reserves in marine worms. *Current Biology*, 23(23), R1031–R1033. https://doi.org/10.1016/j.cub.2013.10.068.

Wu, J., Zhang, Y., & Tang, Y. (2022). Fragmentation of microplastics in the drinking water treatment process: A case study in Yangtze River region, China. *Science of The Total Environment*, 806(Part 1), 150545. https://doi.org/10.1016/j.scitotenv.2021.150545.

Xu, Y., Rillig, M.C., & Waldman, W.R. (2022). A new separation protocol reveals spray painting as a neglected source of microplastics in soils. *Environmental Chemistry Letters*, 20(6), 3363–3369. https://doi.org/10.1007/s10311-022-01500-2.

Yang, J.W., Wu, W., Chung, C.C., Chiang, K.P., Gong, G.C., & Hsieh, C.H. (2018). Predator and prey biodiversity relationship and its consequences on marine ecosystem functioning-interplay between nanoflagellates and bacterioplankton. *ISME Journal*, 12(6), 1532–1542. https://doi.org/10.1038/s41396-018-0111-3.

Yeo, B.G., Mizukawa, K., & Takada, H. (2023). Microplastics in global marine waters and biota: Effectiveness of potential bioindicators in mirroring local pollution levels. *Environmental Monitoring and Contaminants Research*, 3, 4–68. https://doi.org/10.5985/emcr.20230005.

Yin, L., Jiang, C., Wen, X. et al. (2019). Microplastic pollution in surface water of urban lakes in Changsha, China. *International Journal of Environmental Research and Public Health*, 16(10), 1650. https://doi.org/10.3390/ijerph16101650.

Yu, Y., Craig, N., & Su, L. (2023). A hidden pathway for human exposure to micro- and nanoplastics: The mechanical fragmentation of plastic products during daily use. *Toxics*, 11(9), 774. https://doi.org/10.3390/toxics11090774.

Zainuddin, A.H., Aris, A.Z., Zaki, M.R.M., Yusoff, F.M., & Wee, S.Y. (2022). Occurrence, potential sources, and ecological risk estimation of microplastic towards coastal and estuarine zones in Malaysia. *Marine Pollution Bulletin*, 174, 113282. https://doi.org/10.1016/j.marpolbul.2021.113282.

Zambrano, M.C., Pawlak, J.J., Daystar, J., Ankeny, M., Cheng, J.J., Venditti, R.A., & Law, S. (2019). Microfibers generated from the laundering of cotton, rayon and polyester based fabrics and their aquatic biodegradation. *Marine Pollution Bulletin*, 142, 394–407. https://doi.org/10.1016/j.marpolbul.2019.03.058.

Ziani, K., Ioniţă-Mîndrican, C.B., Mititelu, M., Neacşu, S.M., Negrei, C., Moroşan, E., & Preda, O.T. (2023). Microplastics: A real global threat for environment and food safety: A state of the art review. *Nutrients*, 15(3), 617. https://doi.org/10.3390/nu15030617.

7 | Toxicokinetic Effects of Micro/Nano Plastics on Human Health

Ashkhen Hovhannisyan* **and Shushanik Kazaryan**

Department of Medical Biochemistry and Biotechnology,
Russian-Armenian University, Yerevan 0051, Armenia

Abstract: Plastic, due to its diverse properties, finds applications in various fields such as agriculture, packaging, pharmaceuticals, textiles, and construction, to name just a few. However, excessive use of plastic leads to the accumulation of a large amount of plastic waste, and improper management of this waste results in serious environmental problems. During the COVID-19 pandemic, plastic production reached unprecedented levels. It is estimated that by 2024, around 70 million tons of plastic will end up in nature due to the imbalance between the volumes of plastic consumed and the capabilities for managing plastic at the end of its life cycle. A comprehensive understanding of environmental pollution should include not only the assessment of the amount of plastic in landfills but also the more dangerous degradation products, such as micro- and nanoplastics (MNPs) in the hydrosphere, biosphere, lithosphere, and atmosphere. Microplastics (MPs) are plastic particles ranging in size from 1 μm to 5 mm, while nanoplastics (NPs) are particles up to 100 nm in size. MNPs pose a greater threat to the environment than their larger counterparts. Due to their small size and potential toxicity to humans, MNPs are of particular concern. Epidemiological studies indicate that disorders such as lung nodules, asthma, and blood clots may be caused or exacerbated by exposure to MNPs. Although direct evidence of MNPs' impact on human health remains limited, research is needed to assess the health risks posed by MNPs. This chapter discusses issues related to toxicokinetics, mechanisms of toxicity, and methods for detecting MNPs in the human body. Nevertheless, the potential of environmentally derived MNPs to harm human health remains insufficiently studied, and further research is needed to fill the existing knowledge gaps in this area and to develop more effective strategies for the safe management of MNPs.

*Corresponding author: ashkhen.hovhannisyan@rau.am

1. Statistics on Environmental Plastic

Plastics, as synthetic organic polymers composed of long-chain molecules with high molecular weight, have been developed since the mid-nineteenth century. Significant progress in plastic production was made in the early 20th century (Law, 2017), with exponential growth during World War II. Over the past 30 years, plastic consumption has quadrupled due to increasing demand in emerging markets, reaching unprecedented levels during the COVID-19 pandemic (460 million metric tons according to OECD, 2022), driven by the production of essential medical kits, personal protective equipment, and more. With the widespread use of plastic products and limited control over plastic waste management, plastic waste is increasingly being discharged into aquatic environments, buried, or incinerated, and is becoming an ever more serious problem for the environment and human health (Syberg et al., 2021; van Emmerik & Schwarz, 2020; Vaseashta et al., 2021).

Over time, plastic waste undergoes degradation (photodegradation, oxidation, hydrolytic degradation, biodegradation) and mechanical abrasion due to exposure to sunlight, temperature, mechanical impact, and other factors, breaking down into smaller fragments, particularly microplastics (MPs, ranging from 1 μm–5 mm) and nanoplastics (NPs, up to 100 nm). These smaller particles pose a greater threat to the environment and human health due to their size compared to their larger counterparts. It is estimated that around 500 kilotons of plastic enter the World's oceans each year, with projections suggesting that by 2060, this amount could reach between 509 million metric tons and 1,014 million metric tons, depending on the scenario (OECD, 2022). Currently, it is estimated that there are between 75–199 million tons of plastic waste in the world's oceans, and despite existing strategies to increase the proportion of recycled plastic (from 9% in 2019 to 17% by 2060), the majority of produced plastic will continue to be incinerated, buried, or end up in aquatic environments.

Calculations show that about 32% of all produced plastics potentially remain in continental lands/soils due to composting, sewage, irrigation with plastic-contaminated water, plastic mulching, the use of fertilizers containing plastic, and other factors (Zhu et al., 2023; Joos & De Tender, 2022; Xu et al., 2020; He et al., 2018). For instance, China, Japan, and South Korea account for about 80% of the world's consumption of plastic mulching film, increasing its usage from 0.64 million tons to 2.60 million tons per year (Hou et al., 2021; Xu et al., 2020). In Europe, this practice is also on the rise, with an annual increase of 5–10% in usage (He et al., 2018). Moreover, soil contamination with plastic can also be caused by flooding, atmospheric precipitation, disposal of industrial and consumer waste, and so on (Joos & De Tender, 2022; Baho et al., 2021). Although the use of plastic in agriculture accounts for about 3–4% of the total human consumption of plastic, agroplastics remain a significant source of plastic pollution in terrestrial environments (Shafea et al., 2023; Zhou et al., 2021). Additionally, during surface runoff, plastic can be trapped by plant stems, shoots, and leaves (Han et al., 2022), and in primary and secondary roots. Increasing soil porosity, allows less dense plastic particles to rise to the surface. Soil animals (e.g., springtails, earthworms, burrowing mammals) also contribute to the transfer of MPs in the soil (Dissanayake et al., 2022; Bouaicha et al., 2022; Lwanga et al., 2016, 2017).

The threat of micro- and nanoplastics (MNPs) to human health was first reported by parliament members in 2004 (Zhang et al., 2019), and since then, this issue has been the focus of attention for the WHO, many scientists, ecologists, toxicologists, and others. MNPs are present in the air, water, and food, which increases their bioavailability, particularly in the human body. MNPs have been found not only in marine and freshwater (0.02–2.19 particles/m^3) but also in bottled water, where one liter contains an average of about 240,000 tiny plastic pieces, 90% of which are NPs. MNPs make up the majority of plastic pollutants and are ubiquitous in the environment. MPs and NPs inevitably enter the human body through drinking water. Currently, the removal efficiency of MPs at drinking water treatment plants is around 85–95%, with most residual MPs being smaller than 20 μm (Na et al., 2021). Lee et al. (2023) detected NPs in tap water, primarily ranging from 58–255 nm in size and at concentrations of 1.67–2.08 μg/L (Lee et al., 2023).

1.1. MNPs in the Air

An additional environmental and health hazard is the ability of MNPs to be present in the air. The levels of MNPs in the air vary significantly depending on the detection method, but it is well established that both indoor and outdoor air contain MNPs of different sizes. Although the abundance of MNPs varied across studies depending on location and sampling methods, several studies found that indoor air contained particles ranging from 0.2–15.9 particles/m^3/day, while outdoor air ranged up to 3424.7 particles/m^3 (Catarino et al., 2018; Gaston et al., 2020; Perera et al., 2023; Yao et al., 2022; Zhang et al., 2020). The measured sizes of MNPs ranged from 5 to 4950 μm in indoor air and from 2–9555 μm in outdoor air, with studies conducted in California, USA (Gaston et al., 2020). Perera and co-authors conducted corresponding experiments in Australia, where the concentration of MNPs was 1.2–13 times higher in indoor air (Perera et al., 2023). However, in New Jersey, USA, outdoor concentrations were 1.5–3 times higher than indoor concentrations, which may depend on many factors (Yao et al., 2022).

1.2. MNPs in Water

MNPs are present in aquatic environments, including freshwater (Bordós et al., 2019; Mintenig et al., 2019; Yuan et al., 2019), drinking water (Shruti et al., 2020; Vaseashta et al., 2021), seawater (Covernton et al., 2019; Cui et al., 2022; Diez-Pérez et al., 2023), and sediments (Kor et al., 2020; Zhang et al., 2020). MNPs were detected at concentrations as low as 0.02–2.19 particles/m^3 in freshwater, indicating their presence even at minimal levels, with a maximum deposition of

10,500 particles/kg in sediment, highlighting significant accumulation (Egessa et al., 2020; Leslie et al., 2017). In drinking water, MNPs are the smallest, ranging from 1–150 μm, compared to MNPs found in freshwater, seawater, and sediments, which are larger by 10–30 times, respectively.

1.3. MNPs in Food

MNPs enter food products through food chains. As they penetrate the bodies of animals, MNPs become an integral part of the food chain and are found in food products such as seafood (Wootton et al., 2021), vegetables and fruits (Oliveri Conti et al., 2020), dairy products, meat (Kedzierski et al., 2020), and beverages (Pham et al., 2023). In seafood, MNPs are present in concentrations up to 19.9 particles/g, in fruits and vegetables from 26,375–307,750 particles/g, in dairy products from 6.5–53 particles/L, in meat from 0.001–0.019 particles/g, and in beverages up to 117 particles/L. Given the challenges in detecting nano-sized plastics, it is assumed that the number of undetectable NPs could be much higher.

1.4. MNPs in Cosmetics and Personal Care Products

As mentioned earlier, due to their small size, MNPs can penetrate the human body through the skin and mucous membranes. Despite this, MNPs are used in cosmetics as exfoliating agents. For example, polyethylene beads are often added to facial scrubs and exfoliants to mechanically remove dead skin cells and sometimes as abrasive agents in toothpaste or in some sunscreens to enhance absorption and ensure even application on the skin. It has been shown that personal care products can contain up to 50,391 MP/g of product. About 7% of facial cleansers available on the market in China contain 25.04 ± 10.69 mg/g of MP, while 2.2% of shower gels contain up to 17.08 ± 7.50 mg/g of MP (Lei et al., 2017). In addition to potential direct effects on the human body, MNPs enter wastewater, contaminating it and re-entering various environmental distribution chains (Duis et al. 2016).

2. Detection Methods

There is a pressing need to develop a comprehensive set of detection methods for MNPs that are effective, convenient, and accurate. Such methods are essential for effective environmental management and protecting human health from the impact of these particles. Achieving this goal will enable toxicologists to precisely detect micro- and nanometer-sized particles in biological systems, facilitate the establishment of clear standards, and allow for proactive notification of individuals at high risk of exposure. It is known that individuals with pre-existing conditions, such as metabolic syndromes, are more susceptible to the effects of engineered MNPs compared to healthy individuals (Kobos et al., 2020; Alqahtani et al., 2022; Xia et al., 2022). While the removal of NPs released into the environment presents a significant challenge, potential methods for reducing the volume of these released MNPs exist and warrant further investigation. In the current context, it is necessary to implement new measures to reduce the risk of MNP exposure to humans and the environment.

These measures should not only focus on regulatory actions (such as enhancing control over the use and disposal of plastic materials and setting standards for microplastic content in food and water) and the development of effective methods for cleaning water and air from microplastic particles but also on public education and awareness about the risks and ways to minimize MNP exposure. Most importantly, there is a need for research and innovation to gain a more thorough understanding of the mechanisms of MNP impact and to develop new materials that are less harmful to health.

The development of more sensitive detection methods for a wide range of contaminants is crucial, as is the creation of models to assess cumulative risks from multiple pollutants. However, these measures will be impossible to achieve without the advancement of innovative, sustainable, and cost-effective cleaning technologies, as well as closer international collaboration focused on monitoring, researching, and developing policies and strategies to address transboundary pollution issues. Addressing the problem of pollution requires an interdisciplinary approach that integrates scientific research, technological innovation, regulatory measures, and public engagement to ensure water safety and protect ecosystems and human health.

3. Toxicokinetics of MNP

Since MNPs represent a potential hazard to humans, their elimination has become an important issue (Gupta et al., 2024). While rigorous standards are in place to regulate food-contact substances in terms of migration into food (EFSA, 2011), it is less clear how these guidelines offer protection once the plastics themselves have been discarded into the environment. With limited information available about degradation and fragmentation rates, chemical leaching into environmental matrices, and entry into the food chain, it is almost impossible to estimate the cumulative risks of chronic exposure to plastics and their additives.

Many studies on the toxicology of MNPs focus on the study of toxicokinetics (absorption, distribution, and excretion), as well as the biochemical and molecular mechanisms underlying toxicity and the impact of various physicochemical properties of MNPs, such as size, shape, core material, and their interaction with other compounds, on toxicity. Humans are chronically exposed to low concentrations of NPs (Jeong et al., 2024; Prüst et al., 2020; Stabnikova et al., 2021), and while all three exposure routes—ingestion, inhalation, and skin contact—contribute to the overall presence of MPs and NPs in the human body, the risk of exposure is highest when consuming seafood and environmental sources.

In this chapter, we discuss the potential health risks to humans following the entry of MNPs into the body through the food chain, skin, and respiratory systems, without addressing the risks associated with plastic-related chemicals (such as plasticizers). Due to their small size, MNPs can penetrate the human body through the skin, lungs, and gastrointestinal tract (Cheung & Fok, 2016; Ibrahim et al., 2020; Rist & Hartmann, 2018; Syberg et al., 2018). MNPs have been found in colon tissues, and also in placental tissues. Additionally, recent studies have detected the presence of MNPs in human blood and lungs (Jenner et al., 2022; Leslie et al., 2022). As shown by Hoang & Felix-Kim (2020), smaller particles are more prone to absorption, while other studies have demonstrated that fish can absorb PET fibers up to 700 μm in size after 2 hours of exposure (Santana et al., 2021). Moreover, polyethylene (PE) particles up to 850 μm in size were also found to be absorbed by young mummichogs and red sea bream (Ohkubo et al., 2020). Given the transfer of MNPs between different trophic levels of the food chain, some studies have identified an indirect route of entry into the body (da Costa Araujo et al., 2021; Mattsson et al., 2017). MNPs can penetrate the lymphatic system (<150 μm), portal vein (110 μm), and internal organs (20 μm). Once inside the human body, particles, especially those in the submicron range, can cross cellular membranes and overcome various biological barriers, including the blood-brain barrier (BBB) and placental barrier (up to 7%), leading to toxic effects (Braun et al., 2021; Kopatz et al., 2023; Yuan et al., 2022; Wick et al., 2010).

MNPs in the body can accumulate in various organs. Studies on C57BL/6 mice have shown that MNPs ranging from 0.5–5 μm, administered via oral gavage, are capable of accumulating in the liver, spleen, kidneys, heart, and lungs, depending on their size (Zhang et al., 2024). It has also been reported that MNPs were found in the hepatocytes of mice exposed to 35.5 μm PE (60 mg/L) both through the food chain and via gavage (da Costa Araujo et al., 2021). Additionally, MNPs measuring 293 nm can cross the BBB and accumulate in the brain of mice after 2 hours of exposure (Kopatz et al., 2023).

Beyond these mechanisms, it is important to consider that once MNPs penetrate biological environments, they interact with biological macromolecules such as proteins and lipids, resulting in the formation of soft and hard coronas around the MNPs. This alters the physicochemical properties of MNPs, potentially enhancing their toxicity and promoting their penetration and movement within cells (Bashiri et al., 2023; Cao et al., 2022).

As early as 1979, it was shown that intracellular uptake of MNPs smaller than 700 nm occurs through receptor-mediated endocytosis (Goldstein et al., 1979), while larger particles are taken up via phagocytosis (Aderem & Underhill, 1999). Later studies have identified the molecular mechanisms of MNP entry into cells: particles of 50 nm are internalized through clathrin-mediated, caveolin-mediated pathways, and micropinocytosis; particles of 500 nm enter via macropinocytosis; and MNPs of 5 μm in size are not internalized due to their large size and weak Brownian motion (Liu et al., 2021).

The elimination of MNPs from the body also depends on the size of the particles. Plastic particles ranging from 1 μm–5 mm have been found in human stool samples (Zhang et al., 2021). In Wistar rats, 94% of MNPs (108 μm,

1160 μm, and <0.1 μm) were detected in feces within 7 days of exposure, with estimated half-lives of 19.9 hours, 23.7 hours, and 36.9 hours, respectively (Peng et al., 2022). In fish, 95% of MNPs were eliminated within 25 hours (Ohkubo et al., 2020).

3.1. Toxicology of MNPs

Currently, the potential mechanisms of MNP toxicity include the intrinsic toxicity of plastic particles (physical damage following contact with cells), toxicity related to their chemical composition (leaching), and their ability to act as carriers of environmental pollutants (Bradney et al., 2019; Llorca & Farre, 2021; Revel et al., 2018; Vaseashta et al., 2021). However, studies on the toxicity of these particles in mammals remain limited, and their possible impact on humans is poorly understood, often relying on model cell line systems.

MNPs cause significant toxicity across various biological systems, including macromolecules (DNA and gene damage, altered gene expression, and protein transcription changes), cellular organelles (altered cell division, cytotoxicity, apoptosis, oxidative stress response, metabolism, and ion homeostasis disruption), tissues (bone osteolysis, inflammation, and fibrosis), specific organs (carcinogenesis, altered nutrition, changes in metabolic demands, and redistribution of energy reserves) (Kaushik et al., 2024; Hua et al., 2022; Leslie et al., 2022; Liu et al., 2020; Pontecorvi et al., 2023; Rosellini et al., 2023), as well as populations, including animals and humans (reduced fertility, slowed growth, and population decline).

The internalization of MNPs into cells depends on their size; smaller particles (up to 5 nm) can bypass all biological barriers, while larger ones attach to the cell surface through receptor-ligand interactions, and are internalized via endocytosis, forming an endosome or phagosome, which then fuses with lysosomes. Experimental models, including cell lines, organoids, and animal systems, have shown that MNPs initiate the generation of reactive oxygen species (ROS), such as superoxide anions, hydroxyl radicals, and hydrogen peroxide. MNPs can act as electron donors capable of producing superoxide radicals (Hu et al., 2020; Hellack et al., 2017). When reacting with metals like iron, copper, etc., hydrogen peroxide can be converted into hydroxyl radicals (Fenton reaction) or also superoxide radicals (Haber-Weiss reaction) (Hrycay et al., 2015). MNPs can cause both extracellular and intracellular generation of ROS, with extracellular ROS generation closely associated with the degree of 'weathering'. Plastics in the environment undergo aging and weathering through several mechanisms, including photo- and thermal oxidation, as well as UV radiation. These processes lead to chemical changes on the surface of plastics. Specifically, photo-oxidation and UV radiation trigger crosslinking reactions, in which a hydrogen atom is removed from a biomolecule, creating an active site that can bind to another chain. After this, free radicals may react with oxygen in the air, leading to the formation of primary polymeric peroxyl radicals, and subsequently secondary polymeric alkyl radicals. Extracellular free radicals may be responsible for a significant increase in ROS levels after cells uptake weathered plastic particles (Jeon et al., 2021; Andrady et al., 2022; Yousif et al., 2013; Bracco et al., 2018).

The formation of reactive nitrogen species (RNS) in MNP-treated cells is less well-documented compared to ROS formation. However, studies suggest that RNS induction is a result of the high reactivity of nitrogen in the organic structure of some plastics such as polyamide. Photoaging of nitrogen-containing MPs is thought to trigger the formation of NO-containing radicals, which could theoretically promote RNS formation (Zhu et al., 2021).

Various types of cells tested *in vitro*, including immune, corneal, conjunctival, kidney, lung, colon, leukemia, and testicular cell lines, regularly demonstrate ROS formation in response to MNPs exposure. Smaller particle sizes and higher concentrations generally correlate with increased ROS production. In addition to common factors such as concentration, size, and exposure duration, structural changes in microplastics have become important determinants influencing ROS formation (Das, 2023). ROS, in turn, can disrupt membrane structure, impair mitochondrial and other organelle functions, and biomolecules, and cause genetic material damage, ultimately initiating cell death pathways (Das, 2023; Mahmud et al., 2024). This can lead to cellular dysfunction, senescence, and apoptosis, increasing the risk of diseases such as neurodegenerative disorders, cardiovascular diseases, and aging (Aslam et al., 2021; Domenech et al., 2023; Wang et al., 2024; Wu et al., 2023).

Additionally, exposure to microplastics significantly reduces the activity of plasma glutathione peroxidase (GSH-Px), catalase (CAT), and superoxide dismutase (SOD), all of which are associated with oxidative stress, induces ROS production, and causes a Redox imbalance. *In vitro* studies on lung cells, demonstrated that MNPs potentially contribute to the development of chronic obstructive pulmonary disease (COPD) (Dong et al., 2020; Xia et al., 2008). Similarly, liver, stomach, intestine, human bone marrow mesenchymal stem cells (BMMSCs), and lung cells exposed to MNP, undergoing oxidative stress, show significant changes in metabolic and proliferative activity (Domenech et al., 2021; Sun et al., 2024; Goodman et al., 2022; Shen et al., 2022).

Oxidative stress is also dependent on mitochondria. The exposure of MNPs to mitochondria can lead to damage and dysfunction of these organelles, ultimately altering membrane polarity. Mitochondrial membrane depolarization leads to the modification and opening of megachannels (MPTP), voltage-dependent anion channels (VDAC), and MPTP, which include adenine nucleotide translocator (ANT) and cyclophilin D (CyD). The thiol groups of ANT are subjected to modification due to ROS exposure. Moreover, pro-apoptotic proteins Bax and Bak interact with VDAC, leading to an increase in pore size and facilitating the release of cytochrome c, thereby initiating the caspase 3-dependent apoptotic pathway (Yarjanli et al., 2017; Zhang et al., 2023). Furthermore, free radicals not only affect the oxidation of mitochondrial membranes but also increase the levels of toxic products of lipid peroxidation, including malondialdehyde (MDA) and 4-hydroxy-

2-nonenal. The presence of ROS leads to mitochondrial DNA damage, disrupting the function of critical mitochondrial enzymes such as SOD1, SOD2, pyruvate dehydrogenase, aconitase, and L-ketoglutarate dehydrogenase. Elevated levels of free radicals and mitochondrial DNA oxidation have been identified in patients with Alzheimer's disease (Guo et al., 2013; D'Errico et al., 2021). Moreover, it has been observed that ROS enhances the activity of secretase enzymes involved in the production of amyloid, an aberrant protein associated with Alzheimer's disease. Excessive presence of this protein exacerbates neuronal dysfunction, neurodegeneration, and cognitive impairments in people with Alzheimer's disease. Oxidative stress in mitochondria damages the subunits of the electron transport chain, ultimately leading to a reduction in ATP production – an energy collapse (Sharma et al., 2021).

The direct interaction between MNPs and immune cells or the indirect result of cell damage mediated by oxidative stress can lead to an increased production of pro-inflammatory cytokines under various MNPs exposure conditions. Exposure to microplastics activates humoral and cellular immune responses. It induces the expression of pro-inflammatory cytokines and histamine in human cell lines, with the expression of genes IL6, IL8, IL1β being observed. The mechanisms involved in the induction of cytokine production upon MNP exposure are not yet fully studied. The observed reactions may be associated with oxidative stress and the breakdown of mitochondrial and lysosomal membranes, as previously discussed (da Silva Brito et al., 2023; Mattioda et al., 2023; Alijagic et al., 2024; Maharjan et al., 2023).

In vivo studies investigating the toxicity of MNPs, significant increases in pro-inflammatory cytokines, immunoglobulin A (IgA), and neutrophils were observed in the blood samples of rodents, highlighting the involvement of the immune system in response to MNP exposure (Li et al., 2020; Lim et al., 2019). As early as 2001, Brown et al. (2001) demonstrated that polystyrene particles ranging from 202–535 nm could enhance IL-8 expression in human lung cells (A549) (Brown et al., 2001). NPs sized 20–44 nm have also been found to increase the expression of IL-6 and IL-8, contributing to inflammation in various types of human cancers (Prietl et al., 2014; Forte et al., 2016). Moreover, polyethylene particles from plastic prosthetic implants have been shown to cause liver inflammation and periprosthetic bone resorption (Devane et al., 1995; Nich & Goodman, 2014), as well as increase the secretion of IL-6, IL-1β, and TNF in murine macrophages (Green et al., 1998).

The presence of MPs has been detected in human lungs (Amato-Lourenco et al., 2021; Jenner et al., 2022) and in the sputum of patients with respiratory diseases (Huang et al., 2022). MPs are inhaled into the respiratory tract and can be expelled with sputum, with personal smoking habits and invasive tracheal procedures being linked to MP exposure (Huang et al., 2022). However, research on the impact of microplastics deposited in the lungs, especially in patients with respiratory conditions, is still in its early stages. Potential effects and mechanisms of MNPs impact on the respiratory system include interference with the biophysical function of pulmonary surfactants, disruption of alveolar structures, and impairment of airway barrier functions. Exposure to MPs in lung epithelial cells suppresses cell proliferation, induces apoptosis, and stimulates the production of pro-inflammatory cytokines (TNF-α, IL-6, IL-1β, and TGF-β) by lung epithelial cells.

3.2. Effects of MNP on Organ Systems

MNPs can enter the body through various routes. Overall, the bioavailability of MNPs is influenced by various factors, including the physical and chemical characteristics of plastic particles. Further research is needed to understand the risks, including their toxicity, distribution, and accumulation in the body (Hu et al., 2022; Dawson et al., 2018). MNPs enter the body through the airways, in the respiratory system they cause inflammation and damage to lung tissue, increasing the risk of asthma, chronic obstructive pulmonary disease (COPD), and other respiratory disorders (Huang et al., 2022).

When MNPs are ingested, most of them pass through the digestive tract and are excreted from the body. *In vitro* studies have examined the translocation of nanoparticles. The results showed that the size of nanoparticles and the chemical nature of their surface affect translocation, with 50 nm nanoparticles exhibiting a higher translocation rate (up to 7.8%) compared to 100 nm nanoparticles (0.8%) (Walczak et al., 2015). However, a small portion of the microplastic remains in the intestines for several days. In the intestines, MNPs can cause damage, inflammatory diseases, disruption of microbiota, and absorption of toxic substances. MNPs are absorbed in the intestines and enter the bloodstream and lymphatic systems, spreading to other tissues and persisting for long periods of time (Ma et al., 2021). *In vivo* and *in vitro* studies have shown that after exposure, bioavailable particles that enter the bloodstream can move to secondary organs, where they can accumulate to levels that may lead to adverse consequences at the cellular level (Ali et al., 2024).

MNPs can have an adverse impact on the cardiovascular system. MNPs migrate through the bloodstream to the heart (Zhu et al., 2020; Liang et al., 2021). Translocation to the heart depends on the particle size and surface modification (Walczak et al., 2015; Yang et al., 2023). MNPs can lead to impaired circulation and heart function – they reduce heart rate, disrupt heart rate variability, induce inflammation, and cause direct or indirect oxidative damage, leading to apoptosis and pyroptosis of cardiomyocytes, and heart fibrosis through activation of the Wnt/β-catenin pathway, increasing levels of creatine kinase-MB and cardiac troponin I (Wei et al., 2021; Zhu et al., 2020; Liang et al., 2021; Li et al., 2020). MNPs can induce microvascular toxicity through hemolysis, thrombosis, blood clotting, and endothelial cell damage (Wei et al., 2021; Oslakovic et al., 2012).

The liver is the primary organ affected by various xenobiotic substances, including MNPs (Boey and Ho 2020). Studies have shown that chronic liver diseases may be a key factor in the accumulation of MNPs in the liver. *In vitro* studies of the effects of polystyrene (PS-MP) MNPs on Hep G2 cells have shown that they lead to a decrease in the expression of the genes for glyceraldehyde-3-phosphate dehydrogenase (GAPDH), which is involved in glycolysis, as well as CAT and SOD2, thereby reducing the potential of antioxidant enzymes in detoxifying ROS (Goodman et al., 2022; Cheng et al., 2022). The results demonstrated that PS-MPs, even at the lowest concentrations, can cause hepatotoxicity and lipotoxicity through cytotoxicity (increased apoptotic cells and decreased viable cells) and alter molecular markers (increased AST, ALT, and LDH activity in supernatants, reduced GST, GSH, and SOD activity in liver organoids, and increased MDA levels, indicating oxidative stress). PS-MPs induced lipid accumulation decreased ATP production, increased ROS production, and the release of inflammatory markers IL-6 and COL1A1. PS-MPs also increase the expression of HNF4A and CYP2E1 in the liver, raising the risk of steatosis and liver fibrosis (Cheng et al., 2022; Deng et al., 2017; Braeuning, 2019), and may cause damage to nuclear and mitochondrial DNA (Shen et al., 2022), triggering activation of the STING pathway, which leads to liver fibrosis (Shen et al., 2022). However, the exact role of MNPs in liver diseases and their association with liver fibrogenesis and cirrhosis development remain unclear (Horvatits et al., 2022). Most studies have focused on inflammation and oxidative stress-mediated liver damage caused by MNPs. Therefore, future research should investigate other potential mechanisms of liver damage induced by MNPs.

In the study, MPs (PS, PVC, PP, and PE, with sizes ranging from 4 to 15 μm) were found in human urine, suggesting that they may pass through the gastrointestinal tract and be excreted via the kidneys. MNPs are primarily eliminated by the liver through phagocytosis or biliary excretion with feces, and by the kidneys through biotransformation products in urine (Horvatits et al., 2022; Prata, 2022).

The human central nervous system is highly sensitive to environmental pollutants (Prata & Barone, 2022; Goodman et al., 2021). MNPs, by crossing the blood-brain barrier (BBB), accumulate in the nervous system, altering the secretion of neuroinflammatory cytokines and chemokines, as well as the levels of neurotransmitters (Mak et al., 2019). This can lead to excessive accumulation of acetylcholine, causing neurological disorders (Thorat et al., 2018; Ferreira et al., 2012).

Exposure to MNPs can trigger local or systemic immune responses, contributing to immunosuppression or autoimmune diseases (Prata et al., 2020). When exposed to MPs, immune cells cause strong modulation at the transcriptional level of enzymes for cytokine release (IL-8, TNFα, and IL-1β) (Lehner et al., 2020), showing high toxicity and autophagic cell death through the Akt/mTOR and AMPK pathways (Han, S. et al., 2020). However, depending on the exposure time, size, and concentration of MNPs, they may lead to a stronger immune response (Han, S. et al., 2020; Hwang et al., 2019) through the increased expression of TLR4, AP-1, and IRF5 (Li et al., 2020) in innate immunity. The effect of MNPs on adaptive immune responses remains poorly studied.

Components found in MPs, such as bisphenol A and phthalates, are considered endocrine-disrupting compounds that can potentially interfere with normal endocrine function and affect hormonal balance as agonists or antagonists, causing neuroendocrine effects at critical stages of development (Li et al., 2020). In an *in vivo* study, a single dose of BPA (10 μg/kg) led to a rapid increase in plasma insulin levels and consequently, a decrease in blood glucose (Everett et al., 2007), which may increase the risk of developing type 2 diabetes, lower testosterone levels, cause testicular inflammation, and disrupt the blood-testis barrier (Jin et al., 2021).

3.3. Synergistic Effects with Allergens

There are also synergistic effects between MNP exposure and allergens, exacerbating the impact on respiratory health. MNPs increase allergen permeability by disrupting the alveolar barrier function. In mice with asthma, MNPs exacerbate allergen-induced airway inflammation, leading to increased expression of MALT1 in lung tissue (Lu et al., 2021). MALT1 is widely expressed in lymphoid, mast, and endothelial cells, functioning as both a scaffolding protein and a substrate-specific protease. The MALT1 protease plays a dual role in allergic diseases, mediating IgE-dependent cytokine production in mast cells and histamine-induced endothelial permeability (Alfano et al., 2020). This suggests that the combined exposure to microplastics and allergens may synergistically worsen allergic conditions. Studies examining the side effects of MPs have used various chemical materials, particle sizes, or concentrations. Among these, polystyrene MPs are the most frequently used and are also one of the primary air pollutants. However, airborne MNPs coexist with other pollutants and may even have organic substances or metal ions adsorbed on their surfaces, leading to a synergistic effect. Therefore, in-depth research is needed to explore the effects of combined exposure to multiple MNPs on respiratory health. Although the concentration of MPs in the air is very low, the long-term exposure and its link to respiratory diseases require further clinical studies and animal model experiments. Finally, it should be noted that inhaled MNPs can be transported to other organs through the bloodstream after passing through the respiratory tract (Yang et al., 2024).

The liver, in particular, is highly susceptible to potential MNP toxicity. Adverse outcomes related to MNP-induced hepatotoxicity include liver dysfunction, inflammation, metabolic disturbances, and liver fibrosis (Cheng et al., 2022; Ge et al., 2023; Tsochatzis et al., 2024). There has been an observed increase in the activity of liver damage biomarkers such as ALT and AST. Hepatocyte cell lines can serve as a model for testing the toxic effects of MNPs (Tsochatzis et al., 2024). *In vivo* studies on MNP reproductive toxicity have revealed changes in sperm count, morphology, and motility in Wistar rats and Balb/c mice, as well as alterations in the morphology of the fallopian tubes and in follicle formation and growth in ICR mice (Amereh et al., 2020; An et al., 2021; Park et al., 2020; Xia et al., 2020).

MNPs enter the intestines during digestion through the gastrointestinal tract (GIT). Existing data from various studies indicate that the accumulation of MNPs in mammalian models and human cells can lead to adverse outcomes in terms of epithelial toxicity, immunotoxicity, and gut microbiota disruption (Covello et al., 2024). Research has been conducted using model systems (Abdelkhaliq et al., 2018; Magrì et al., 2018; Hesler et al., 2019; Schwarzfischer et al., 2022; Bredeck et al., 2022). An increase in the activity of intestinal biomarkers such as diamine oxidase and D-lactate has been observed (Hirt et al., 2020; Covello et al., 2024). Polystyrene particles of 5 μm and 20 μm cause gut microbiota dysbiosis, disrupt the intestinal barrier, and negatively affect neurotransmission (Deng et al., 2017; Luo et al., 2019a; Jin et al., 2019), and 30 nm particles suppress vesicle movement and cytokinesis overall (Xia et al., 2020). Carboxylated polystyrene particles (20 nm) can activate basolateral K^+ channels, leading to Cl^- and HCO_3^+ ion efflux (McCarthy et al., 2011). NPs have demonstrated increased macrophage activation in the gut compared to MPs (Yang et al., 2023). The study showed that NPs caused lysosomal damage, triggering the reprogramming of macrophages to produce IL-1. Notably, IL-1 signaling from the gut impacted brain immunity, leading to microglial activation and Th17 differentiation, which was associated with cognitive impairment in mice (Yang et al., 2023). Mice exposed to NPs via nasal inhalation exhibited altered behavior and neurotoxic effects, as evidenced by decreased acetylcholinesterase (AChE) activity. This study highlights the role of physicochemical properties in the deposition of organic NPs in the mammalian brain (Liu et al., 2022; Liu et al., 2020). It underscores the complex relationship between the gut and the brain, as well as the need to address MNP pollution globally.

MNPs, by directly crossing the BBB, can influence the secretion of neuroinflammatory cytokines, chemokines, transporters, and receptor markers. The neurotoxicity, neuroinflammation, and brain damage that may result from this are significant concerns, potentially leading to central nervous system (CNS) dysfunction. Neurotoxic responses are often assessed using indicators like acetylcholinesterase activity (Prüst et al., 2020).

MNPs induce dysfunction in nerve cells, leading to either an increase or decrease in levels of relevant biomarkers. For example, in astrocytes, MNPs cause a decrease in synaptic function regulation, proteostasis, energy metabolism, and mitochondrial function. Similarly, in endothelial cells, there is a decline in synaptic function regulation, proteostasis, neurite formation regulation, energy metabolism, and mitochondrial function. In microglia, there is an increase in cell signaling and anti-inflammatory response but a decrease in synaptic function regulation and proteostasis. Neurons exhibit reduced energy metabolism, mitochondrial function, and proteostasis. Lastly, in oligodendrocytes, MNP exposure results in decreased synaptic function regulation, proteostasis, neurite formation regulation, energy metabolism, and mitochondrial function (Kaushik et al., 2024).

The shape, size, chemical composition of the corona, surface modification, and cell type significantly influence the effects of MNPs. For instance, smaller polystyrene particles are more readily internalized by macrophages through phagocytosis compared to larger ones, leading to the induction of inflammatory gene expression (Olivier et al., 2004). MNPs made of polyethyleneimine with a polypeptide corona increase cellular internalization in K562 lymphoblast cells, while polyethylene NP damages the structure of the cell membrane, alters its fluidity, and ultimately initiates cell death (Rossi et al., 2014).

Molecular mechanisms of NP cytotoxicity are often linked to disruptions in mitochondrial, endoplasmic reticulum, and lysosomal function, resulting in metabolic and functional disorders (Armstrong, 2007). MNPs alter mitochondrial function by increasing oxygen consumption and generating reactive oxygen species (ROS) (Pitt et al., 2018). Additionally, MNPs are associated with the anti-apoptotic Bcl-2-caspase 8 signaling pathway (Qu et al., 2019). In *Sterechinus neumayeri*, MNPs led to elevated antioxidant activity, including catalase, superoxide dismutase, metallothionein, and anti-apoptotic signaling through Bcl-2-caspase 8 (Bergami et al., 2019). Several studies have observed increased levels of antioxidant system markers (superoxide dismutase, catalase, glutathione peroxidase, glutathione-S-transferase, glutathione, glutathione reductase, and malondialdehyde) in response to MNP exposure. NPs induce oxidative stress (Kazaryan et al., 2022) and stress-related autophagy pathways in the endoplasmic reticulum, marked by increased expression of Grp78 and Grp170 (Bergami et al., 2019), and trigger phosphorylation of MAPK14 and upregulation of XBP1 (Qu et al., 2019; Yang et al., 2020).

Several studies have found that MNPs are internalized and accumulate in lysosomes, leading to their dysfunction, altering pH levels, and disrupting membrane integrity. This process activates the EB transcription factor, which enhances lysosome-autophagosome fusion and the clearance of autophagic cargo (Song et al., 2015; Saftig & Haas, 2016), potentially leading to cell death (Wang et al., 2018). Additionally, MNPs with a corona have been reported to cause damage due to the degradation of the surface corona in lysosomes (Wang et al., 2013).

NPs up to 50 nm in size penetrate the nucleus, causing structural changes and single- and double-strand DNA damage. Exposure to NPs from transparent polypropylene food containers has also been positively associated with DNA strand breaks [i.e., a net induction of 0.10 lesions/10^6 base pairs (95% CI: –0.04; 0.23 lesions/10^6 base pairs)] at the highest concentration. Moreover, oxidative-reductive activation of the β-catenin/YAP cascade suggests a new toxic mechanism for MNPs and implies potential carcinogenic effects (Ding et al., 2024; Roursgaard et al., 2022). Statistical analysis shows that specially modified polystyrene (PS) particles are genotoxic, with immune cells appearing to be more sensitive to genotoxicity than other cell types, such as epithelial cells. In contrast, there is no clear link between statistically significant effects in genotoxicity tests and the primary size of PS particles (i.e., smaller or larger than 100 nm) or between the type of genotoxic endpoint (i.e., repairable or permanent DNA damage). Three studies on the effects of PS particles in animals showed increased levels of DNA strand breaks in leukocytes and prefrontal cortex cells. Nanoplastics made from polyethylene, polypropylene, polyvinyl chloride, and polyethylene terephthalate have been studied in very few studies, and it is currently impossible to conclude their genotoxic risk. In summary, there is some evidence suggesting that PS particles may be genotoxic in mammalian cells (Møller et al., 2023). The toxic effects of MNPs can be broadly categorized into the following types of impacts (Figure 1).

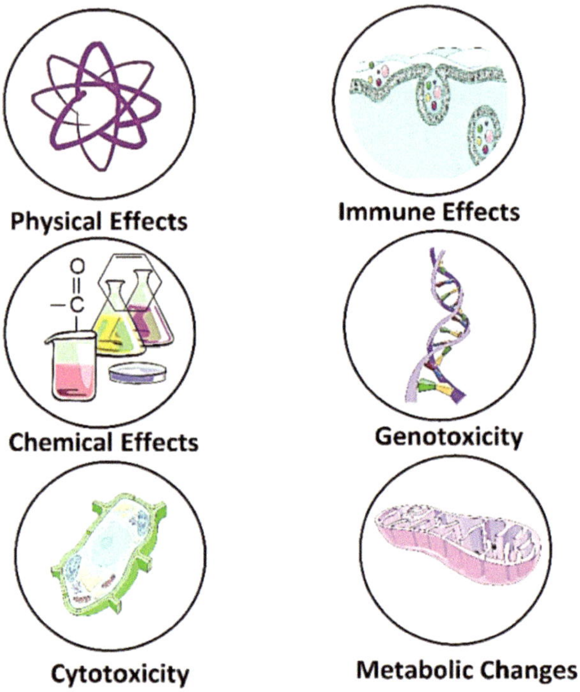

Figure 1. Main types of MNPs toxic impact.

- **Physical Effects:** Accumulation and mechanical damage—Sharp particles can physically damage tissues, particularly in the respiratory system, gastrointestinal tract, and central nervous system (CNS), causing obstructions, and blockages of small vessels, bronchioles, and microvilli.
- **Chemical Effects:** MNPs can adsorb various chemicals on their surface, such as heavy metals and organic pollutants, which may be released into the body. Additionally, plasticizers like phthalates and bisphenol can leach out during degradation, leading to toxic effects.
- **Immune Effects:** MNPs entering the body can cause inflammation in tissues and organs, leading to immune dysregulation and autoimmune reactions.
- **Cytotoxicity:** MNPs can induce apoptosis and autophagy in cells.

- **Genotoxicity:** MNPs may cause DNA damage, increasing the risk of developing cancer.
- **Metabolic Changes:** Alterations in redox reactions, energy balance, and ionic homeostasis.

4. Conclusions and Path Forward

MNPs are widespread in the environment and are more than just pollutants. This raises concerns about their potential impact, and cumulative effects on human health, animals, and the ecosystem as a whole. The constant and overwhelming presence of MNPs in the environment makes them more accessible for absorption by skin cells through direct contact, lung cells through inhalation, and stomach cells through ingestion of food and water. The primary mechanism of MNP action is the generation of ROS (Reactive Oxygen Species), reactive nitrogen species (RNS), and the initiation of an oxidative stress cascade. MNPs cause inflammatory and immune reactions, accelerate aging processes, lead to dysfunction of humoral status, changes in cellular homeostasis and energy metabolism, tissue degeneration, abnormal development and dysfunction of various organs, alterations in biochemical parameters, and even may cause genotoxicity with pathological consequences. Although numerous studies in animals and cell cultures have pointed to the primary mechanisms of MNP action, it is still unclear which exposure pathway represents the greatest risk to humans in relation to MNPs. Furthermore, there is insufficient information about the process of MNP elimination from the human body, and whether MNPs have an affinity for interacting with specific chemicals or microorganisms during their impact on the human body. The toxicity mechanisms of MNPs are not fully understood, including their interaction with other pollutants, and the long-term exposure and lasting effects of MNPs on human health have not been assessed. Therefore, the risk MNPs assessment for human health is limited and complex. Moreover, it is necessary to study the susceptibility of MNPs to diseases, and it is crucial to quantitatively assess the impact of MNPs on human health and their pathogenesis. Currently, standardized methods for detecting and identifying MNPs are lacking. Therefore, it is necessary to establish unique methods for identifying and classifying plastic waste based on several criteria, such as size, shape, physical and chemical properties, 'aging' processes, and sources. Solving the problems associated with MNPs will require an interdisciplinary approach, involving ecologists, toxicologists, health specialists, and policymakers to protect public health and the environment, and further research is needed to fill existing knowledge gaps in this area and to develop more effective strategies for monitoring and safely managing MNPs.

References

Abdelkhaliq, A., van der Zande, M., Punt, A., Helsdingen, R., Boeren, S., Vervoort, J.J.M. et al. (2018). Impact of nanoparticle surface functionalization on the protein corona and cellular adhesion, uptake, and transport. *J. Nanobiotechnology*, 16, 70. doi: 10.1186/s12951-018-0394.

Aderem, A., & Underhill, D.M. (1999). Mechanisms of phagocytosis in macrophages. *Annu. Rev. Immunol.*, 17, 593–623. doi: 10.1146/annurev.immunol.17.1.593. PMID: 10358769.

Alfano, D.N., Klei, L.R., Klei, H.B., Trotta, M., Gough, P.J., Foley, K.P. et al. (2020). MALT1 protease plays a dual role in the allergic response by acting in both mast cells and endothelial cells. *J. Immunol.*, 204(9), 2337–2348 doi: 10.4049/jimmunol.1900281. Epub 2020 Mar 25. PMID: 32213560.

Ali, N., Katsouli J., Marczylo, E., Gant, T., Wright, S., & Bernardino de la Serna, J. (2024). The potential impacts of micro- and nanoplastics on various organ systems in humans. *eBioMedicine*, 99, 104901.

Alijagic, A., Kotlyar, O., Larsson, M., Salihovic, S., Hedbrant, A., Eriksson, U. et al. (2024). Immunotoxic, genotoxic, and endocrine disrupting impacts of polyamide microplastic particles and chemicals. *Environ. Int.*, 183, 108412. doi: 10.1016/j.envint.2023.108412.

Alqahtani, S., Xia, L., & Shannahan, J.H. (2022). Enhanced silver nanoparticle-induced pulmonary inflammation in a metabolic syndrome mouse model and resolving D1 treatment. *Part Fibre Toxicol.*, 19(1), 54. doi: 10.1186/s12989-022-00495-6. PMID: 35933425; PMCID: PMC9356467.

Amato-Lourenço, L.F., Carvalho-Oliveira, R., Júnior, G.R., Dos Santos, Galvão, L., Ando, R.A., & Mauad, T. (2021). Presence of airborne microplastics in human lung tissue. *J. Hazard. Mater.*, 416, 126124. doi: 10.1016/j.jhazmat.2021.126124. Epub 2021 May 24. PMID: 34492918.

Amereh, F., Babaei, M., Eslami, A., Fazelipour, S. & Rafiee, M. (2020). The emerging risk of exposure to nano(micro) plastics on endocrine disturbance and reproductive toxicity: From a hypothetical scenario to a global public health challenge. *Environ. Pollut.*, 261, 114158. doi: 10.1016/j.envpol.2020.114158. Epub 2020 Feb. 10. PMID: 32088433.

An, R., Wang, X., Yang, L., Zhang, J., Wang, N., Xu, F. et al. (2020). Polystyrene microplastics cause granulosa cells apoptosis and fibrosis in ovary through oxidative stress in rats. *Toxicology*, 449, 152665. doi: 10.1016/j.tox.2020.152665. Epub 2020 Dec. 24. Erratum in: *Toxicology*, 2022 478, 153291. doi: 10.1016/j.tox.2022.153291. PMID: 33359712.

Andrady, A., Barnes, P., Bornman, J., Gouin, T., Madronich, S., White, C. et al. (2022). Oxidation and fragmentation of plastics in a changing environment: From UV-radiation to biological degradation. *Sci. Total Environ.*, 851, 158022. doi: 10.1016/j.scitotenv.2022.158022.

Armstrong, J.S. (2007). Mitochondrial medicine: Pharmacological targeting of mitochondria in disease. *Br. J. Pharmacol.*, 151(8),1154–1165. doi: 10.1038/sj.bjp.0707288. Epub 2007 May, 21. PMID: 17519949; PMCID: PMC2189819.

Aslam, M., Ali, M.H., Erum, W., Shehzadi, S., & Khalid, Z. (2021). Effect of environmental pollutants on neurological disorders. *Markhor* (*J. Zool.*), 2, 9–13. doi: 10.54393/mjz.v2i2.18.

Baho, D.L., Bundschuh, M., & Futter, M.N. (2021). Microplastics in terrestrial ecosystems: Moving beyond the state of the art to minimize the risk of ecological surprise. *Glob. Chang. Biol.*, 27(17), 3969–3986. doi: 10.1111/gcb.15724. Epub 2021 Jun., 13. PMID: 34042229.

Bashiri, G., Padilla, M.S., Swingle, K.L., Shepherd, S.J., Mitchell, M.J., & Wang, K. (2023). Nanoparticle protein corona: From structure and function to therapeutic targeting. *Lab Chip.*, 23(6),1432–1466. doi: 10.1039/d2lc00799a. PMID: 36655824; PMCID: PMC10013352.

Bergami, E., Krupinski Emerenciano, A., González-Aravena, M., Cárdenas, C., Hernández, P., Silva, J. et al. (2019). Polystyrene nanoparticles affect the innate immune system of the Antarctic sea urchin Sterechinus neumayeri. *Polar Biol.*, 42, 743–757. doi:10.1007/s00300-019-02468-6.

Boey, A., & Ho, H.K. (2020). All roads lead to the liver: Metal nanoparticles and their implications for liver health. *Small*, 16, Article 2000153.

Bordós, G., Urbányi, B., Micsinai, A., Kriszt, B., Palotai, Z., Szabó, I. et al. (2019). Identification of microplastics in fish ponds and natural freshwater environments of the Carpathian basin, Europe. *Chemosphere*, 216,110–116. doi: 10.1016/j.chemosphere.2018.10.110. Epub 2018 Oct 19. PMID: 30359912.

Bouaicha, O., Mimmo, T., Tiziani, R., Praeg, N., Polidori, C., Lucini, L. &, Borruso, L. (2022). Microplastics make their way into the soil and rhizosphere: A review of the ecological consequences. *Rhizosphere*, Article 100542.

Bracco, P., Costa, L., Luda, M.P., & Billingham, N. (2018). A review of experimental studies of the role of free-radicals in polyethylene oxidation. *Polym. Degrad. Stab.*, 155, 67–83. doi: 10.1016/j.polymdegradstab.2018.07.011.

Bradney, L., Wijesekara, H., Palansooriya, K.N, Obadamudalige, N., Bolan, N.S, Ok, Y.S. et al. (2019). Particulate plastics as a vector for toxic trace-element uptake by aquatic and terrestrial organisms and human health risk. *Environ. Int.*, 131, 104937. doi: 10.1016/j.envint.2019.104937. Epub 2019 July 5. PMID: 31284110.

Braeuning, A. (2019). Uptake of microplastics and related health effects: A critical discussion of Deng et al. *Scientific Reports*, 7, 46687, *Arch Toxicol.*, 93 (2017), pp. 219-220.

Braun, T., Ehrlich, L., Henrich, W., Koeppel, S., Lomako, I., Schwabl, P. et al. (2021). Detection of microplastic in human placenta and meconium in a clinical setting. *Pharmaceutics*, 13(7), 921. doi: 10.3390/pharmaceutics13070921. PMID: 34206212; PMCID: PMC8308544.

Bredeck, G., Halamoda-Kenzaoui, B., Bogni, A., Lipsa, D., & Bremer-Hoffmann, S. (2022). Tiered testing of micro- and nanoplastics using intestinal *in vitro* models to support hazard assessments. *Environ. Int.*, 158, 106921. doi: 10.1016/j.envint.2021.106921. Epub 2021 Oct 8. PMID: 34634620.

Brown, D.M., Wilson, M.R., MacNee, W., Stone, V., & Donaldson, K. (2001). Size-dependent proinflammatory effects of ultrafine polystyrene particles: A role for surface area and oxidative stress in the enhanced activity of ultrafines. *Toxicol. Appl. Pharmacol.*, 175(3), 191–199. doi: 10.1006/taap.2001.9240. PMID: 11559017.

Cao, J., Yang, Q., Jiang, J., Dalu, T., Kadushkin, A., Singh, J. et al. (2022). Coronas of micro/nano plastics: A key determinant in their risk assessments. *Part. Fibre Toxicol.*, 19(1), 55. doi: 10.1186/s12989-022-00492-9. PMID: 35933442; PMCID: PMC9356472.

Catarino, A.I., Macchia, V., Sanderson, W.G., Thompson, R.C., & Henry, T.B. (2018). Low levels of microplastics (MP) in wild mussels indicate that MP ingestion by humans is minimal compared to exposure via household fibres fallout during a meal. *Environ. Pollut.*, 237, 675–684. doi: 10.1016/j.envpol.2018.02.069. Epub 2018 Mar 29. PMID: 29604577.

Cheng, W., Li, X., Zhou, Y., Yu, H., Xie, H., Guo, H. et al. (2022). Polystyrene microplastics induce hepatotoxicity and disrupt lipid metabolism in the liver organoids. *Science of The Total Environment*, 806,150328. https://doi.org/10.1016/j.scitotenv.2021.150328.

Cheung, P.K., & Fok, L. (2016). Evidence of microbeads from personal care products contaminating the sea. *Mar. Pollut. Bull.*, 109, 582–585. doi: 10.1016/j.marpolbul.2016.05.046.

Covello, C., Di Vincenzo, F., Cammarota, G., & Pizzoferrato, M. (2024). Micro(nano) plastics and their potential impact on human gut health: A narrative review. *Curr. Issues Mol. Biol.*, 46(3), 2658–2677. doi: 10.3390/cimb46030168. PMID: 38534784; PMCID: PMC10968954.

Covernton, G.A., Pearce, C.M., Gurney-Smith, H.J., Chastain, S.G., Ross, P.S., Dower, J.F., & Dudas, S.E. (2019). Size and shape matter: A preliminary analysis of microplastic sampling technique in seawater studies with implications for ecological risk assessment. *Sci. Total Environ.*, 1, 667, 124–132. doi: 10.1016/j.scitotenv.2019.02.346. Epub 2019 Feb 24. PMID: 30826673.

Cui, Y., Liu, M., Selvam, S., Ding, Y., Wu, Q., Pitchaimani, V.S. et al. (2022). Microplastics in the surface waters of the South China sea and the western Pacific Ocean: Different size classes reflecting various sources and transport. *Chemosphere*, 299, 134456. doi: 10.1016/j.chemosphere.2022.134456. Epub 2022 Mar 29. PMID: 35364074.

D'Errico, M., Parlanti, E., Pascucci, B., Filomeni, G., Mastroberardino, P.G., & Dogliotti, E. (2021). The interplay between mitochondrial functionality and genome integrity in the prevention of human neurologic diseases. *Arch. Biochem. Biophys.*, 710, 108977. doi: 10.1016/j.abb.2021.108977.

da Costa Araújo, A.P., & Malafaia, G. (2021). Microplastic ingestion induces behavioral disorders in mice: A preliminary study on the trophic transfer effects via tadpoles and fish. *J. Hazard. Mater.*, 401, 123263. doi: 10.1016/j.jhazmat.2020.123263. Epub 2020 Jun 22. PMID: 32629346.

da Silva Brito, W.A., Singer, D., Miebach, L., Saadati, F., Wende, K., Schmidt, A., & Bekeschus, S. (2023). Comprehensive *in vitro* polymer type, concentration, and size correlation analysis to microplastic toxicity and inflammation. *Sci. Total Environ.*, 854, 158731. doi: 10.1016/j.scitotenv.2022.158731.

Das, A. (2023). The emerging role of microplastics in systemic toxicity: Involvement of reactive oxygen species (ROS). *Sci. Total Environ.*, 895, 165076. doi: 10.1016/j.scitotenv.2023.165076.

Dawson, A.L., Kawaguchi, S., King, C.K. et al. (2018). Turning microplastics into nanoplastics through digestive fragmentation by Antarctic krill. *Nat. Commun.*, 9, 1001.

Deng, Y., Zhang, Y., Lemos, B., & Ren, H. (2017). Tissue accumulation of microplastics in mice and biomarker responses suggest widespread health risks of exposure. *Sci. Rep.*, 7, 46687. doi: 10.1038/srep46687. PMID: 28436478; PMCID: PMC5402289.

Devane, P.A., Bourne, R.B., Rorabeck, C.H., MacDonald, S., & Robinson, E.J. (1995). Measurement of polyethylene wear in metal-backed acetabular cups. II. Clinical application. *Clin. Orthop. Relat. Res.*, 319, 317–326. PMID: 7554645.

Diez-Pérez, D.B., Arenas, I., Maidana, E., López-Rosales, A., Andrade, J.M., & Muniategui-

Lorenzo, S. (2023). Microplastics in surface water of the Bay of Asunción, Paraguay. *Mar. Pollut. Bull.*, 192, 115075. doi: 10.1016/j.marpolbul.2023.115075. Epub 2023 Jun 2. PMID: 37269701.

Ding, R., Chen, Y., Shi, X., Li, Y., Yu, Y., Sun, Z., & Duan, J. (2024). Size-dependent toxicity of polystyrene microplastics on the gastrointestinal tract: Oxidative stress related-DNA damage and potential carcinogenicity. *Sci. Total Environ.*, 912, 169514. doi: 10.1016/j.scitotenv.2023.169514. Epub 2023 Dec 20. PMID: 38135073.

Dissanayake, P.D., Kim, S., Sarkar, B., Oleszczuk, P., Sang, M.K., Haque, M.N. et al. (2022). Effects of microplastics on the terrestrial environment: A critical review. *Environ. Res.*, 209, 112734. doi: 10.1016/j.envres.2022.112734. Epub 2022 Jan 21.PMID: 35065936.

Domenech, J., de Britto, M., Velázquez, A., Pastor, S., Hernández, A., Marcos, R., & Cortés, C. (2021). Long-term effects of polystyrene nanoplastics in human intestinal CaCo-2 cells. *Biomolecules*, 11, 1442. doi: 10.3390/biom11101442.

Domenech, J., Annangi, B., Marcos, R., Hernández, A., & Catalán, J. (2023). Insights into the potential carcinogenicity of micro- and nano-plastics. *Mutat. Res./Rev. Mutat. Res.*, 791, 108453. doi: 10.1016/j.mrrev.2023.108453.

Dong, X., Liu, X., Hou, Q., & Wang Z. (2023). From natural environment to animal tissues: A review of microplastics(nanoplastics) translocation and hazards studies. *Sci. Total Environ.*, 855, 158686. doi: 10.1016/j.scitotenv.2022.158686. Epub 2022 Sep 10. PMID: 36099943.

Duis, K., & Coors, A. (2016). Microplastics in the aquatic and terrestrial environment: Sources (with a specific focus on personal care products), fate, and effects. *Environ. Sci. Eur.*, 28(1), 1–25. doi: 10.1186/s12302-015-0069-y.

Egessa, R., Nankabirwa, A., Ocaya, H., & Pabire, W.G. (2020). Microplastic pollution in surface water of Lake Victoria. *Sci. Total Environ.*, 741, 140201. doi: 10.1016/j.scitotenv.2020.140201. Epub 2020 Jun 15. PMID: 32603936.

Everett, J., Frithsen, I.L., Diaz, V.A., Koopman, R.J., Simpson Jr., W.M., & Mainous, A.G. III (2007). Association of a polychlorinated dibenzo-p-dioxin, a polychlorinated biphenyl, and DDT with diabetes in the 1999–2002 national health and nutrition examination survey. *Environ. Res.*, 103(2007), Article 413-418.

Ferreira, G.K., Carvalho-Silva, M., Gonçalves, C.L. et al. (2012). L-tyrosine administration increases acetylcholinesterase activity in rats. *Neurochem. Int.*, 61(2012), 1370–1374.

Forte, M., Iachetta, G., Tussellino, M., Carotenuto, R., Prisco, M., De Falco, M. et al. (2016). Polystyrene nanoparticles internalization in human gastric adenocarcinoma cells. *Toxicol. In Vitro*, 126–136. doi: 10.1016/j.tiv.2015.11.006. Epub 2015 Nov 14. PMID: 26585375.

Gaston, E., Woo, M., Steele, C., Sukumaran, S., & Anderson, S. (2020). Microplastics differ between indoor and outdoor air masses: Insights from multiple microscopy methodologies. *Appl. Spectrosc.*, 74(9), 1079–1098. doi: 10.1177/0003702820920652. Epub 2020 Apr 27. PMID: 32233850.

Ge, Y., Yang, S., Zhang, T., Wan, X., Zhu, Y., Yang, F. et al. (2023). The hepatotoxicity assessment of micro/nanoplastics: A preliminary study to apply the adverse outcome pathways. *Sci. Total Environ.*, 902,165659. doi: 10.1016/j.scitotenv.2023.165659. Epub 2023 Jul 29. PMID: 37517720.

Goldstein, J.L., Anderson, R.G., & Brown, M.S. (1979). Coated pits, coated vesicles, and receptor-mediated endocytosis. *Nature*, 279(5715), 679–685. doi: 10.1038/279679a0. PMID: 221835.

Goodman, K..E., Hare, J.T. , Khamis, Z.I., Hua, T., & Sang, Q.-X.A. (2021). Exposure of human lung cells to polystyrene microplastics significantly retards cell proliferation and triggers morphological changes *Chem. Res. Toxicol.*, 34, 1069–1081.

Goodman, K.E., Hua, T., & Sang, Q.-X.A. (2022). Effects of polystyrene microplastics on human kidney and liver cell morphology, cellular proliferation, and metabolism. *ACS Omega*, 7, 34136–34153. doi: 10.1021/acsomega.2c03453.

Green, T.R., Fisher, J., Stone, M., Wroblewski, B.M., & Ingham, E. (1998). Polyethylene particles of a 'critical size' are necessary for the induction of cytokines by macrophages *in vitro*. *Biomaterials*, 19(24), 2297–2302. doi: 10.1016/s0142-9612(98)00140-9. PMID: 9884043.

Guo, C., Sun, L., Chen, X., & Zhang, D. (2013). Oxidative stress, mitochondrial damage, and neurodegenerative diseases. *Neural. Regen. Res.*, 8, 2003–2014. doi: 10.3969/j.issn.1673-5374.2013.21.009.

Gupta, N., Parsai, T., & Kulkarni, H.V. (2024). A review on the fate of micro and nano plastics (MNPs) and their implication in regulating nutrient cycling in constructed wetland systems. *J. Environ. Manage.*, 350, 119559. doi: 10.1016/j.jenvman.2023.119559. Epub 2023 Nov 28. PMID: 38016236.

Han, N., Zhao, Q., Ao, H., Hu, H., & Wu, C. (2022). Horizontal transport of macro- and microplastics on soil surface by rainfall induced surface runoff as affected by vegetations. *Sci. Total Environ.*, 831, 154989. doi: 10.1016/j.scitotenv.2022.154989. Epub 2022 Apr 2. PMID: 35381242.

Han, S., Bang, J.,. Choi, D. et al. (2020). Surface pattern analysis of microplastics and their impact on human-derived cells. *ACS Appl. Polym. Mater.*, 2, 4541–4550.

He, D., Luo, Y., Lu, S., Liu, M., Song, Y., & Lei, L. (2018). Microplastics in soils: Analytical methods, pollution characteristics, and ecological risks. *TrAC Trends Anal. Chem.*, 109, 163–172. doi:10.1016/J.TRAC.2018.10.006.

Hellack, B., Nickel, C., Albrecht, C., Kuhlbusch, T.A., Boland, S., Baeza-Squiban, A. et al. (2017). Analytical methods to assess the oxidative potential of nanoparticles: A review. *Environ. Sci. Nano*. 4, 1920-1934.

Hesler, M., Aengenheister, L., Ellinger, B., Drexel, R., Straskraba, S., Jost, C. et al. (2019). Multi-endpoint toxicological assessment of polystyrene nano- and microparticles in different biological models *in vitro*. *Toxicol. Vitr.*, 61, 104610. doi: 10.1016/j.tiv.2019.104610.

Hirt, N., & Malapel, M. (2020). Body-immunotoxicity and intestinal effects of nano- and microplastics: A review of the literature. *Part. Fibre Toxicol.*, 17(1), 57. doi: 10.1186/s12989-020-00387-7. PMID: 33183327; PMCID: PMC7661204.

Hoang, T.C., & Felix-Kim, M. (2020). Microplastic consumption and excretion by fathead minnows (*Pimephales promelas*): Influence of particles size and body shape of fish. *Sci. Total Environ.*, 704, 135433. doi: 10.1016/j.scitotenv.2019.135433. Epub 2019 Nov 23. PMID: 31896224.

Horvatits, T., Tamminga, M., Liu, B., Motta, O., Giorgini, E., & Montano, L. (2022). Microplastics detected in cirrhotic liver tissue. *eBioMedicine*, 82, Article 104147.

Hou, J., Xu, X., Yu, H., Xi, B., & Tan, W. (2021). Comparing the long-term responses of soil microbial structures and diversities to polyethylene microplastics in different aggregate fractions. *Environ. Int.*, 149, 106398. doi: 10.1016/j.envint.2021.106398. Epub 2021 Jan 24. PMID: 33503554.

Hrycay, E.G., & Bandiera, S.M. (2015). Involvement of cytochrome P450 in reactive oxygen species formation and cancer. *Adv. Pharmacol.*, 74, 35–84. doi: 10.1016/bs.apha.2015.03.003.

Hu, L., Zhou, Y., Wang, Y., Zhang, D., & Pan, X. (2022). Transfer of micro(nano)plastics in animals: A mini-review and future research recommendation. *J. Hazard. Mater. Adv.*, 7(2022), Article 100101.

Hu, M., & Palić, D. (2020). Micro- and nano-plastics activation of oxidative and inflammatory adverse outcome pathways. *Redox. Biol.*, 37, 101620. doi: 10.1016/j.redox.2020.101620.

Hua, T., Kiran, S., Li, Y., & Sang, Q.A. (2022). Microplastics exposure affects neural development of human pluripotent stem cell-derived cortical spheroids. *J. Hazard. Mater.*, 435, 128884. doi: 10.1016/j.jhazmat.2022.128884. Epub 2022 Apr 11. PMID: 35483261.

Huang, S., Huang, X., Bi, R., Guo, Q., Yu, X., Zeng, Q. et al. (2022). Detection and analysis of microplastics in human sputum. *Environ. Sci. Technol.*, 56, 2476–2486. doi: 0.1021/acs.est.1c03859.

Hwang, J., Choi, D., Han, S., Choi, J., & Hong, J. (2019). An assessment of the toxicity of polypropylene microplastics in human derived cells. *Sci. Total Environ.*, 684(2019), 657–669.

Ibrahim, Y.S., Tuan Anuar, S., Azmi, A.A., Wan Mohd Khalik, W.M.A., Lehata, S., Hamzah, S.R. et al. (2020). Detection of microplastics in human colectomy specimens. *JGH Open*, 5(1),116–121. doi: 10.1002/jgh3.12457. PMID: 33490620; PMCID: PMC7812470.

Jaeseong Jeong, Jeongeun Im, & Jinhee Choi. (2024). Integrating aggregate exposure pathway and adverse outcome pathway for micro/nanoplastics: A review on exposure, toxicokinetics, and toxicity studies. *Ecotoxicology and Environmental Safety*, 272, 116022, https://doi.org/10.1016/j.ecoenv.2024.116022.

Jenner, L.C., Rotchell, J.M., Bennett, R.T., Cowen, M., Tentzeris, V., & Sadofsky, L.R. (2022). Detection of microplastics in human lung tissue using μFTIR spectroscopy. *Sci. Total Environ.*, 831, 154907. doi: 10.1016/j.scitotenv.2022.154907. Epub 2022 Mar 29. PMID: 35364151.

Jeon, S., Lee, D.-K., Jeong, J., Yang, S.I., Kim, J.-S., Kim, J., & Cho, W.-S. (2021). The reactive oxygen species as pathogenic factors of fragmented microplastics to macrophages. *Environ. Pollut.*, 281, 117006. doi: 10.1016/j.envpol.2021.117006.

Jin, H., Ma, T., Sha, X. et al. (2021). Polystyrene microplastics induced male reproductive toxicity in mice. *J. Hazard. Mater.*, 401(2021), Article 123430.

Jin, Y., Lu, L., Tu, W., Luo, T., & Fu, Z. (2019). Impacts of polystyrene microplastic on the gut barrier, microbiota, and metabolism of mice. *Sci. Total Environ.*, 649, 308–317. doi: 10.1016/j.scitotenv.2018.08.353. Epub 2018 Aug 28. PMID: 30176444.

Joos, L., & De Tender, C. (2022). Soil under stress: The importance of soil life and how it is influenced by (micro)plastic pollution. *Comput. Struct. Biotechnol. J.*, 20, 1554–1566. doi: 10.1016/j.csbj.2022.03.041. PMID: 35422972; PMCID: PMC8991314.

Kaushik, A., Singh, A.; Kumar Gupta, V., & Mishra, Y.K. (2024). Nano/micro-plastic, an invisible threat getting into the brain. *Chemosphere*, 361, 142380. doi: 10.1016/j.chemosphere.2024.142380. Epub 2024 May 18. PMID: 38763401.

Kazaryan, S., Farsiyan, L., Tumoyan, J., Kirakosyan, G., Ayvazyan, N., Gasparyan, H. et al. (2022). Oxidative stress and histopathological changes in several organs of mice injected with biogenic silver nanoparticles. *Artif. Cells Nanomed. Biotechnol.*, 50(1), 331–342. doi: 10.1080/21691401.2022.2149931. PMID: 36476283.

Kedzierski, M., Lechat, B., Sire, O., Le Maguer, G., Tilly, V., & Bruzaud, S. (2024). Microplastic contamination of packaged meat: Occurrence and associated risks. *Food Packag. Shelf Life*, 24, Article 100489.

Kobos, L., Alqahtani, S., Xia, L., Coltellino, V., Kishman, R., McIlrath, D. et al. (2020). Comparison of silver nanoparticle-induced inflammatory responses between healthy and metabolic syndrome mouse models. *J. Toxicol. Environ. Health A.*, 83(7), 249–268. doi: 10.1080/15287394.2020.1748779. Epub 2020 Apr 12. PMID: 32281499; PMCID: PMC7493428.

Kopatz, V., Wen, K., Kovács, T., Keimowitz, A.S., Pichler, V., Widder, J. et al. (2023). Micro- and nanoplastics breach the blood-brain barrier (BBB): Biomolecular Corona's role revealed. *Nanomaterials (Basel)*. 13(8), 1404. doi: 10.3390/nano13081404. PMID: 37110989; PMCID: PMC10141840.

Kor, K., Ghazilou, A., & Ershadifar, H. (2020). Microplastic pollution in the littoral sediments of the northern part of the Oman Sea. *Mar. Pollut. Bull.*, 155, 111166. doi: 10.1016/j.marpolbul.2020.111166. Epub 2020 Apr 16. PMID: 32310102.

Law, K.L. (2017). Plastics in the marine environment. *Annual Review of Marine Science*, 9, 205–229. doi: 10.1146/annurev-marine-010816-060409.

Lee, J., Wang, J., Oh, Y., & Jeong, S. (2023). Highly efficient microplastics removal from water using in-situ ferrate coagulation: Performance evaluation by micro-Fourier-transformed infrared spectroscopy and coagulation mechanism. *Chem. Eng. J.*, 451(4). doi: 10.1016/j.cej.2022.138556.

Lehner, R., Wohlleben, W., Septiadi, D., Landsiedel, R., Petri-Fink, A., & Rothen-Rutishauser, B. (2020). A novel 3D intestine barrier model to study the immune response upon exposure to microplastics. *Arch. Toxicol.*, 94, 2463–2479.

Lei, K., Qiao, F., Liu, Q., Wei, Z., Qi, H., Cui, S. et al. (2017). Microplastics releasing from personal care and cosmetic products in China. *Mar. Pollut. Bull.*, 123(1–2), 122–126. doi: 10.1016/j.marpolbul.2017.09.016.

Leslie, H.A., Brandsma, S.H., van Velzen, M.J., & Vethaak, A.D. (2017). Microplastics en route: Field measurements in the Dutch River delta and Amsterdam canals, wastewater treatment plants, North Sea sediments, and biota. *Environ. Int.*, 101, 133–142. doi: 10.1016/j.envint.2017.01.018. Epub 2017 Jan., 28. PMID: 28143645.

Leslie, H.A., van Velzen, M.J.M., Brandsma, S.H., Vethaak, A.D., Garcia-Vallejo, J.J., & Lamoree, M.H. (2022). Discovery and quantification of plastic particle pollution in human blood. *Environ Int.*, May, 163,107199. doi: 10.1016/j.envint.2022.107199. Epub 2022 Mar 24. PMID: 35367073.

Li, B., Ding, Y., Cheng, X., Sheng, D., Xu, Z., Rong, Q. et al. (2020). Polyethylene microplastics affect the distribution of gut microbiota and inflammation development in mice. *Chemosphere*, 244, 125492. doi: 10.1016/j.chemosphere.2019.125492. Epub 2019 Nov 27. PMID: 31809927.

Li, C., Zhu, S., Liu, Q. et al. (2020). Polystyrene microplastics cause cardiac fibrosis by activating Wnt/β-catenin signaling pathway and promoting cardiomyocyte apoptosis in rats. *Environ. Pollut.*, 265, Article 115025.

Liang B., Zhong, Y., Huang, Y. et al. (2021). Underestimated health risks: Polystyrene micro- and nanoplastics jointly induce intestinal barrier dysfunction by ROS-mediated epithelial cell apoptosis. *Part. Fibre Toxicol.*, 18, 20.

Lim, S.L., Ng, C.T., Zou, L., Lu, Y., Chen, J., Bay, B.H. et al. (2019). Targeted metabolomics reveals differential biological effects of nanoplastics and nano ZnO in human lung cells. *Nanotoxicology*,13(8), 1117–1132. doi: 10.1080/17435390.2019.1640913. Epub 2019 July 24. PMID: 31272252.

Liu, C., & Liu, C. (2023). Exploring plastic-management policy in China: Status, challenges, and policy insights. *Sustainability*, 15(11), 9087. https://doi.org/10.3390/su15119087.

Liu, L., Xu, K., Zhang, B., Ye, Y., Zhang, Q., & Jiang, W. (2021). Cellular internalization and release of polystyrene microplastics and nanoplastics. *Sci. Total Environ.*, 779, 146523. doi: 10.1016/j.scitotenv.2021.146523. Epub 2021 Mar 18. PMID: 34030247.

Liu, Q., Chen, C., Li, M., Ke, J., Huang, Y., Bian, Y. et al. (2020). Neurodevelopmental toxicity of polystyrene nanoplastics in *Caenorhabditis elegans* and the regulating effect of presenilin. *ACS Omega*, 5(51), 33170–33177. doi: 10.1021/acsomega.0c04830. PMID: 33403278; PMCID: PMC7774258.

Liu, X., Zhao, Y., Dou, J., Hou, Q., Cheng, J., & Jiang, X. (2022). Bioeffects of inhaled nanoplastics on neurons and alteration of animal behaviors through deposition in the brain. *Nano. Lett.*, 22(3), 1091–1099. doi: 10.1021/acs.nanolett.1c04184. Epub 2022 Jan 28. PMID: 35089039.

Llorca, M., & Farré, M. (202). Current insights into potential effects of micro-nanoplastics on human health by *in-vitro* tests. *Front. Toxicol.*, 3, 752140. doi: 10.3389/ftox.2021.752140. PMID: 35295102; PMCID: PMC8915894.

Lu, K., Lai, K., Stoeger, T., Ji, S., Lin, Z., Lin, X. et al. (2021). Detrimental effects of microplastic exposure on normal and asthmatic pulmonary physiology. *J. Hazard. Mat.*, 416, 126069. 10.1016/j.jhazmat.2021.126069.

Luo, T., Wang, C., Pan, Z., Jin, C., Fu, Z., & Jin, Y. (2019a). Maternal polystyrene microplastic exposure during gestation and lactation altered metabolic homeostasis in the dams and their F1 and F2 offspring. *Environ. Sci. Technol.*, 53(18), 10978–10992. doi: 10.1021/acs.est.9b03191. Epub 2019 Sep 4. PMID: 31448906.

Lwanga Huerta, E., Gertsen, H., Gooren, H., Peters, P., Salánki, T., van der Ploeg, M. et al. (2016). Microplastics in the terrestrial ecosystem: Implications for *Lumbricus terrestris* (*Oligochaeta*, Lumbricidae). *Environ. Sci. Technol.*, 50(5), 2685–2691. doi: 10.1021/acs.est.5b05478. Epub 2016 Feb 8. PMID: 26852875.

Lwanga Huerta, E., Gertsen, H., Gooren, H., Peters, P., Salánki, T., van der Ploeg, M. et al. (2017). Incorporation of microplastics from litter into burrows of *Lumbricus terrestris*. *Environ. Pollut.*, 220(Pt A), 523–531. doi: 10.1016/j.envpol.2016.09.096. Epub 2016 Oct 8. PMID: 27726978.

Ma, C., Li, L., Chen, Q., Lee, J.-S., Gong, J., & Shi, H. (2021). Application of internal persistent fluorescent fibers in tracking microplastics *in vivo* processes in aquatic organisms. *J. Hazard. Mater.*, 401, Article 123336.

Magrì, D., Sánchez-Moreno, P., Caputo, G., Gatto, F., Veronesi, M., Bardi, G. et al. (2018). Laser ablation as a versatile tool to mimic polyethylene terephthalate nanoplastic pollutants: Characterization and toxicology assessment. *ACS Nano.*, 12, 7690–7700. doi: 10.1021/acsnano.8b01331.

Maharjan, K.C.P.B., Acharya, A.M., Lee, D., Kusma, S., Gautam, R., Kwon, J.-T. et al. (2023). Polytetrafluorethylene microplastic particles mediated oxidative stress, inflammation, and intracellular signaling pathway alteration in human-derived cell lines. *Sci. Total Environ.*, 897, 165295. doi: 10.1016/j.scitotenv.2023.165295.

Mahmud, F., Sarker, D.B., Jocelyn, J.A., & Sang, Q.A. (2024). Molecular and cellular effects of microplastics and nanoplastics: Focus on inflammation and senescence. *Cells*, 13(21), 1788. doi: 10.3390/cells13211788.

Mak, C.W., Ching-Fong Yeung, K., & Chan, K.M. (2019). Acute toxic effects of polyethylene microplastic on adult zebrafish. *Ecotoxicol. Environ. Saf.*, 182(2019), Article 109442.

Mattioda, V., Benedetti, V., Tessarolo, C., Oberto, F., Favole, A., Gallo, M. et al. (2023). Pro-inflammatory and cytotoxic effects of polystyrene microplastics on human and murine intestinal cell lines. *Biomolecules*, 13, 140. doi: 10.3390/biom13010140.

Mattsson, K., Johnson, E.V., Malmendal, A., Linse, S., Hansson, L.A., & Cedervall, T. (2017). Brain damage and behavioural disorders in fish induced by plastic nanoparticles delivered through the food chain. *Sci. Rep.*, 7(1), 11452. doi: 10.1038/s41598-017-10813-0. PMID: 28904346; PMCID: PMC5597631.

McCarthy, J., Gong, X., Nahirney, D., Duszyk, M., & Radomski, M. (2011). Polystyrene nanoparticles activate ion transport in human airway epithelial cells. *Int. J. Nanomedicine*, 6, 1343–1356. doi: 10.2147/IJN.S21145. Epub 2011 Jun 28. PMID: 21760729; PMCID: PMC3133525.

Mintenig, S.M., Löder, M.G.J., Primpke, S., & Gerdts, G. (2019). Low numbers of microplastics detected in drinking water from ground water sources. *Sci. Total Environ.*, 648, 631–635. doi: 10.1016/j.scitotenv.2018.08.178. Epub 2018 Aug 16. PMID: 30121540.

Møller, P., & Roursgaard, M. (2023). Exposure to nanoplastic particles and DNA damage in mammalian cells. *Mutat. Res. Rev.*, 792, 108468. doi: 10.1016/j.mrrev.2023.108468. Epub 2023 Sep 4. PMID: 37666295.

Na, S.H., Kim, M.J., Kim, J.T., Jeong, S., Lee, S., Chung, J., & Kim, E.J. (2021). Microplastic removal in conventional drinking water treatment processes: Performance, mechanism, and potential risk. *Water Res.*, 202, 117417. doi: 10.1016/j.watres.2021.117417. Epub 2021 Jul 7. PMID: 34271456.

Nich, C., & Goodman, S.B. (2014). Role of macrophages in the biological reaction to wear debris from joint replacements. *J. Long Term Eff. Med. Implants*, 24(4), 259–265. doi: 10.1615/jlongtermeffmedimplants.2014010562. PMID: 25747029; PMCID: PMC4366682.

OECD. (2022). *Global Plastics Outlook Economic Drivers, Environmental Impacts, and Policy Options*. https://doi.org/10.1787/de747aef-en.

Ohkubo, N., Ito, M., Hano, T., Kono, K., & Mochida, K. (2020). Estimation of the uptake and gut retention of microplastics in juvenile marine fish: Mummichogs (*Fundulus heteroclitus*) and red sea breams (*Pagrus major*). *Mar. Pollut. Bull.*, 160, 111630. doi: 10.1016/j.marpolbul.2020.111630. Epub 2020 Sep 7. PMID: 32911116.

Oliveri Conti, G., Ferrante, M., Banni, M., Favara, C., Nicolosi, I., Cristaldi, A. et al. (2020). Micro- and nano-plastics in edible fruit and vegetables. The first diet risks assessment for the general population. *Environ. Res.*, 187, 109677. doi: 10.1016/j.envres.2020.109677. Epub 2020 May 20. PMID: 32454310.

Oslakovic, C., Cedervall, T., Linse, S., & Dahlbäck, B. (2012). Polystyrene nanoparticles affecting blood coagulation. *Nanomedicine Nanotechnol. Biol. Med.*, 8, 981–986.

Park, E.J., Han, J.S., Park, E.J., Seong, E., Lee, G.H., Kim, D.W. et al. (2020). Repeated-oral dose toxicity of polyethylene microplastics and the possible implications on reproduction and development of the next generation. *Toxicol. Lett.*, 324, 75–85. doi: 10.1016/j.toxlet.2020.01.008. Epub 2020 Jan 16. PMID: 31954868.

Peng, C., He, N., Wu, Y., Lu, Y., Sun, H., & Wang, L. (2022). Excretion characteristics of nylon microplastics and absorption risk of nanoplastics in rats. *Ecotoxicol. Environ. Saf.*, 238, 113586. doi: 10.1016/j.ecoenv.2022.113586. Epub 2022 May 2. PMID: 35512478.

Perera, K., Ziajahromi, S., Nash, S.B., & Leusch, F.D.L. (2023). Microplastics in Australian indoor air: Abundance, characteristics, and implications for human exposure. *Sci. Total Environ.*, 889, 164292. doi: 10.1016/j.scitotenv.2023.164292. Epub 2023 May 19. PMID: 37211105.

Pham, D.T., Kim, J., Lee, S.H., Kim, J., Kim, D., Hong, S. et al. (2023). Analysis of microplastics in various foods and assessment of aggregate human exposure via food consumption in Korea. *Environ. Pollut.*, 322, 121153. doi: 10.1016/j.envpol.2023.121153. Epub 2023 Jan 25. PMID: 36709032.

Pitt, J.A., Kozal, J.S., Jayasundara, N., Massarsky, A., Trevisan, R., Geitner, N. et al. (2018). Uptake, tissue distribution, and toxicity of polystyrene nanoparticles in developing zebrafish (Danio rerio). *Aquat. Toxicol.*, 194, 185–194. doi: 10.1016/j.aquatox.2017.11.017. Epub 2017 Nov 24. PMID: 29197232; PMCID: PMC6959514.

Pontecorvi, P., Ceccarelli, S., Cece, F., Camero, S., Lotti, L.V., Niccolai, E. et al. (2023). Assessing the impact of polyethylene nano/microplastic exposure on human vaginal keratinocytes. *Int. J. Mol. Sci.*, 24(14) 11379. doi: 10.3390/ijms241411379. PMID: 37511139; PMCID: PMC10380279.

Prata, J.C., da Costa, J.P., Lopes, I., Duarte, A.C., & Rocha-Santos, T. (2020). Environmental exposure to microplastics: An overview on possible human health effects. *Sci. Total Environ.*, 702, Article 134455.

Prata J.C., & Barone, S. (2022). Microplastics and human health: Integrating pharmacokinetics. *Crit. Rev. Environ. Sci. Technol.*, 53(2022), 1489–1511.

Prietl, B., Meindl, C., Roblegg, E., Pieber, T.R., Lanzer, G., & Fröhlich, E. (2014). Nano-sized and micro-sized polystyrene particles affect phagocyte function. *Cell Biol. Toxicol.*, 30(1), 1–16. doi: 10.1007/s10565-013-9265-y. Epub 2013 Nov 29. PMID: 24292270; PMCID: PMC4434214.

Prüst, M., Meijer, J., & Westerink, R.H.S. (2020). The plastic brain: Neurotoxicity of micro- and nanoplastics. *Part. Fibre Toxicol.*, 17(1), 24. doi: 10.1186/s12989-020-00358-y. PMID: 32513186; PMCID: PMC7282048.

Qu, M., Liu, Y., Xu, K., & Wang, D. (2019). Activation of p38 MAPK signaling-mediated endoplasmic reticulum unfolded protein response by nanopolystyrene particles. *Adv. Biosyst.*, 3(4), e1800325. doi: 10.1002/adbi.201800325. Epub 2019 Mar 14. PMID: 32627431.

Revel, M., Roman, C., & Châtel, A. (2021). Is cell culture a suitable tool for the evaluation of micro- and nanoplastics ecotoxicity? *Ecotoxicology*, 30(3), 421–430. doi: 10.1007/s10646-021-02355-z. Epub 2021 Feb 12. PMID: 33580466.

Rist, S., & Hartmann, N.B. (2017). Aquatic ecotoxicity of microplastics and nanoplastics: Lessons learned from engineered nanomaterials. *In:* Wagner, M., & Lambert, S. (Eds.), *Freshwater Microplastics: Emerging Environmental Contaminants* (pp. 25–49). Springer. (*The Handbook of Environmental Chemistry*, Vol. 58). doi: 10.1007/978-3-319-61615-5_2.

Rosellini, M., Schulze, A., Omer, E.A., Ali, N.T., Marini, F., Küpper, J.H., & Efferth, T. (2023). The effect of plastic-related compounds on transcriptome-wide gene expression on CYP2C19-overexpressing HepG2 cells. *Molecules*, 28(16), 5952. doi: 10.3390/molecules28165952. PMID: 37630204; PMCID: PMC10459118.

Rossi, G., Barnoud, J., & Monticelli, L. (2014). Polystyrene nanoparticles perturb lipid membranes. *J. Phys. Chem. Lett.*, 5(1), 241–246. doi: 10.1021/jz402234c. Epub 2013 Dec 23. PMID: 26276207.

Roursgaard, M., Hezareh Rothmann, M., Schulte, J., Karadimou, I., Marinelli, E., & Møller, P. (2022). Genotoxicity of particles from grinded plastic items in Caco-2 and HepG2 Cells. *Front. Public Health*, 10, 906430. doi: 10.3389/fpubh.2022.906430. PMID: 35875006; PMCID: PMC9298925.

Saftig, P., & Haas, A. (2016). Turn up the lysosome. *Nat. Cell. Biol.*, 18(10), 1025–1027. doi: 10.1038/ncb3409. PMID: 27684505.

Santana, M.F.M., Dawson, A.L., Motti, C.A., van Herwerden, L., Lefevre, C. & Kroon, F.J. (2021). Ingestion and depuration of microplastics by a Planktivorous Coral Reef

Fish (*Pomacentrus amboinensis*). *Front. Environ. Sci.*, 9, 641135. doi: 10.3389/fenvs.2021.641135.

Schwarzfischer, M., & Rogler, G. (2022). The intestinal barrier-shielding the body from nano- and microparticles in our diet. *Metabolites*, 12(3), 223. doi: 10.3390/metabo12030223. PMID: 35323666; PMCID: PMC8952728.

Shafea, L., Yap, J., Beriot, N., Felde, V.J., Okoffo, E.D., Enyoh, C.E., & Peth, S. (2023). Microplastics in agroecosystems: A review of effects on soil biota and key soil functions. *J. Plant Nutr. Soil Sci.*, 186(1), 5–22. doi: 10.1002/jpln.202200136.

Sharma, C., Kim, S., Nam, Y., Jung, U.J., & Kim, S.R. (2021). Mitochondrial dysfunction as a driver of cognitive impairment in Alzheimer's Disease. *Int. J. Mol. Sci.*, 2, 4850. doi: 10.3390/ijms22094850.

Shen, F., Li, D., Guo, J., & Chen J. (2022). Mechanistic toxicity assessment of differently sized and charged polystyrene nanoparticles based on human placental cells. *Water Res.*, 223, 118960. doi: 10.1016/j.watres.2022.118960.

Shen, R., Yang, K., Cheng, X. et al. *(2022).* Accumulation of polystyrene microplastics induces liver fibrosis by activating cGAS/STING pathway. *Environ. Pollut.*, 300(2022), Article 118986.

Shruti, V.C., Pérez-Guevara, F., & Kutralam-Muniasamy, G. (2020). Metro station free drinking water fountain: A potential "microplastics hotspot" for human consumption. *Environ Pollut.*, 261, 114227. doi: 10.1016/j.envpol.2020.114227. Epub 2020 Feb 20. PMID: 32113111.

Song, W., Popp, L., Yang, J., Kumar, A., Gangoli, V.S., & Segatori, L. (2015). The autophagic response to polystyrene nanoparticles is mediated by transcription factor EB and depends on surface charge. *J. Nanobiotechnology*, 13, 87. doi: 10.1186/s12951-015-0149-6. PMID: 26596266; PMCID: PMC4657241.

Stabnikova, O., Stabnikov, V., Marinin, A., Klavins, M., Klavins, L., & Vaseashta, A. (2021). Microbial life on the surface of microplastics in natural waters. *Applied Sciences*, 11(24), 11692. https://doi.org/10.3390/app112411692.

Sun, R., Liu, M., Xiong, F., Xu, K., Huang, J., Liu, J. et al. (2024). Polystyrene micro- and nanoplastics induce gastric toxicity through ROS mediated oxidative stress and P62/Keap1/Nrf2 pathway. *Sci. Total Environ.*, 912, 169228. doi: 10.1016/j.scitotenv.2023.169228.

Syberg, K., Hansen, S.F., Christensen, T.B., & Khan, F.R. (2018). Risk perception of plastic pollution: Importance of stakeholder involvement and citizen science. *In:* Wagner, M., & Lambert, S. (Eds.), *Freshwater Microplastics: The Handbook of Environmental Chemistry* (58 pp.). Springer, Cham. https://doi.org/10.1007/978-3-319-61615-5_10.

Syberg, K., Nielsen, M.B., Westergaard Clausen, L.P., Calster, G., Wezel, A. et al. (2021). Regulation of plastic from a circular economy perspective. *Current Opinion in Green and Sustainable Chemistry*, 29, 100462. doi: 10.1016/j.cogsc.2021.100462.

Thorat, K., Pandey, S., Chandrashekharappa, S. et al. (2018). Prevention of pesticide-induced neuronal dysfunction and mortality with nucleophilic poly-Oxime topical gel. *Sci. Adv.*, 4, Article eaau1780.

Tsochatzis, E.D., Gika, H., Theodoridis, G., Maragou, N., Thomaidis, N., & Corredig, M. (2024). Microplastics and nanoplastics: Exposure and toxicological effects require important analysis considerations. *Heliyon*, 10(11), e32261. doi: 10.1016/j.heliyon.2024.e32261. PMID: 38882323; PMCID: PMC11180319.

van Emmerik, T., & Schwarz, A. (2020). Plastic debris in rivers. *Wiley Interdisciplinary Reviews: Water*. 7(1), 1–24. e1398. doi: 10.1002/wat2.1398.

Vaseashta, A., Ivanov, V., Stabnikov, V., & Marinin, A. (2021). Environmental safety and security investigations of neustonic microplastic aggregates near water-air interphase. *Polish Journal of Environmental Studies*, 30(4), 3457–3469. https://doi.org/10.15244/pjoes/131947

Vaseashta, A., Duca, G., & Travin, S.O. (Eds.). (2022). *Handbook of Research on Water Sciences and Society*. Engineering Science Reference. DOI: 10.4018/978-1-7998-7356-3.

Walczak, A.P., Kramer, E., Hendriksen, P.J.M. et al. (2015). Translocation of differently sized and charged polystyrene nanoparticles in *in vitro* intestinal cell models of increasing complexity. *Nanotoxicology*, 9, 453–461.

Wang, F., Yu, L., Monopoli, M.P., Sandin, P., Mahon, E., Salvati, A., & Dawson, K.A. (2013). The biomolecular corona is retained during nanoparticle uptake and protects the cells from the damage induced by cationic nanoparticles until degraded in the lysosomes. *Nanomedicine*, 9(8), 1159–1168. doi: 10.1016/j.nano.2013.04.010. Epub 2013 May 7. PMID: 23660460.

Wang, F., Salvati, A., & Boya, P. (2018). Lysosome-dependent cell death and deregulated autophagy induced by amine-modified polystyrene nanoparticles. *Open Biol.*, 8(4), 170271. doi:10.1098/rsob.170271.

Wang, K., Du, Y., Li, P., Guan, C., Zhou, M., Wu, L. et al. (2024). Nanoplastics cause heart aging/myocardial cell senescence through the Ca^{2+}/mtDNA/cGAS-STING signaling cascade. *J. Nanobiotechnol.*, 22, 96. doi: 10.1186/s12951-024-02375-x.

Wei, J., Wang, X., Liu, Q. et al. (2021). The impact of polystyrene microplastics on cardiomyocytes pyroptosis through NLRP3/Caspase-1 signaling pathway and oxidative stress in Wistar rats. *Environ. Toxicol.*, 36, 935–944.

Wick, P., Malek, A., Manser, P. et al. (2010). Barrier capacity of human placenta for nanosized materials. *Environ. Health Perspect.*, 118, 432–436.

Wootton, N., Reis-Santos, P., Dowsett, N., Turnbull, A., & Gillanders, B.M. (2021). Low abundance of microplastics in commercially caught fish across southern Australia. *Environ. Pollut.*, 2021, 290. doi: 118030, 10.1016/j.envpol.2021.118030.

Wu, D., Zhang, M., Bao, T.T., & Lan, H. (2023). Long-term exposure to polystyrene microplastics triggers premature testicular aging. *Part. Fibre Toxicol.*, 20, 35. doi: 10.1186/s12989-023-00546-6.

Xia, L., Alqahtani, S., Ferreira, C.R., Aryal, U.K., Biggs, K., & Shannahan, J.H. (2022). Modulation of pulmonary toxicity in metabolic syndrome due to variations in iron oxide nanoparticle-bio corona composition. *Nanomaterials* (*Basel*). 12(12), doi: 10.3390/nano12122022. PMID: 35745361; PMCID: PMC9230893.

Xia, T., Kovochich, M., Liong, M., Zink, J.I., & Nel, A.E. (2008). Cationic polystyrene nanosphere toxicity depends on cell-specific endocytic and mitochondrial injury pathways. *ACS Nano.*, 2(1), 85–96. doi: 10.1021/nn700256c. PMID: 19206551.

Xia, W., Rao, Q., Deng, X., Chen, J., & Xie, P. (2020). Rainfall is a significant environmental factor of microplastic pollution in inland waters. *Sci. Total Environ.*, 732, 139065. doi: 10.1016/j.scitotenv.2020.139065. Epub 2020 Apr 29. PMID: 32422477.

Xie, X., Deng, T., Duan, J., Xie, J., Yuan, J., &, Chen, M. (2020). Exposure to polystyrene microplastics causes reproductive toxicity through oxidative stress and activation of the p38 MAPK signaling pathway. *Ecotoxicol. Environ. Saf.*, 190, 110133. doi: 10.1016/j.ecoenv.2019.110133. Epub 2019 Dec 30. PMID: 31896473.

Xu, B., Liu, F., Cryder, Z., Huang, D., Lu, Z., He, Y., & Xu, J. (2020). Microplastics in the soil environment: Occurrence, risks, interactions, and fate: A review. *Crit. Rev.*

Environ. Sci. Technol., 50(21), 2175–2222. https://doi.org/10.1080/10643389.2019.1694822.

Yang, Q., Dai, H., Cheng, Y., Wang, B., Xu, J., Zhang, Y. et al. (2023). Oral feeding of nanoplastics affects brain function of mice by inducing macrophage IL-1 signal in the intestine. *Cell Rep.*, 42(4), 112346. doi: 10.1016/j.celrep.2023.112346. Epub 2023 Apr 4. PMID: 37022934.

Yang, S., Zhang, T., Ge, Y., Yin, L., Pu, Y., & Liang, G. (2024). Inhalation exposure to polystyrene nanoplastics induces chronic obstructive pulmonary disease-like lung injury in mice through multidimensional assessment. *Environ. Pollut.*, 347, 123633. doi: 10.1016/j.envpol.2024.123633. Epub 2024 Feb 27. PMID: 38423272.

Yang, Y., Xie, E., Du, Z. et al. (2023). Detection of various microplastics in patients undergoing cardiac surgery. *Environ. Sci. Technol.*, 57, 10911–10918.

Yao, Y., Glamoclija, M., Murphy, A., & Gao, Y. (2022). Characterization of microplastics in indoor and ambient air in northern New Jersey. *Environ. Res.*, 207, 112142. doi: 10.1016/j.envres.2021.112142. Epub 2021 Sep 29. PMID: 34597660.

Yarjanli, Z., Ghaedi, K., Esmaeili, A., Rahgozar, S., & Zarrabi, A. (2017). Iron oxide nanoparticles may damage to the neural tissue through iron accumulation, oxidative stress, and protein aggregation. *BMC Neurosci.*, 18, 51. doi: 10.1186/s12868-017-0369-9.

Yong, C.Q.Y., Valiyaveettil. S., & Tang, B.L. (2020). Toxicity of microplastics and nanoplastics in mammalian systems. *Int. J. Environ. Res. Public Health*, 17(5), 1509. doi: 10.3390/ijerph17051509. PMID: 32111046; PMCID: PMC7084551.

Yousif, E., & Haddad, R. (2013). Photodegradation and photostabilization of polymers, especially polystyrene. *SpringerPlus*, 2, 398. doi: 10.1186/2193-1801-2-398.

Yuan, W., Liu, X., Wang, W., Di, M., & Wang, J. (2019). Microplastic abundance, distribution, and composition in water, sediments, and wild fish from Poyang Lake, China. *Ecotoxicol. Environ. Saf.*, 170, 180–187. doi: 10.1016/j.ecoenv.2018.11.126. Epub 2018 Dec 5. PMID: 30529617.

Yuan, Z., Nag, R., & Cummins, E. (2022). Human health concerns regarding microplastics in the aquatic environment: From marine to food systems. *Sci. Total Environ.*, 823, 153730. doi: 10.1016/j.scitotenv.2022.153730. Epub 2022 Feb 7. PMID: 35143789.

Zhang, D., Liu, X., Huang, W., Li, J., Wang, C., Zhang, D., & Zhang, C. (2020). Microplastic pollution in deep-sea sediments and organisms of the Western Pacific Ocean. *Environ. Pollut.*, 259, 113948. doi: 10.1016/j.envpol.2020.113948. Epub 2020 Jan 10. PMID: 32023798.

Zhang, S., Wang, J., Liu, X., Qu, F., Wang, X., Wang, X. et al. (2019). Microplastics in the environment: A review of analytical methods, distribution, and biological effects. *TrAC Trends Anal. Chem.*, 111, 62–72. doi:10.1016/j.trac.2018.12.002

Zhang, Z., Chen, W., Chan, H., Peng, J., Zhu, P., Li, J. et al. (2024). Polystyrene microplastics induce size-dependent multi-organ damage in mice: Insights into gut microbiota and fecal metabolites. *J. Hazard. Mater.*, 461, 132503. doi: 10.1016/j.jhazmat.2023.132503. Epub 2023 Sep 6. PMID: 37717443.

Zhang, Z., Huang, Q., Zhao, D., Lian, F., Li, X., & Qi, W. (2023). The impact of oxidative stress-induced mitochondrial dysfunction on diabetic microvascular complications. *Front. Endocrinol.*, 14, 1112363. doi: 10.3389/fendo.2023.1112363.

Zhou, J., Wen, Y., Marshall, M.R., Zhao, J., Gui, H., Yang, Y., & Zang, H. (2021). Microplastics as an emerging threat to plant and soil health in agroecosystems. *Sci. Total Environ.*, 787. doi: 10.1016/j.scitotenv.2021.147444.

Zhu, J., Dong, G., Feng, F., Ye, J., Liao, C.H., Wu, C.H., & Chen, S.C. (2023). Microplastics in the soil environment: Focusing on the sources, its transformation, and change in morphology. *Sci. Total Environ.*, 896, 165291. doi: 10.1016/j.scitotenv.2023.165291. Epub 2023 July 3. PMID: 37406689.

Zhu, K., Jia, H., Jiang, W., Sun, Y., Zhang, C., Liu, Z. et al. (2021). The first observation of the formation of persistent aminoxyl radicals and reactive nitrogen species on photo-irradiated nitrogen-containing microplastics. *Environ. Sci. Technol.*, 56, 779–789. doi: 10.1021/acs.est.1c05650.

8 | Micro Plastic Hazards in Agricultural Ecosystems: Management and Policies for Sustainable Agriculture

Manjistha Baruah[1], Rahul Gogoi[2], Madhurjya Ranjan Sarmah[1], Oliva Saha[1,2] and Madhumita Barooah[1,2*(0000-0002-3768-5336)]

[1] DBT – Northeast Centre for Agricultural Biotechnology, Assam Agricultural University, Jorhat – 785013, Assam, India

[2] Department of Agricultural Biotechnology, Assam Agricultural University, Jorhat – 785013, Assam, India

Abstract: The advent of industrialization goes hand in hand with environmental degradation. A major element that has managed to penetrate all aspects of the ecosystem is plastic. The release of toxic fumes as by-products coupled with their non-biodegradable nature adds to their tenacity in the environment. It has been reported that about 8 million tonnes of plastic are deposited in oceans every year and by 2025 over 250 million tonnes of plastics are estimated to be accumulated. Terrestrial environments are an even larger sink with 4–23 times more plastic buildup. These plastic wastes are decomposed or disintegrated into smaller fragments by mechanical abrasion, ultraviolet degradation, oxidation, and biotical action. Micro- and nano plastics (MNPs) are broken-down plastics with sizes 100 nm–5 mm, and <100 nm, respectively. They accumulate in the terrestrial environment through littering, atmospheric deposition, irrigation and flooding, agricultural mulching, and compost and sewage sludge application which ultimately disrupt the congruency in a soil-plant system. This sets off a series of adverse effects on soil in relation to its water-holding capacity and bulk density properties, triggers oxidative stress on plants affecting root development and crop biomass, and finally causes an imbalance in the abundance, diversity and activity of microbes associated with it. The current chapter is an attempt to

*Corresponding author: madhumita.barooah@aau.ac.in

understand the nature of MNPs and their transportation in the soil provoking an avalanche of detrimental effects scarring soil health, causing disparity in crop growth and development, and toppling the very composition of soil microflora.

1. Introduction

Plastics have revolutionized human activities due to their affordability and versatility, replacing traditional materials like wood and metals. The creation of synthetic plastic by Leo Baekeland in 1907 marked a pivotal moment, which saw a meteoric surge in plastic production and usage. The world, at present, produces over 450 million tons of plastic annually which finds uses in various spheres of our lives. Out of this, more than 280 million tons are short-lived plastics which become waste and contaminate our environment. Various sectors such as packaging, construction, fishing, automotive, electronics, and agriculture significantly contribute to plastic depositions. The recent COVID-19 pandemic witnessed a higher usage of single-use plastics including protective kits, facemasks, and gloves leading to exacerbating this issue (Wang et al., 2022). Industrialization has introduced plastics into every facet of the ecosystem, causing widespread environmental degradation. Concerns arise from the release of toxic by-products and the non-biodegradable nature of plastics, resulting in approximately 8 million tons deposited in oceans annually and an estimated accumulation of over 250 million tons of plastics by 2025. Terrestrial environments, acting as significant sinks, accumulate 4–23 times more plastic. The breakdown of plastic waste into smaller fragments, including micro- and nano-plastics (MNPs), poses additional challenges. Despite their invisible nature, they infiltrate air, soil, water, and food, to cause ecotoxicological effects. Their miniscule size along with varying chemical composition facilitates their transportation to terrestrial ecosystems and aquatic systems and therefore they have been placed on the list of ten emergent pollutants by the United Nations Environment Programme (UNEP) in 2014 (Lam et al., 2018; Hartmann et al., 2019; Karasik et al., 2020). Microplastics (MPs), particles smaller than 5 mm in dimension, have become ubiquitous pollutants, contaminating diverse environments including oceans, rivers, lakes, soils, and even the atmosphere (Horton et al., 2017; Tang, 2021). Particularly alarming is the substantial release of plastics into agricultural soils, estimated to be 4–23 times higher than in aquatic environments, and European farmland receiving a significant amount ranging from 125–850 tons (Nizzetto et al., 2016).

These MPs enter farmland soils through various pathways, originating as primary particles from sources like seed and agrochemical coating, or as secondary particles arising out from the weathering of larger plastic items such as agricultural materials (Wu et al., 2019). Once introduced into the soil, MPs exhibit remarkable persistence, enduring for decades due to their resistance to degradation. They can move through soil layers via agricultural practices and soil biota, potentially disrupting soil health and function (Fei et al., 2020). MNPs in soil profoundly disrupt soil nutrient cycling, impair plant nutrient acquisition, and alter soil microbial composition, resulting in adverse effects on crop yield and quality, proving to be hazardous to human health. MPs have been observed to attach to seed coat and root surfaces or accumulate in their vascular systems, hindering the uptake of water and essential nutrients by crops (Ju et al., 2021). Additionally, MPs induce oxidative damage to crops, disrupting metabolic activities necessary for crop growth (de Souza Machado et al., 2019). Chemical additives released from MPs can further damage by inducing cytotoxicity and genotoxicity in crops (Chang et al., 2022). MPs influence the availability of water and nutrients deemed essential for crop growth through the modification of the biotic and abiotic environments of the soil (Fei et al., 2020). This impacts the enzymatic profile of the soil, altering the interactions involving plants and soil. Production of crops can be hampered by the combined toxicity of MPs and other prevalent pollutants that exist in agricultural soil, which poses significant barriers to agricultural productivity and food security (Qi et al., 2018). Furthermore, various shapes and sizes along with different polymer compositions, have unwanted effects on soil conditions and crop health (Chang et al., 2022; Wang et al., 2022).

The investigation into the impact of MPs on soil is relatively nascent, with studies emerging around 2012. Existing research suggests that MPs have the potential to modify soil properties and possibly influence crop health and yield (Rillig, 2012; Boots et al., 2019; de Souza Machado et al., 2019; Wang et al., 2019; Lehmann et al., 2020; Zhao et al., 2021). However, the research findings are often limiting and conflicting in nature. Given the critical role of agroecosystems in food production and ecosystem services, comprehending the ecological implications of MPs in agricultural soils is imperative for ensuring future sustainability (Rillig & Lehmann, 2020; Wang et al., 2022). This chapter discusses the sources of MPs in an agroecosystem, their distribution and transportation patterns, and how they impact soil health, crop growth, and changes in the soil microbial community. Additionally, it also brings into focus some of the regulations and policies regarding the usage of MPs to promote sustainable agriculture.

2. Source of MPs in Agricultural Soils

The usage of plastic has become an integral part of modern farming practices. Within the agricultural value chains, 80% of plastic is consumed during crop cultivation and livestock, 18% and 2% in aquaculture and forestry. The agricultural value chains and food packaging consume annually another 12.5 Mt and 37.5 Mt of plastics, respectively (FAO, 2021). The production of agricultural and horticultural crops and livestock makes extensive use of plastic as protective films during fumigation, silage preparation, bale wrap films, protective films for mulching, nurseries, wind tunnels, greenhouses, direct cover, and non-woven floating cover, nets (ant-hail, anti-bird, wind-breaking, and shading nets), twine, and pipes for irrigation and drainage. Plastics polymers such as polystyrene (PS), ethylene-vinyl acetate copolymer, and polyethylene terephthalate (PET) find frequent uses in the agricultural sectors, followed by polyethylene (PEs) of both low- and high-density, polypropylene (PP), and PVC (Sarkar et al., 2018). Over time, the combined effects of wind, water, weathering, and UV light break down these plastic materials to various sizes, leading to the formation of MPs, and MNPs. The primary sources of MPs and MNPs stem from the common practice of using biosolids (processed sewage sludge) and compost, plastic mulching films, water pipes, plastic greenhouse covers, polymer-based fertilizers, and pesticides (Koutink et al., 2021; Corradini et al., 2019; Steinmetz et al., 2016; Wang et al., 2019). Plastic films are often used as mulching covering to conserve soil moisture, regulate soil temperature, and suppress weed growth. The plastic film, viz., polyvinyl chloride (PVC), has improved water use efficiency and agricultural growth output. However, the use of thin film and mechanized cultivation results in higher residual levels of plastic mulch, which interfere with the movement of nutrients and moisture, slow down root growth, decrease seed germination, cause salinization, and build up toxic chemicals like ketones, aldehydes, and phthalate esters in soils (Liu et al., 2014). Agroecosystems' varied responses may be adversely affected when MPs accumulate in farmed areas. Although various factors such as wind and water reduce plastic residues from the soil surface, a significant number of MPs remain and subsequently penetrate to the soil layers (Rillig et al., 2017). In Europe and North America, high (50%) usage of processed sewage for farming has alarmingly increased the load of MPs amounting to 63,000–430,000 tons per year. Plastic contaminants can enter agricultural production systems through non-agricultural sources such as contaminated water, air, and waste, as well as broken or deteriorating agricultural plastic products. The use of treated wastewater to irrigate farmlands is another source of MPs in farmlands. About 99% or more of MP contaminants can be removed from the existing wastewater treatment (WWT) facilities (Corradini et al., 2019). However, the recovered MPs remain in sludge and can contaminate agricultural soils (Li et al., 2018). Waste management sludge is often spread directly onto agricultural soils in several countries across the globe. It is reported that 87% of the biosolids generated by WWT facilities in Portugal are used in agricultural practices (Alvarenga et al., 2016). The amount of sludge applied in the farmed land is positively correlated to the amount of MP burden in the soil and thus, MP pollution from urban wastewater to agricultural ecosystems *via* biosolids needs attention (Nizzetto et al., 2016). Plastic packaging and wrappers often find their way into the compost pits and despite picking up the large and medium-sized plastic contaminants from composting, the probability of secondary MPs being created during the processing still exists (Bolan & Bradney, 2019). The extensive usage of plastic on pastures and agricultural land is giving rise to secondary MPs which have become the main cause of residual plastic in farmlands. Apart from plastic mulch films, polymer coatings for seeds to enhance germination and for fertilizers to boost nutrient utilization efficiency and lower the risk of runoff and emissions are the primary sources of MPs in modern agriculture (Accinelli et al., 2016; Gil-Ortiz et al., 2020).

Plastic materials offer easier storage, greater safety, and transport of bulk agrochemicals. They are also used as packaging material to transport and preserve the quality of agricultural produce leading them to find their way into agricultural soils. The extensive and continuous usage of polymer-coated seeds and fertilizers will eventually impact the soil's physical, chemical, and biological properties. Several factors including, the physical and chemical characteristics of MPs, prevailing environmental conditions, topography, and the transport of MPs influence the MP burden (He et al., 2018; Karbalaei et al., 2018; Oliviero et al., 2019). High levels of synthetic fibers ranging from 1.25–3.50 picogram were reported in farmlands following a one to five-year sludge amendment, indicating their persistence in the soils (Zubris & Richards, 2005; Corradini et al., 2019). The major paths of plastic pollution can vary depending on the location. Most agricultural plastics are single-use products with short shelf lives that break down into garbage after a year. Inadequate waste management and disposal will lead to an increase in secondary MPs pollution of the soil. Depending on the sampling sites, the climate, and other environmental conditions including the usage of plastic (agricultural intensification), the number of MPs and MNPs may change. Risk and toxicological assessments are essential due to the variety of polymers and additives used for various applications. There is a wide range of variations in the toxicological characteristics and dangers of MPs to plants and animals due to the many types of polymers and additives used in agricultural plastics (Sarkar et al., 2018).

3. Distribution and Transport of MPs in Agricultural Land

MPs have been found to occur in different niches of ecosystems ranging from farmlands, urbanized landscapes, rural areas, high-altitude mountains, and remote areas and pose a serious and detrimental pollutant (Chen et al., 2020). The adoption and extensive use of plastic in modern farming systems have led to their increased abundance in soil and contamination in agricultural lands (Steinmetz et al., 2016; Zhang et al., 2018). Primary plastics such as mulching material, coated fertilizers, and pesticides, irrigation application of biosolids as organic amendments, silage, and bale wrap films are extensively used in agro-horticultural crop and livestock production. These primary forms of plastics get converted to secondary forms of plastics due to environment and microbial action resulting in widespread distribution and persistence in farmlands including, crop rhizosphere (Bala et al., 2023). A recent survey conducted by Yoon et al. (2024) in Korea to determine the distribution of MPs in different land use patterns reported the highest load (5047 particles kg^{-1}) in agricultural lands, indicating that present agricultural practices are one of the major sources of MPs pollution. A higher abundance of MPs in soils of intensively cultivated agricultural soil (~1883.33 number of MP particles per kg) was also reported in China (Wang et al., 2024). The majority of MPs found in soil occur on the top 30 cm, where they are influenced by tillage and harvesting techniques.

The amount of soil moisture, surface and colloid characteristics, and soil type all affect MP diffusivity, which can vary greatly. Soil texture, saturation, porosity, and ion strength function in concert to cause electrostatic interactions between soils and MPs (Wu et al., 2020). Moreover, environmental factors like rain and runoff affect MP transport. Rillig et al. (2017) reported that repeated wetting and drying cycles sped up PE and PP pristine pellets' migration downhill. The transportation of MPs is influenced by their size and shape. Larger particles may accumulate near the soil surface unless preferential flow paths or macropores are present, while smaller particles and irregularly shaped particles may entangle in the soil matrix to influence their transport behavior (Bläsing & Amelung, 2018). Additionally, particle shape influences transport, with fibrous MPs being retained more in soils as compared to spherical particles. Conceptual transport models serve as the foundation for understanding MPs exchange between terrestrial and aquatic ecosystems (Raza et al., 2022). The characteristics of MPs, including density, surface area, shape, size, and composition, affect behavior and may influence the soil environment. Earthworms and microarthropods create preferential flow paths, accelerating MP transport vertically into deeper soil layers. These pathways, coupled with weather conditions like wet-dry cycles, can facilitate MP transport even in the absence of biological activity (O'Connor et al., 2019). Soil organic matter enhances MP transport by promoting attachment onto mobile organic particles or by increasing overall MP mobility (Alimi et al., 2021). The characteristics of the organism as well as the particle size of MPs influence how they will be transported by soil organisms. Smaller organisms facilitate the transport of small particles more effectively. Microbial activity on MP surfaces may lead to degradation and secondary MPs generation, contributing to further transport into soil and potentially groundwater (Danso et al., 2019). Electrostatic attraction among particles plays a role in transportation. Aged plastics exhibit higher electrostatic attraction among particles facilitating their aggregation and subsequent colonization by microorganisms which facilitates their movement. Ingestion by various soil invertebrates including earthworms, snails, larvae, nematodes, and mites is another predominant transport mechanism observed in soil. Additionally, earthworms mechanically introduce and transport MP into soil layers. Despite limited documentation, terrestrial vertebrates contribute to MP transport. Chickens and sheep ingest MP from contaminated environments, transferring them through defecation. Ungulates, like deer, carry MPs in their feces, while birds incorporate plastics into nests. Vultures transport MP from rubbish dumps to nesting sites. Percolating water plays an important part in the transport of MPs through soil profiles. Preferential flow pathways, such as macropores, facilitate rapid and deep MP transport. Leaching is another way via which MPs are released into groundwater (O'Connor et al., 2019). Several studies are reporting the movement of MPs and MNP particles through the soil by attaching mobile organic particles present in the soil, which not only facilitates but also expedites their transport (Rodriguez-Seijo et al., 2019). Plastic mobility in agricultural soils may exhibit variations under different conditions. Agricultural landscapes, susceptible to erosion, may serve as significant sources of MP to downstream ecosystems. However, research on the magnitude, timing, and drivers of plastic transport in these systems is lacking. Wind erosion facilitates the transport of lightweight plastic particles across terrestrial environments, particularly in arid regions. Wind action can disperse plastics over long distances, posing threats to remote ecosystems. Agricultural practices such as tillage can redistribute MPs within soils, affecting soil structure and composition, which, in turn, influences MPs transport dynamics (Dong et al., 2021a; Lozano et al., 2021; Wang et al., 2020). Chemical conditions in soils including pH and ionic strength also impact MP transport and retention, soils with higher pH and lower ionic strength increase the mobility of the MP (Wu et al., 2020).

4. Effect of MPs

4.1. Soil

MPs and MNPs have been reported to infiltrate agricultural soil and subsequently influence its physicochemical and biological properties (Lozano et al., 2021). The intensity of the effects on soil structure depends on the type of MPs and MNPs, aggregate size fraction, and plant presence. They can reach up to a concentration of 40,000 particles kg^{-1} in soil, thereby affecting soil bulk density, water-stable aggregates, and soil structure (Zhang et al., 2018). Polyester MP fibers (PES-MPs) and MP (PP-MPs) fibers were reported to significantly decrease the soil bulk density in soils grown with different crops (Zhou et al., 2023; de Souza Machado et al., 2018). The size of MP particles can modulate soil pore space and interactions, potentially impacting water retention capacity and hydraulic conductivity. Small particles may affect micropores, increasing water retention, while large particles tend to occupy small and medium pore spaces, enhancing hydraulic conductivity and water infiltration. Studies have indicated that MPs decrease saturated hydraulic conductivity and water-holding capacity, posing long-term threats to soil quality, microbial composition, and plant health (Guo et al., 2022). In contrast, MPs have been demonstrated to have a favorable impact on soil aeration. This frequently enhances plant growth by decreasing resistance to root penetration and enhancing the capacity to retain water. Additionally, they have the potential to modify soil aggregates, influencing infiltration, porosity, and the soil's capacity to facilitate the movement of gases and liquids (Islam et al., 2023). MPs have been observed to absorb and discharge harmful substances, accelerating soil pollution and changing soil properties. This could impact the arrangement of heavy metal types and the soil's ability to absorb soluble substances and retain water consequently affecting the creation of soil aggregates and their specific gravity (Wen et al., 2022).

The release of chemical components from MPs into the soil could additionally modify soil organisms and pH (Kim et al., 2020; Waldman & Rillig 2020). Soil composition, in conjunction with the presence of certain plant species, might also regulate the influence of MPs on soil pH (Lozano et al., 2021). Application of MPs, including low-density polypropylene and biodegradable materials, initially increased soil pH before decreasing over time, as observed by Qi et al. (2019). Similarly, de Souza Machado et al. (2019) and Lozano et al. (2021) reported that applying MP foams and fragments resulted in a rise in soil pH, primarily because it enhances soil aeration and porosity. The absorption of (PS-MPs) particles in soils is significantly affected by soil physicochemical attributes, with organic carbon content playing a crucial role (Hüffer et al., 2018). These interactions also affect soil organic carbon (SOC) estimations and the global carbon cycle (Kim et al., 2021). In agricultural contexts, plastic mulch films and clothing fibers have shown varied effects on soil organic matter (SOM), with some increasing and others decreasing SOM concentrations (Seth & Misra, 2014; Agnihotri & Seth, 2020; Boots et al., 2019).

The impact of MPs on soil extracellular enzyme activities has been extensively revealing significant variability in their effects. According to some reports, MPs can obstruct the activities of extracellular enzymes while others propose that they may enhance them, with different types of MPs exerting diverse effects on soil enzyme activities (Yu et al., 2020; Oladele et al., 2023). The accumulation of plastic film residues notably hinders soil enzyme activities, leading to reductions in fluorescein diacetate hydrolysis (FDA) and dehydrogenase activity by varying percentages (Wang et al., 2016). An increase in MP load in the soil resulted in significant reductions in urease (URE) and β-glucosidase (GLU) activities (Qian et al., 2018). Similarly, the activities of several enzymes involved in nitrogen, phosphorus, and carbon cycles were found to decrease significantly following exposure to MPs (Awet et al., 2018). Contrary to these reports, Liu et al. (2017) observed increased activities of FDA and phenol oxidase (PO) in the presence of elevated levels of MPs, while low levels did not show any distinct effect. Different soil enzyme activities exhibit differential responses to MPs, with certain MPs inhibiting some activities while stimulating others. For example, in acid soils, PE and polyvinyl chloride inhibit FDA activity but enhance the activities of URE and acid phosphatase (Fei et al., 2020). Moreover, the properties of the polymers may have specific effects on extracellular enzyme activities, with polypropylene and polyester fibers shown to inhibit FDA activities (De Souza Machado et al., 2018). Despite numerous studies, the precise effects of MPs on soil extracellular enzyme activities remain undetermined due to variations in MP characteristics, incubation time, and soil type. The inhibition in enzyme activities might occur either due to changes in soil nutritional substrates and physicochemical properties or through adsorption mechanisms. MPs competed with soil microorganisms for physicochemical niches, leading to reduced microbial activity and consequently decreased extracellular enzyme activity (Yu et al., 2020). Furthermore, enzyme activities in different aggregate-size fractions exhibited diverse responses to

MP exposure. Specifically, catalase activity was notably affected in the coarse particulate fraction, while phenol oxidase and GLU activities showed significant responses in the micro-aggregate fraction. Conversely, urease, manganese peroxidase, and laccase activities were most affected in the non-aggregated silt and clay fraction. These variations are attributed to differences in critical factors influencing enzyme activities across distinct aggregate-size fractions.

4.2. Vegetations

Seed germination, considered as the initial stage of plant growth, is affected by stress, particularly MP stress (Wang & Tanveer, 2023). MPs could potentially block the pores within the seed capsule, reducing water absorption and imbibition, and ultimately hindering seed germination (Bosker et al., 2019). Decreases in germination index, root length, and sprout length were found in Rapeseed (*Brassica campestris* L.) seeds when co-exposed to polymethyl methacrylate (PMMA) MPs and As (V) (Dong et al., 2022) although, De Silva et al. (2022) thought that it may not be dependable to assess the direct effects by considering only the traditional metrics. Therefore, it is imperative to investigate the effects of MP stress on seed germination beyond the traditional metrics of germination, such as germination rate and germination viability. Low and moderate concentrations of MPs in soil were found to inhibit seed germination, whereas high concentrations of MPs in soil promoted the germination of wheat seeds (Lian et al., 2019). The diverse particle sizes, agglomeration, and charge on MP particles could influence the ability of seeds to sprout and grow under MP stress conditions (Ziajahromi et al., 2018). Roots rank second among plant components experiencing direct and mechanical damage due to MP stress. The MPs present in soil and aquatic environments may attach to the root surface of the plants while the MNP particles suspended in the air may deposit on the plant leaves, flowers, and fruits. Kalčíková et al. (2017) reported the presence of MNP (30–600 μm) in freshwater on the root surface of duckweed which in turn affected root cell viability and root length. PE microbeads (3 μm) amassing on the root surface of maize led to the formation of only a few lateral roots (Urbina et al., 2020). MP is taken up by the plant roots, mainly by the lateral root cap and apical meristem where cell division occurs actively. The active cell division within the apical meristem creates a porous structure, allowing MPs captured by the root cap to enter the roots. Upon entry into the roots, MPs can be transported to other plant organs through the xylem vessels (Li et al., 2020). The accumulation of MPs (with sizes ranging from 100–700 nm) in the interstitial tissue of roots was observed to subsequently transfer to stems, leaves, flowers, and fruits, as reported by Li et al. (2021).

In a recent investigation, Urbina et al. (2020) observed that compared to control plants, the bioaccumulation of MPs in the maize rhizosphere led to a 50% reduction in plant biomass and height, accompanied by decreases in transpiration rates and nitrogen content. Additionally, nano PS-induced stress in the roots by inhibiting their growth. (Giorgetti et al., 2020). Changes in various important physiological functions, such as photosynthesis, redox regulation, ionic homeostasis, and hormone regulation, occur concurrently with the slowdown in physical growth caused by MP. These processes are susceptible to stressful situations, and any changes to them result in a decrease in crop development. Plants use photosynthesis, a crucial process that produces sugars and oxygen. This process is influenced by a variety of elements, such as the manufacture of pigments used in photosynthetic reactions, the exchange of gases between leaves and chlorophyll, ionic homeostasis, and redox control. Plant photosynthesis is decreased because of MP stress's negative regulation of these variables. MPs induced stress on the photosynthetic process, resulting in reduced shoot dry weight, shoot height, and leaf area in lettuce, along with decreased levels of carotenoids, chlorophyll a, and b by 12.5%, 9.1%, and 8.7%, respectively (Lian et al., 2021). Similarly, the application of PAN MPs exhibited a negative association with chlorophyll a and b contents in leaves, thereby inhibiting the growth of various plant species, including cucumber, tomato, cabbage, lettuce, and bok choy (Ahammed et al., 2012). However, Mondal et al. (2022) noted a higher chlorophyll (a/b) ratio, potentially due to the suppression of chlorophyll 'b' synthesis under MP stress conditions. This indicates that the decline in photosynthesis under MP stress may be attributed to a decrease in the synthesis of chlorophyll b (Pflugmacher et al., 2021). Chai et al. (2023) suggested that the reduced photosynthesis in mangroves induced by PVC, PP, and PE MPs resulted from a decrease in the concentration of carotenoids and chlorophyll a, as well as the efficiency of the electron transport chain (ETC), rather than affecting leaf gas exchange or chlorophyll b levels. The variability in outcomes could be attributed to the dosage of MPs and the plant species involved. Transpiration pull plays a significant role in the movement of MPs within plants; hence, higher transpiration rates could lead to increased uptake of MPs in plants (Li et al., 2020). Studies have shown that algae can accumulate MP on their surface which in turn induces detrimental effects on the organism (Zhang et al., 2021). Various effects of MPs have been documented in photosynthetic organisms inhabiting both aquatic and terrestrial environments (Meng et al., 2021). It can superficially attach organisms and influence photosynthesis directly. They can also enter plant systems with other pollutants changing the bioavailability of the pollutants and affecting photosynthesis indirectly (Dong et al., 2021).

MPs have been identified as inhibitory agents affecting the photosynthetic pigment content in aquatic photoautotrophs (Wu et al., 2019). Wang et al. (2020) demonstrated that PVC decreased chlorophyll content in three different types of algae, with the inhibitory effect directly correlating with the concentration of MPs. When plants receive external stimuli, they may initiate a stress response that disrupts normal cellular activities, including photosynthesis. Consequently, parameters such as intercellular carbon dioxide concentration, photosynthetic pigment content, and stomatal conductance may be impacted. In response, plants may undergo self-regulation, such as increasing leaf area, to mitigate the adverse effects of reduced chlorophyll content (An et al., 2021). Plant hormones like jasmonate, along with auxin, play a role in stimulating nectar secretion and synthesis. Research indicates that MPs treatment leads to a significant decrease in auxin levels in barley leaves. MPs also impact phytohormone levels in the roots and leaves of other crops such as barley, tomato, and cucumber. Furthermore, studies have demonstrated that exposure to MPs tends to reduce fatty acid levels and influence amino acid profiles, potentially altering nectar chemistry (Bouaicha et al., 2022). Consequently, the disturbance of biochemical and phytohormonal profiles by MPs could affect nectar production, consequently impacting pollination biology and potentially disrupting or delaying interactions between plants and pollinators. The impact of MPs on terrestrial plants is mostly concerned with grain crops such as wheat, corn and cash crops such as lettuce and cucumber. MPs formed by polylactic acid (PLA) were found to reduce chlorophyll content in maize leaves (Wang et al., 2020). The relative chlorophyll content of common bean (*Phaseolus vulgaris* L.) leaves was drastically reduced after exposure to low-density polyethylene (LDPE) (Meng et al., 2021). However, few studies have reported an increase in photosynthetic pigment content in the leaves of terrestrial photoautotrophs under the influence of MPs. Ren et al. (2021) observed an increase in chlorophyll upon the addition of (PS) to soil in flowering Chinese cabbage. On the other hand, plants exposed to MPs for prolonged periods showed severe disruption in different pigment levels including carotenoids. Hence, MPs, akin to other abiotic stress factors like heavy metals, influence the quantity and production of plant carotenoids by disrupting chlorophyll pathways through the inhibition of enzymes engaged in carotenoid biosynthesis (Li et al., 2020). In an aquatic environment, MPs, attached to the surface of aquatic phototrophs, influence the access to light energy causing physical disruption to cells. Some small particles of MPs can enter the cell, leading to the generation of reactive oxygen species (ROS) and a cascade of chain of oxidative stress reactions. The ROS generated damaged cell organelles affecting normal physiological processes. MPs are found in several habitats ranging from lakes, rivers, oceans, and farmlands to the atmosphere and have emerged as a serious environmental issue with a potentially detrimental effect on vegetation (Chen et al., 2020). Uptake of MP by plants can cause inhibition of their growth and development and affect their reproduction, immunity, and genetic stability (Anbumani & Kakkar, 2018).

4.3. Soil Microflora

Microplastics in soil change the physio-chemical properties of the soil resulting in alteration of soil microbial population and structure (Huang et al., 2019; Aralappanavar et al., 2024). The loosely bound compositional chemicals involved in making the MPs interact and disrupt the equilibrium between the microbial diversity and their activity (Wang et al., 2016). Wan et al. (2019) reported an increase in the phylum Gemmatimonadetes due to a 1% reduction in moisture content induced by the addition of MPs to soil. MPs can cause considerable changes in soil bacterial and fungal community diversity and structure (Hou et al., 2021). Microbes can adapt quickly to their surroundings by evolving their genetic makeup and metabolism and thus microbes with the ability to degrade plastic residues and use them as a source of carbon have been reported (Hadad et al., 2005). Fungal genera like *Fusarium, Penicillium, Aspergillus* that can use complex sources of carbon derived from MPs were reported to be higher in abundance and diversity in MP-contaminated soils (Accinelli et al., 2020). The effect of MPs depends on type, size, polymer composition, concentration, and exposure time (Zhou et al., 2020). Ren et al. (2020) reported that 5% larger PE (150 μm) in clay soil reduced bacterial and fungal abundance and diversity while smaller MPs (13 μm) tend to enhance bacterial and fungal populations. A 5% increase of both PVC and PE in acidic soil decreased soil alpha diversity while the addition of 1% PVC does not affect soil bacterial diversity (Fei et al., 2020). However, application of 1% MPs did not significantly change the bacterial population (Judy et al., 2019). Similar findings reported insignificant changes in soil bacterial diversity and community composition when a low concentration of MPs was applied for a short duration (30–60 days) (Qin et al., 2023). Plasticizing agents such as phthalates that are desorbed from plastic may damage fauna dwelling in the soil. Certain soil fauna may absorb MPs leading to changes in their related microbiota. The MPs may also change the composition of gut microbes in organisms such as arthropods (Zhu et al., 2018). Alteration in the feeding habits of the earthworm fed with polystyrene beads not only changed the gut microbiota but also negatively impacted their survival (Huerta et al., 2016).

MPs play an important role in shaping the microbial community composition of soil. Actinobacteria, a known hydrocarbon degrader, is found to degrade synthetic polymers in MP polluted soil. Similarly, Proteobacteria and Bacteroidetes can degrade complex organic matter (Yi et al., 2021). Studies have shown that MPs can facilitate the enrichment of plastic-degrading species such as Actinobacteria, Bacteroidetes, and Proteobacteria in both MP surface and MP-amended soils (Qiu et al., 2023). An increase in the relative abundance of Actinobacteria (31–36%) over a 30-day incubation resulted in the application of large PE MPs (150 μm) (Ren et al., 2020). Pathogenic genera like *Nocardia*, *Aeromicrobium*, and *Amycolaptosis* can adapt and survive under extreme MP-polluted soil (Yi et al., 2021). Study shows that application of PVC and PE at 2% (w/w) for 1 year increased the relative abundance of *Nocardia* in silty loam soils (Yi et al., 2021). Pathogenic organisms appear on the surface of MPs and act as a vector assisting in the transfer of the microbes to organs and tissues of plants and animals resulting in tissue damage and infection (Khan et al., 2020). Thus, factors including plastic amount, plastic size, plastic type, and incubation period influenced the effect of plastics on soil microbial communities.

MPs have a profound effect on soil microbial activity and soil biogeochemical cycles. It can alter soil DOC (Dissolved organic carbon), MBC (Microbial biomass carbon), soil CO_2, and CH_4 emission thus interfering with soil C turnover (Wan et al., 2023). For instance, PE MPs (<150 μm) at 5% w/w application in soil for 30 days increased soil CO_2 emission by 24–28.67 % compared with control soils (Gao et al., 2021). Long-term exposure to PA, PS, PET, and PVC increased the DOC content of silty loam soils (Zhang et al., 2023). MPs have a significant impact on soil microbial/enzymatic activity. Alteration in microbial communities due to the presence of MPs has a significant effect on soil enzymatic function. Han et al. (2022b) demonstrated that the application of PE, PAN, and hydrochar at a concentration of 0.5% w/w over 5 months led to increased expression of genes (mcrA) associated with methane production. Feng et al. (2022) found that applying PE, PS, PA, PLA, PBS, and PHB (with particle sizes ranging from 39–80 μm) at concentrations of 0.2% and 2% (w/w) in sandy loam soil for 120 days resulted in significant changes in the relative abundance of various bacteria, including N2-fixing bacteria (Nocardia), nitrifiers (Nitrospirae), denitrifiers (Gemmatimonadetes), and P-solubilizers (Streptomyces and Sphingomonas). The use of MPs hydrochar in soil significantly boosted the expression of nitrogen-fixation genes (nifD, nifH, and nifK), while PAN-hydrochar notably increased the expression of nitrate-reduction genes (nirB, nirD, nrfA, and nrfH). Zhao et al. (2021) observed that MPs increased acid phosphatase activity while decreasing β-glucosidase activity in sandy loam soil after 31 days of incubation. Additionally, Li et al. (2024) found that the incubation of PE and PLA for 2 months significantly reduced the abundance of genes involved in soil methane oxidation (Mox), P-solubilization (gcd), and thiosulfate reduction (soxB, soxY, soxZ, and soxC). Thus, the interaction between MPs and soil nutrients alters the MP-microbe interaction, subsequently affecting the functional characteristics of the soil microflora.

5. MPs as Carriers of Other Contaminants

The MPs besides being known pollutants on their own, are also, considered potential carriers of various other contaminants, including organic pollutants such as agrochemicals, hydrocarbons, plasticizers, detergents, oils, and inorganic compounds e.g., nitrates, phosphates, and heavy metals (Caruso et al., 2019; Wan, 2020; Vaseashta et al., 2021; Dube & Okuthe, 2023). Polycyclic aromatic hydrocarbons (PAHs), drugs, 17β-estradiol, and triclosan are among the other organic contaminants that MPs can co-transfer. The transportation of MPs coupled with the sorbed contaminants in agricultural soils is a major source of worry as immobile pollutants have a strong inclination to interact with soil matrices. MPs possess a substantial surface area and exhibit effective adsorption capabilities, enabling the absorption of pollutants and facilitating their widespread transport into the environment. The primary sorption mechanism for organic contaminants on MPs is hydrophobic interaction, which is followed by electrostatic interactions, pore filling, hydrogen bonding, halogen bonding, and π-π interactions. Polymers, viz., PET, PS, PP, and PE used in the manufacture of plastics are hydrophobic in nature and thus attract other pollutants materials through hydrophobicity that drives MPs as vectors for a range of contaminants (Torres et al., 2021). Besides this, the specific surface area, porosity in it, and the quantity of adsorption sites upon the surface affect the initial adsorption of contaminants by MPs. Both the chemical and physical characteristics of the soil, especially pH, ionic strength, or just salinity, texture, and metal cation concentration, also influence these factors (Rai et al., 2022). Important variables that influence the flow of organic pollutants alongside MPs include the type of contaminants, the amount of particles, the movement of the particles in relation to the particular contaminants, and the pace at which contaminants degrade from the MPs during the MPs' transit (Maddela et al., 2022). Notwithstanding the fulfillment of the initial three prerequisites, the contaminant's rate of desorption plays a pivotal role in the MP-mediated transportation of contaminants. Additionally, MPs and MNPs have different durations of contamination within the soil matrix that a particle carries. The UV light exposure of PE MPs increases their carbonyl index (CI) by almost ninefold ranging from 0.07–0.62. Exposure of MPs to UV produces free oxygen radicals, which can stimulate the development of oxidative functional groups and enhance the carrier capacity of MPs and MNPs. Exposing PE MPs to UV light boosted their CI ranging from 0.07–0.62, an almost ninefold rise. In general, UV exposure to MPs produces free oxygen radicals, which can stimulate the development of oxidative functional groups. Thus, creating such functional groups improves MNPs' ability to serve as vectors. MPs are important vectors in the build-up and movement of artificial nanomaterials, organic pollutants, and heavy metals (HMs) (Cheng et al., 2021). Chemical processes such as precipitation, surface complexity, and electrostatic attraction causes HMs to be adsorbed on the surface of MNPs. By using electrostatic adsorption, Guan et al. (2020) discovered 819.90 mg $kg^{-}1$ Cu^{2+} on PS-MS The Pb^{2+} adsorption rates with various MPs vary with polymethylmethacrylate > PE > PP. The maximum Pb^{2+} absorption by polymethylmethacrylate was 4.21 mg g^{-1}. The *in vitro* investigations using aquatic matrices showed that salinity along with pH are important factors in the HM adsorption process by MPs. Higher salinity inhibits MPs' ability to adsorb HMs, higher pH increases Cd's adsorption onto HDPE, while an acidic pH promotes Cr(VI) adsorption. The metals' partition coefficient, ion exchange capacity, and reactivity are accountable for the variance in the uptake rates of heavy metals by MNPs (Khalid et al., 2021). Many inorganic pollutants, like heavy metals (HMs), are adsorbed to the surface of metal NPs, which leads to the co-transfer of pollutants to the soil. Glassy polymers like PS, PVC, and PA, along with secondary MP particles, particularly absorb pollutants through pore-filling interactions. This process entails contaminants physically becoming enclosed within the micro or nanopores of MPs, effectively acting as carriers for other contaminants (Hüffer et al., 2016; Zang et al., 2018). The sorption of pollutants by MPs is influenced by various physicochemical properties of the MPs, such as their size, structure, and functional groups, as well as environmental parameters like pH, temperature, and salinity (Kinigopoulou et al., 2022). The MPs serve as a favorable host for bacteria including several pathogenic bacteria due to their hydrophobic nature, greater surface area, and pit-forming nature (Bowley et al., 2021). MPss have also been associated with harboring antibiotic-resistant bacteria such as Mycobacterium, *Vibrio* sp., *Escherichia coli, Pseudomonas* sp., presenting a substantial risk to the existing healthcare system (Dong et al., 2021; Zhao et al., 2021).

6. Policies and Interventions to Reduce Microplastic Pollution

The increased use of plastic products in agricultural operations and inappropriate disposal of plastic waste have made agricultural fields a major pollution sink for various plastic wastes and MNPs. They are highly stable and are imperishable in the environment due to which they accumulate in agricultural soil posing a major threat to the ecosystem. The widespread distribution and contamination of MPs in the environment along with their potential risk has drawn attention from several quarters. Remediation of MPs contaminated agroecosystems is challenging as various sources contribute to their build-up in farmlands. The initial step in the proposed framework measures would be to identify the sources of MPs that find their way into the Agro-ecosystem. Sewage sludge applied as a source of fertilizers for crop production is regarded as one of the major contributors. Improved technologies to process sustainable biosolids through hyperthermophilic composting (HTC) have improved solid waste treatments (Chen et al., 2021). The existing regulatory guidelines in Western countries have set limits on the presence of heavy metals and organic substances in the sludge. However, the limits of MNPs contamination in the European (EU 86/278/EEC) and U.S. (Code 503) guidelines have yet to be included. Plastic film mulching, extensively used in agricultural fields, contributes a substantial proportion of the MNPs in the agricultural field (Huang et al., 2020). Incorporating biodegradable organic mulching materials, like the residue of crops, tree leaves, rice straws, husks, wood dust, and water hyacinths along with the development of innovative biodegradable polymers prepared from plants and microbes for use in such usage can lead to significant reduction (Mancinelli et al., 2015). Scientific classification of garbage is vital to materialize recycling and proper disposal of plastic waste. This will help reduce the accumulation of plastics in agroecosystems during landfill or stacking. Converting various types of MP waste into viable raw materials represents an effective strategy for addressing environmental pollution resulting from improper disposal of plastic waste. Furthermore, establishing a comprehensive industrial framework for recycling and utilizing plastic waste is crucial for mitigating pollution (Yu et al., 2022). Institutes of higher learning and scientific research can play a proactive role in the identification of other potential sources of MPs, their distribution and pattern of transportation along with their anticipated impacts on agroecosystems and human health. Such studies can generate data and help policymakers make informed decisions on formulating policies to address the issues of reducing MP pollution. Monitoring agencies can initiate comprehensive surveys and studies to understand the sources, distribution, and abundance of MPs in agro-ecosystems, and collect data on the pollution levels which will serve as a foundation for evidence-based policymaking and targeted interventions. Effectively managing the ecological and human health risks linked to MP pollution requires the implementation of robust management strategies aimed at alleviating the adverse effects of MPs on the environment and public health. They can enforce regulations and standards aimed at reducing MP pollution. An effective surveillance system and early warning mechanisms to detect emerging trends and hotspots of MP pollution developed by the monitoring programs will go a long way in mitigating the ill effects of MPs.

7. Regulatory Initiatives and Guidelines on the Control of Microplastic Pollution

Developing regulatory standards for handling MPs in various situations is a challenging task, given that classification and quantification of the intentional or accidental release of MPs are yet to be clearly defined. Certain MPs are not intentionally included in the products but are generated through the decomposition of polymers. In such situations, establishing restrictions on unintentionally released MPs is difficult, and such MPs cannot be controlled under existing chemical legislation. Conversely, specific MPs, such as those found in fertilizer coatings, phytosanitary products, cosmetics, household and industrial detergents, cleaning products, paints, as well as items utilized in the oil and gas industries, are intentionally produced and deliberately incorporated during the manufacturing process of various products. Legal and administrative measures are being developed to address the serious issue of MPs. In 2015, the United Nations Environment Programme initiated the phasing out of the use of MPs in cosmetic industries. The European Commission also declared MP pollution as a key concern while China has promulgated provisions on "Plastic Restriction Order" and "Opinions on Further Strengthening the Control of Plastic Pollution" in 2008 and 2020, respectively. These policies primarily aim to decrease plastic pollution by limiting their source of production, use, and recycling process. The Indian Government has taken the initiative to promulgate laws to manage plastic usage. The Ministry of Environment, Forest and Climate Change, Government of India amended the Plastic Waste Management Rules, 2016 and notified the Plastic Waste Management (Amendment) Rules, 2024 which came into force on 14-3-2024. The rules prohibit the use, sale, stocking, and distribution of single-use plastics and also brought in the local government like the Panchayat at the District level under the provision and placed responsibilities on the local government to enforce the provisions of the law. The European Chemical Agency (ECHA) has recognized that the intentional release of MPs creates a significant risk to ecosystems if not properly controlled and accordingly, in January 2019, the ECHA established limits on MPs in products placed on EU/EEA markets to prevent the discharge of 500,000 Tons of MPs over a 20-year period which presently stands at 145,000 Tons usage. In September 2022, the Registration, Evaluation, Authorization, and Restriction of Chemicals (REACH) Committee met for the first time to discuss these laws with Member States. Some of the proposed regulatory approaches to control the release of purposely added MPs are to restrict the introduction of MP-containing products into the market, prohibit the usage of MPs in natural/biodegradable polymers, promote the use of products with no predicted MPs release, and implement a system of mandatory reporting for the identification, description of usage, quantity, and release of MNPs (Allan et al., 2021). In Canada, plastic-manufactured objects have been placed on the list of dangerous substances under the "Canadian Environment Protection Act's Schedule 1" regulation. The Canadian government issued proposed laws to outlaw single-use plastics (such as grocery/plastic bags, stir sticks, beverage six-pack holders, cutlery/plates, straws, and food packaging materials) in December 2021 in accordance with this legislation. The United Nations Environmental Assembly (UNEA) conducted a meeting with 175 countries to develop a legally enforceable resolution to address plastic pollution by 2024, called the "International Plastic Treaty". The draft is expected to be completed by the end of 2024 (Metzler & Simbeck, 2022). According to the Microbead-Free Waters Act (MFWA) of 2015, the United States established legislation that forbids the manufacturing, packaging, and distribution of cosmetics containing plastic microbeads considered a subset of MPs. This new regulation applies to cosmetics as well as non-prescription medications such as toothpaste. The few rules in place to regulate MPs in the environment are directed at plastic goods that are widely utilized in industries other than agriculture. The MPs are currently not subject to any limitations in the agricultural sector. This suggests that future research is necessary to learn more about MPs and their behavior in the soil environment, particularly in the agricultural ecosystem. The regulations and laws for the prevention and control of agricultural soil MP pollution still need to be improved.

8. Conclusion

The detrimental effects of MPs are well documented in numerous studies across various ecosystems. Agro-ecosystem is closely associated with anthropogenic activities which is the primary source of MP pollution. The MPs seep through the soil system affecting its fertility, altering microbial activity and plant growth and yield. Recent research findings point towards a hazardous environment with serious health implications. It is of utmost importance to adopt practices and implement judicious use of plastics, promoting sustainable practices, and improving waste management practices to safeguard human health. Future research can explore the influence of additional factors on the global-scale accumulation of MPs in soil with the help of numerical and spatial modeling, aiming to uncover long-term trends and patterns, with a special focus on food security and human health. Policymakers with the help of research-based studies can formulate informed decisions to enact and enforce laws aimed at addressing this critical issue.

References

Accinelli, C., Abbas, H.K., Bruno, V., Nissen, L., Vicari, A., Bellaloui, N. et al. (2020). Persistence in soil of microplastic films from ultra-thin compostable plastic bags and implications on soil *Aspergillus flavus* population. *Waste Management*, 113, 312–318.

Accinelli, C., Abbas, H.K., Little, N.S., Kotowicz, J.K., Mencarelli, M., & Shier, W.T. (2016). A liquid bioplastic formulation for film coating of agronomic seeds. *Crop Protection*, 89, 123–128.

Agnihotri, A., & Seth, C.S. (2020). Does jasmonic acid regulate photosynthesis, clastogenicity, and phytochelatins in *Brassica juncea* L. in response to Pb-subcellular distribution? *Chemosphere*, 243, 125361.

Ahammed, G.J., Wang, M.M., Zhou, Y.H., Xia, X.J., Mao, W.H., Shi, K. et al. 2012. The growth, photosynthesis and antioxidant defense responses of five vegetable crops to phenanthrene stress. *Ecotoxicology and environmental safety,* 80, 132-139.

Allan, J., Belz, S., Hoeveler, A., Hugas, M., Okuda, H., Patri, A. et al. (2021). Regulatory landscape of nanotechnology and nanoplastics from a global perspective. *Regulatory Toxicology and Pharmacology*, 122, 104885.

Alvarenga, P., Farto, M., Mourinha, C., & Palma, P. (2016). Beneficial use of dewatered and composted sewage sludge as soil amendments: Behavior of metals in soils and their uptake by plants. *Waste and Biomass Valorization*, 7, 1189–1201.

An, J., Liu, H., Zheng, Y., Cheng, J., & Song, C. (2021). Effects of soil microplastics residue on soybean seedlings growth and the physiological and biochemical characteristics. *J. Sichuan Agric. Univ.*, 39, 41–46.

Anbumani, S., & Kakkar, P. (2018). Ecotoxicological effects of microplastics on biota: A review. *Environmental Science and Pollution Research*, 25, 14373–14396.

Aralappanavar, V.K., Mukhopadhyay, R., Yu, Y., Liu, J., Bhatnagar, A., Praveena, S.M. et al. (2024). Effects of microplastics on soil microorganisms and microbial functions in nutrients and carbon cycling: A review. *Science of The Total Environment*, 171435.

Atugoda, T., Vithanage, M., Wijesekara, H., Bolan, N., Sarmah, A.K., Bank, M.S. et al. (2021). Interactions between microplastics, pharmaceuticals and personal care products: Implications for vector transport. *Environment International*, 149, 106367.

Awet, T.T., Kohl, Y., Meier, F., Straskraba, S., Grün, A.L., Ruf, T. et al. (2018). Effects of polystyrene nanoparticles on the microbiota and functional diversity of enzymes in soil. *Environmental Sciences Europe*, 30, 1–10.

Bala, S., Garg, D., Sridhar, K., Inbaraj, B.S., Singh, R., Kamma, S. et al. (2023). Transformation of agro-waste into value-added bioproducts and bioactive compounds: Micro/nano formulations and application in the agri-food-pharma sector. *Bioengineering*, 10(2), 152.

Bläsing, M., & Amelung, W. (2018). Plastics in soil: Analytical methods and possible sources. *Science of The Total Environment*, 612, 422–435.

Bolan, N., & Bradney, L. (2019). Particulate plastics in soils. *Chemistry in Australia*, 18–20.

Boots, B., Russell, C.W., & Green, D.S. (2019). Effects of microplastics in soil ecosystems: Above and below ground. *Environmental Science & Technology*, 53(19), 11496–11506.

Bosker, T., Bouwman, L.J., Brun, N.R., Behrens, P., & Vijver, M.G. (2019). Microplastics accumulate on pores in seed capsule and delay germination and root growth of the terrestrial vascular plant *Lepidium sativum*. *Chemosphere*, 226, 774–781.

Bouaicha, O., Tiziani, R., Maver, M., Lucini, L., Miras-Moreno, B., Zhang, L. et al. (2022). Plant species-specific impact of polyethylene microspheres on seedling growth and the metabolome. *Science of The Total Environment*, 840, 156678.

Bowley, J., Baker-Austin, C., Porter, A., Hartnell, R., & Lewis, C. (2021). Oceanic hitchhikers: Assessing pathogen risks from marine microplastic. *Trends in Microbiology*, 29(2), 107–116.

Caruso, G. (2019). Microplastics as vectors of contaminants. *Marine Pollution Bulletin*, 146, 921–924.

Chang, H., Brahney, J., Carvill, S., Granek, E.F., Hunter, N., & Wolfand, J. (2022, December). A social-ecological-technological systems approach to understanding microplastic pollution at the river basin scale. In: *AGU Fall Meeting Abstracts* (Vol. 2022, pp. H56D–01).

Chen, G., Feng, Q., & Wang, J. (2020). Mini review of microplastics in the atmosphere and their risks to humans. *Science of The Total Environment*, 703, 135504.

Chen, Z., Xing, R., Yang, X., Zhao, Z., Liao, H., & Zhou, S. (2021). Enhanced in situ Pb (II) passivation by biotransformation into chloropyromorphite during sludge composting. *Journal of Hazardous Materials*, 408, 124973.

Cheng, F., Zhang, T., Liu, Y., Zhang, Y., & Qu, J. (2021). Non-negligible effects of UV irradiation on transformation and environmental risks of microplastics in the water environment. *Journal of Xenobiotics*, 12(1), 1–12.

Corradini, F., Meza, P., Eguiluz, R., Casado, F., Huerta-Lwanga, E., & Geissen, V., 2019. Evidence of microplastic accumulation in agricultural soils from sewage sludge disposal. *Science of The Total Environment*, 671, 411–420.

Danso, D., Chow, J., & Streit, W.R. (2019). Plastics: Environmental and biotechnological perspectives on microbial degradation. *Applied and Environmental Microbiology*, 85(19), e01095-01119.

De Silva, Y.S.K., Rajagopalan, U.M., Kadono, H., & Li, D. (2022). Effects of microplastics on lentil (*Lens culinaris*) seed germination and seedling growth. *Chemosphere*, 303, 135162.

de Souza Machado, A.A., Kloas, W., Zarfl, C., Hempel, S., & Rillig, M.C. (2018). Microplastics as an emerging threat to terrestrial ecosystems. *Global Change Biology*, 24(4), 1405–1416.

de Souza Machado, A.A., Lau, C.W., Kloas, W., Bergmann, J., Bachelier, J.B., Faltin, E. et al. (2019). Microplastics can change soil properties and affect plant performance. *Environmental Science & Technology*, 53(10), 6044–6052.

Dong, H., Chen, Y., Wang, J., Zhang, Y., Zhang, P., Li, X. et al. (2021). Interactions of microplastics and antibiotic resistance genes and their effects on aquaculture environments. *Journal of Hazardous Materials*, 403, 123961.

Dong, R., Liu, R., Xu, Y., Liu, W., Wang, L., Liang, X. et al. (2022). Single and joint toxicity of polymethyl methacrylate microplastics and As (V) on rapeseed (*Brassia campestris* L.). *Chemosphere*, 291, 133066.

Dong, Y., Gao, M., Qiu, W., & Song, Z. (2021). Uptake of microplastics by carrots in the presence of As (III): Combined toxic effects. *Journal of Hazardous Materials*, 411, 125055.

Dube, E., & Okuthe, G.E. (2023). Plastics and micro/nano-plastics (MNPs) in the environment: Occurrence, impact, and toxicity. *International Journal of Environmental Research and Public Health*, 20(17), 6667.

FAO. (2021). The Food and Agriculture Organization (FAO), the United Nations, Assessment of Agricultural Plastics and their Sustainability, a Call for Action. Rome, Italy. https://doi.org/10.4060/cb7856en

Fei, Y., Huang, S., Zhang, H., Tong, Y., Wen, D., Xia, X. et al. (2020). Response of soil enzyme activities and bacterial communities to the accumulation of microplastics in an acid cropped soil. *Science of the Total Environment,* 707, p.135634

Feng, X., Wang, Q., Sun, Y., Zhang, S., & Wang, F. (2022). Microplastics change soil properties, heavy metal availability, and bacterial community in a Pb-Zn-contaminated soil. *Journal of Hazardous Materials*, 424, 127364.

Gao, B., Yao, H., Li, Y., & Zhu, Y. (2021). Microplastic addition alters the microbial community structure and stimulates soil carbon dioxide emissions in vegetable-growing soil. *Environmental Toxicology and Chemistry*, 40(2), 352–365.

Gil-Ortiz, R., Naranjo, M.Á., Ruiz-Navarro, A., Caballero-Molada, M., Atares, S., García, C., & Vicente, O. (2020). New eco-friendly polymeric-coated urea fertilizers enhanced crop yield in wheat. *Agronomy*, 10(3), 438.

Giorgetti, L., Spanò, C., Muccifora, S., Bottega, S., Barbieri, F., Bellani, L., & Castiglione, M.R. (2020). Exploring the interaction between polystyrene nanoplastics and *Allium cepa* during germination: Internalization in root cells, induction of toxicity, and oxidative stress. *Plant Physiology and Biochemistry*, 149, 170–177.

Guan, J., Qi, K., Wang, J., Wang, W., Wang, Z., Lu, N., & Qu, J. (2020). Microplastics as an emerging anthropogenic vector of trace metals in freshwater: Significance of biofilms and comparison with natural substrates. *Water Research*, 184, 116205.

Guo, M., Zhao, F., Tian, L., Ni, K., Lu, Y., & Borah, P. (2022). Effects of polystyrene microplastics on the seed germination of herbaceous ornamental plants. *Science of The Total Environment*, 809, 151100.

Hadad, D., Geresh, S., & Sivan, A. (2005). Biodegradation of polyethylene by the thermophilic bacterium *Brevibacillus borstelensis*. *Journal of Applied Microbiology*, 98(5), 1093–1100.

Han, L., Zhang, B., Li, D., Chen, L., Feng, Y., Xue, L. et al. (2022). The co-occurrence of microplastics and hydrochar stimulated the methane emission but suppressed nitrous oxide emission from rice paddy soil. *Journal of Cleaner Production*, 337, 130504.

Hartmann, N.B., Huffer, T., Thompson, R.C., Hassellöv, M., Verschoor, A., Daugaard, A.E. et al. (2019). Are we speaking the same language? Recommendations for a definition and categorization framework for plastic debris. *Environ. Sci. Technol.*, 53(3), 1039–1047.

He, D., Luo, Y., Lu, S., Liu, M., Song, Y., & Lei, L. (2018). Microplastics in soils: Analytical methods, pollution characteristics and ecological risks. *TrAC Trends in Analytical Chemistry*, 109, 163–172.

Horton, A.A., Walton, A., Spurgeon, D.J., Lahive, E., & Svendsen, C. (2017). Microplastics in freshwater and terrestrial environments: Evaluating the current understanding to identify the knowledge gaps and future research priorities. *Science of The Total Environment*, 586, 127–141.

Hou, J., Xu, X., Yu, H., Xi, B., & Tan, W. (2021). Comparing the long-term responses of soil microbial structures and diversities to polyethylene microplastics in different aggregate fractions. *Environment International*, 149, 106398.

Huang, Y., Liu, Q., Jia, W., Yan, C., & Wang, J. (2020). Agricultural plastic mulching as a source of microplastics in the terrestrial environment. *Environmental Pollution*, 260, 114096.

Huang, Y., Zhao, Y., Wang, J., Zhang, M., Jia, W., & Qin, X. (2019). LDPE microplastic films alter microbial community composition and enzymatic activities in soil. *Environmental Pollution*, 254, 112983.

Huerta Lwanga, E., Gertsen, H., Gooren, H., Peters, P., Salánki, T., Van Der Ploeg, M. et al. (2016). Microplastics in the terrestrial ecosystem: Implications for *Lumbricus terrestris* (*Oligochaeta*, Lumbricidae). *Environmental Science & Technology*, 50(5), 2685–2691.

Hüffer, T., & Hofmann, T. (2016). Sorption of non-polar organic compounds by micro-sized plastic particles in aqueous solution. *Environmental Pollution*, 214, 194–201.

Hüffer, T., Weniger, A.K., & Hofmann, T. (2018). Sorption of organic compounds by aged polystyrene microplastic particles. *Environmental Pollution*, 236, 218-225

Islam, M.S., Islam, Z., Jamal, A.S.I.M., Momtaz, N., & Beauty, S.A. (2023). Removal efficiencies of microplastics of the three largest drinking water treatment plants in Bangladesh. *Science of The Total Environment*, 895, 165155.

Ju, S., Shin, G., Lee, M., Koo, J.M., Jeon, H., Ok, Y.S. et al. (2021). Biodegradable chito-beads replacing non-biodegradable microplastics for cosmetics. *Green Chemistry*, 23(18), 6953–6965.

Judy, J.D., Williams, M., Gregg, A., Oliver, D., Kumar, A., Kookana, R., & Kirby, J.K. (2019). Microplastics in municipal mixed-waste organic outputs induce minimal short- to long-term toxicity in key terrestrial biota. *Environmental Pollution*, 252, 522–531.

Kalčíková, G., Gotvajn, A.Ž., Kladnik, A., & Jemec, A. (2017). Impact of polyethylene microbeads on the floating freshwater plant duckweed Lemna minor. *Environmental Pollution*, 230, 1108–1115.

Karasik, R., Vegh, T., Diana, Z., Bering, J., Caldas, J., Pickle, A. et al. (2020). 20 years of government responses to the global plastic pollution problem: The plastics policy inventory. NI X, 20(05). Durham, NC: Duke University.

Karbalaei, S., Hanachi, P., Walker, T.R., & Cole, M. 2018. Occurrence, sources, human health impacts and mitigation of microplastic pollution. *Environmental Science and Pollution Research*, 25, 36046–36063.

Khalid, N., Aqeel, M., Noman, A., Khan, S.M., & Akhter, N. (2021). Interactions and effects of microplastics with heavy metals in aquatic and terrestrial environments. *Environmental Pollution*, 290, 118104.

Khan, S., Ali, S., Muhammad, S., Khan, B., Ali, A., Hesham, E.L., & Begum, S. (2020). Bacterial contamination in drinking water of urban Peshawar: A comparative study at the sources and user points of tube wells. *Desalination and Water Treatment*, 181, 221–227.

Kim, S.W., Jeong, S.W., & An, Y.J. (2021). Microplastics disrupt accurate soil organic carbon measurement based on chemical oxidation method. *Chemosphere*, 276, 130178.

Kim, S.W., Waldman, W.R., Kim, T.Y., & Rillig, M.C. 2020. Effects of different microplastics on nematodes in the soil environment: Tracking the extractable additives using an ecotoxicological approach. *Environmental Science & Technology*, 54(21), 13868–13878.

Kinigopoulou, V., Pashalidis, I., Kalderis, D., & Anastopoulos, I. (2022). Microplastics as carriers of inorganic and organic contaminants in the environment: A review of recent progress. *Journal of Molecular Liquids*, 350, 118580.

Koutnik, V.S., Alkidim, S., Leonard, J., DePrima, F., Cao, S., Hoek, E.M., & Mohanty, S.K. (2021). Unaccounted microplastics in wastewater sludge: Where do they go? *ACS ES&T Water*, 1(5), 1086–1097.

Lam, C.S., Ramanathan, S., Carbery, M., Gray, K., Vanka, K.S., Maurin, C. et al. (2018). A comprehensive analysis of plastics and microplastic legislation worldwide. *Water, Air, & Soil Pollution*, 229, 1–19.

Lehmann, A., Leifheit, E.F., Gerdawischke, M., & Rillig, M.C. (2020). Microplastics have shape- and polymer-dependent effects on soil processes. *BioRxiv*, 2020-06.

Li, L., Luo, Y., Li, R., Zhou, Q., Peijnenburg, W.J., Yin, N. et al. (2020). Effective uptake of sub-micrometre plastics by crop plants via a crack-entry mode. *Nature Sustainability*, 3(11), 929–937.

Li, X., Chen, L., Mei, Q., Dong, B., Dai, X., Ding, G., & Zeng, E.Y. (2018). Microplastics in sewage sludge from the wastewater treatment plants in China. *Water Research*, 142, 75–85.

Li, Y., Shi, X., Qin, P., Zeng, M., Fu, M., Chen, Y. et al. (2024). Effects of polyethylene microplastics and heavy metals on soil-plant microbial dynamics. *Environmental Pollution*, 341, 123000.

Li, Z., Li, R., Li, Q., Zhou, J., & Wang, G. (2020). Physiological response of cucumber (*Cucumis sativus* L.) leaves to polystyrene nanoplastics pollution. *Chemosphere*, 255, 127041.

Li, Z., Li, Q., Li, R., Zhou, J., & Wang, G. (2021). The distribution and impact of polystyrene nanoplastics on cucumber plants. *Environmental Science and Pollution Research*, 28, 16042–16053.

Lian, J., Liu, W., Meng, L., Wu, J., Chao, L., Zeb, A., & Sun, Y. (2021). Foliar-applied polystyrene nanoplastics (PSNPs) reduce the growth and nutritional quality of lettuce (*Lactuca sativa* L.). *Environmental Pollution*, 280, 116978.

Lian, J., Shen, M., & Liu, W. (2019). Effects of microplastics on wheat seed germination and seedling growth. *Journal of Agro-Environment Science*, 38(4), 737–745.

Liu, E.K., He, W.Q., & Yan, C.R. (2014). 'White revolution' to 'white pollution': Agricultural plastic film mulch in China. *Environmental Research Letters*, 9(9), 091001.

Liu, H., Yang, X., Liu, G., Liang, C., Xue, S., Chen, H. et al. (2017). Response of soil dissolved organic matter to microplastic addition in Chinese loess soil. *Chemosphere*, 185, 907–917.

Lozano, Y.M., Lehnert, T., Lehmann, A., & Rillig, M.C. (2021). Microplastic shape, polymer type, and concentration affect soil properties and plant biomass. *Frontiers in Plant Science*, 12, 616645.

Maddela, N.R., Ramakrishnan, B., Kakarla, D., Venkateswarlu, K., & Megharaj, M. (2022). Major contaminants of emerging concern in soils: A perspective on potential health risks. *RSC Advances*, 12(20), 12396–12415.

Mancinelli, R., Marinari, S., Brunetti, P., Radicetti, E., & Campiglia, E. (2015). Organic mulching, irrigation and fertilization affect soil CO_2 emission, and C storage in tomato crop in the Mediterranean environment. *Soil and Tillage Research*, 152, 39–51.

Meng, F., Yang, X., Riksen, M., Xu, M., & Geissen, V. (2021). Response of common bean (*Phaseolus vulgaris* L.) growth to soil contaminated with microplastics. *Science of The Total Environment*, 755, 142516.

Metzler, D., & Simbeck, S. (2022). Microplastics: Global buzz and concern spur increased regulation. *Haley & Aldrich*, 70.

Mondal, N.K., Kundu, S., Debnath, P., Mondal, A., & Sen, K. (2022). Effects of polyethylene terephthalate microplastic on germination, biochemistry and phytotoxicity of Cicer arietinum L. and cytotoxicity study on Allium cepa L. *Environmental toxicology and pharmacology*, 94, 103908.

Nizzetto, L., Bussi, G., Futter, M.N., Butterfield, D., & Whitehead, P.G. (2016). A theoretical assessment of microplastic transport in river catchments and their retention by soils and river sediments. *Environmental Science: Processes & Impacts*, 18(8), 1050–1059.

Nizzetto, L., Futter, M., & Langaas, S. (2016). Are agricultural soils dumps for microplastics of urban origin? *Environ. Sc. Tech.*, 50, 10777–10779.

O'Connor, D., Pan, S., Shen, Z., Song, Y., Jin, Y., Wu, W.M., & Hou, D. (2019). Microplastics undergo accelerated vertical migration in sand soil due to their small size and wet-dry cycles. *Environmental Pollution*, 249, 527–534.

Oladele, S.O., Ojokole, W., & Oladele, B.B. (2023). Microplastics in agricultural soil: Polystyrene fragments inhibit soil microbial and enzymatic activities but promote nutrient concentration of Cowpea (*Vigna unguiculata*). *Journal of Hazardous Materials Advances*, 10, 100263.

Oliviero, M., Tato, T., Schiavo, S., Fernández, V., Manzo, S., & Beiras, R. (2019). Leachates of micronized plastic toys provoke embryotoxic effects upon sea urchin *Paracentrotus lividus*. *Environmental Pollution*, 247, 706–715.

Qi, R., Jones, D.L., Li, Z., Liu, Q., & Yan, C. (2020). Behavior of microplastics and plastic film residues in the soil environment: A critical review. *Science of The Total Environment*, 703, 134722.

Qi, Y., Yang, X., Pelaez, A.M., Lwanga, E.H., Beriot, N., Gertsen, H. et al. (2018). Macro- and micro-plastics in soil-plant system: Effects of plastic mulch film residues on wheat (*Triticum aestivum*) growth. *Science of The Total Environment*, 645, 1048–1056.

Qian, H., Zhang, M., Liu, G., Lu, T., Qu, Q., Du, B., & Pan, X. (2018). Effects of soil residual plastic film on soil microbial community structure and fertility. *Water, Air, & Soil Pollution*, 229, 1–11.

Qin, P., Li, T., Cui, Z., Zhang, H., Hu, X., Wei, G., & Chen, C. (2023). Responses of bacterial communities to microplastics: More sensitive in less fertile soils. *Science of The Total Environment*, 857, 159440.

Qiu, G., Wang, Q., Wang, Q., Wang, T., Kang, Z., Zeng, Y. et al. (2023). Effects of polyethylene microplastics on properties, enzyme activities, and the succession of microbial community in Mollisol: At the aggregate level. *Environmental Research*, 237, 116976.

Pflugmacher, S., Sulek, A., Mader, H., Heo, J., Noh, J.H., Penttinen, O.P. et al. (2020). The influence of new and artificial aged microplastic and leachates on the germination of Lepidium sativum L. *Plants*, 9(3), 339.

Rai, P.K., Sonne, C., Brown, R.J., Younis, S.A., & Kim, K.H. (2022). Adsorption of environmental contaminants on micro- and nano-scale plastic polymers and the influence of weathering processes on their adsorptive attributes. *Journal of Hazardous Materials*, 427, 127903.

Raza, M., Lee, J.Y., & Cha, J. (2022). Microplastics in soil and freshwater: Understanding sources, distribution, potential impacts, and regulations for management. *Science Progress*, 105(3), 00368504221126676.

Ren, X., Tang, J., Liu, X., & Liu, Q. (2020). Effects of microplastics on greenhouse gas emissions and the microbial community in fertilized soil. *Environmental Pollution*, 256, 113347.

Ren, X., Tang, J., Wang, L., & Liu, Q. (2021). Microplastics in soil-plant system: Effects of nano/microplastics on plant photosynthesis, rhizosphere microbes, and soil properties in soil with different residues. *Plant and Soil*, 462, 561–576.

Rillig, M.C. (2012). Microplastic in terrestrial ecosystems and the soil? *Environ. Sc. & Tech.*, 46(12), 6453–6454.

Rillig, M.C., Ingraffia, R., & de Souza Machado, A.A. (2017). Microplastic incorporation into soil in agroecosystems. *Frontiers in Plant Science*, 8, 305719.

Rillig, M.C., Ziersch, L., & Hempel, S. (2017). Microplastic transport in soil by earthworms. *Scientific Reports*, 7(1), 1362.

Rillig, M.C., & Lehmann, A. (2020). Microplastic in terrestrial ecosystems. *Science*, 368(6498), 1430–1431.

Rodríguez-Seijo, A., Santos, B., da Silva, E.F., Cachada, A., & Pereira, R. (2018). Low-density polyethylene microplastics as a source and carriers of agrochemicals to soil and earthworms. *Environmental Chemistry*, 16(1), 8–17.

Sarkar, D.J., Barman, M., Bera, T., De, M., & Chatterjee, D. (2018). Agriculture: Polymers in crop production mulch and fertilizer. *Encyclopedia of Polymer Applications*, 28–47.

Seth, C.S., & Misra, V. (2014). Changes in C–N metabolism under elevated CO_2 and temperature in Indian mustard (*Brassica juncea* L.): An adaptation strategy under climate change scenarios. *Journal of Plant Research*, 127, 793–802.

Seth, C.S. (2014). A review on the effects of climate change on plants and ecosystems and certain approaches for plant response studies under climate change scenarios with a specific focus on FACE. *Journal of Food and Nutritional Disorder*, 4(1), 1–9.

Steinmetz, Z., Wollmann, C., Schaefer, M., Buchmann, C., David, J., Tröger, J. et al. (2016). Plastic mulching in agriculture. Trading short-term agronomic benefits for long-term soil degradation? *Science of The Total Environment*, 550, 690–705.

Tang, Y., Liu, Y., Chen, Y., Zhang, W., Zhao, J., He, S. et al. (2021). A review: Research progress on microplastic pollutants in aquatic environments. *Science of The Total Environment*, 766, 142572.

Torres, F.G., Dioses-Salinas, D.C., Pizarro-Ortega, C.I., & De-la-Torre, G.E. (2021). Sorption of chemical contaminants on degradable and non-degradable microplastics: Recent progress and research trends. *Science of The Total Environment*, 757, 143875.

Urbina, M.A., Correa, F., Aburto, F., & Ferrio, J.P. (2020). Adsorption of polyethylene microbeads and physiological effects on hydroponic maize. *Science of The Total Environment*, 741, 140216.

Vaseashta, A., Ivanov, V., Stabnikov, V., & Marinin, A. (2021). Environmental safety and security investigations of neustonic microplastic aggregates near water-air interphase. *Pol. J. Environ. Stud.*, 30, 3457–3469. https://doi.org/10.15244/pjoes/131947.

Waldman, W.R., & Rillig, M.C. (2020). Microplastic research should embrace the complexity of secondary particles. *Env. Sc. & Tech.*, 54(13), 7751–7753.

Wan, L., Cheng, H., Liu, Y., Shen, Y., Liu, G., & Su, X. (2023). Global meta-analysis reveals differential effects of microplastics on soil ecosystem. *Science of the Total Environment*, 867, 161403.

Wan, Y., Wu, C., Xue, Q., & Hui, X. (2019). Effects of plastic contamination on water evaporation and desiccation cracking in soil. *Science of The Total Environment*, 654, 576–582.

Wang, Chengqian, David O'Connor, Liuwei Wang, Wei-Min Wu, Jian Luo, & Deyi Hou. (2022). Microplastics in urban runoff: Global occurrence and fate. *Water Research*, 225(2022), 119129.

Wang, F., Zhang, X., Zhang, S., Zhang, S., & Sun, Y. (2020). Interactions of microplastics and cadmium on plant growth and arbuscular mycorrhizal fungal communities in agricultural soil. *Chemosphere*, 254, 126791.

Wang, H.T., Ding, J., Xiong, C., Zhu, D., Li, G., Jia, X.Y. et al. (2019). Exposure to microplastics lowers arsenic accumulation and alters gut bacterial communities of earthworm *Metaphire californica*. *Environmental Pollution*, 251, 110–116.

Wang, J., Liu, X., Li, Y., Powell, T., Wang, X., Wang, G., & Zhang, P. (2019). Microplastics as contaminants in the soil environment: A mini review. *Science of The Total Environment*, 691, 848–857.

Wang, J., Lv, S., Zhang, M., Chen, G., Zhu, T., Zhang, S. et al. (2016). Effects of plastic film residues on occurrence of phthalates and microbial activity in soils. *Chemosphere*, 151, 171–177.

Wang, L., & Tanveer, M. (2023). Editorial to the special issue "Eco-Physiological and Molecular Basis of Stress Tolerance in Plants". *Biology*, 12(3), 485.

Wang, S., Wang, Y., Liang, Y., Cao, W., Sun, C., Ju, P., & Zheng, L. (2020). The interactions between microplastic polyvinyl chloride and marine diatoms: Physiological, morphological, and growth effects. *Ecotoxicology and Environmental Safety*, 203, 111000.

Wang, Y., Xv, X., Shao, T., He, Q., Guo, Z., Wang, Y. et al. (2024). A case on source to soil to solutions: Distribution characteristics of microplastics in farmland soil of the largest vegetable base in Northwest China. *Science of The Total Environment*, 907, 167910.

Wen, X., Yin, L., Zhou, Z., Kang, Z., Sun, Q., Zhang, Y. et al. (2022). Microplastics can affect soil properties and chemical speciation of metals in yellow-brown soil. *Ecotoxicology and Environmental Safety*, 243, 113958.

Wu, X., Liu, Y., Yin, S., Xiao, K., Xiong, Q., Bian, S. et al. (2020). Metabolomics revealing the response of rice (*Oryza sativa* L.) exposed to polystyrene microplastics. *Environmental Pollution*, 266, 115159.

Wu, X., Lyu, X., Li, Z., Gao, B., Zeng, X., Wu, J., & Sun, Y. (2020). Transport of polystyrene nanoplastics in natural soils: Effect of soil properties, ionic strength, and cation type. *Science of The Total Environment*, 707, 136065.

Wu, Y., Guo, P., Zhang, X., Zhang, Y., Xie, S., & Deng, J. (2019). Effect of microplastics exposure on the photosynthesis system of freshwater algae. *Journal of Hazardous Materials*, 374, 219–227.

Xiao, M., Luo, Y., Zhang, H., Yu, Y., Yao, H., Zhu, Z. et al. (2022). Microplastics shape microbial communities affecting soil organic matter decomposition in paddy soil. *Journal of Hazardous Materials*, 431, 128589.

Yi, M., Zhou, S., Zhang, L., & Ding, S. (2021). The effects of three different microplastics on enzyme activities and microbial communities in soil. *Water Environment Research*, 93(1), 24–32.

Yoon, J.H., Kim, B.H. & Kim, K.H. (2024). Distribution of microplastics in soil by types of land use in metropolitan area of Seoul. *Applied Biological Chemistry*, 67(1), 15.

Yu, H., Fan, P., Hou, J., Dang, Q., Cui, D., Xi, B., & Tan, W. (2020). Inhibitory effect of microplastics on soil extracellular enzymatic activities by changing soil properties and direct adsorption: An investigation at the aggregate-fraction level. *Environmental Pollution*, 267, 115544.

Yu, H., Zhang, Y., Tan, W., & Zhang, Z. (2022). Microplastics as an emerging environmental pollutant in agricultural soils: Effects on ecosystems and human health. *Frontiers in Environmental Science*, 10, 855292.

Zhang, H., Wang, J., Zhou, B., Zhou, Y., Dai, Z., Zhou, Q. et al. (2018). Enhanced adsorption of oxytetracycline to weathered microplastic polystyrene: Kinetics, isotherms, and influencing factors. *Environmental Pollution*, 243, 1550–1557.

Zhang, Q., Zhao, M., Meng, F., Xiao, Y., Dai, W., & Luan, Y. (2021). Effect of polystyrene microplastics on rice seed germination and antioxidant enzyme activity. *Toxics*, 9(8), 179.

Zhang, S., Yang, X., Gertsen, H., Peters, P., Salánki, T., & Geissen, V. (2018). A simple method for the extraction and identification of light density microplastics from soil. *Science of The Total Environment*, 616, 1056–1065.

Zhang, W., Sun, S., Du, X., Han, Y., Tang, Y., Zhou, W. et al. (2022). Toxic impacts of microplastics and tetrabromobisphenol A on the motility of marine microalgae and potential mechanisms of action. *Gondwana Research*, 108, 158–170.

Zhang, X., Li, Y., Lei, J., Li, Z., Tan, Q., Xie, L. et al. (2023). Time-dependent effects of microplastics on soil bacteriome. *Journal of Hazardous Materials*, 447, 130762.

Zhao, T., Lozano, Y.M., & Rillig, M.C. (2021). Microplastics increase soil pH and decrease microbial activities as a function of microplastic shape, polymer type, and exposure time. *Frontiers in Environmental Science*, 9, 675803.

Zhou, G., Wu, Q., Tang, P., Chen, C., Cheng, X., Wei, X.F. et al. (2023). How many microplastics do we ingest when using disposable drink cups? *Journal of Hazardous Materials*, 441, 129982.

Zhou, Y., Wang, J., Zou, M., Jia, Z., Zhou, S., & Li, Y. (2020). Microplastics in soils: A review of methods, occurrence, fate, transport, ecological, and environmental risks. *Science of The Total Environment*, 748, 141368.

Zhu, D., Chen, Q.L., An, X.L., Yang, X.R., Christie, P., Ke, X. et al. (2018). Exposure of soil collembolans to microplastics perturbs their gut microbiota and alters their isotopic composition. *Soil Biology and Biochemistry*, 116, 302–310.

Ziajahromi, S., Kumar, A., Neale, P.A., & Leusch, F.D. (2018). Environmentally relevant concentrations of polyethylene microplastics negatively impact the survival, growth and emergence of sediment-dwelling invertebrates. *Environmental Pollution,* 236, 425-431.

Zubris, K.A.V., & Richards, B.K. (2005). Synthetic fibers as an indicator of land application of sludge. *Environmental Pollution*, 138(2), 201–211.

9 | Biodegradation Prerequisites for Microplastic Pollution in Freshwater Lakes in Armenia

Lusine Hambaryan[1,2*], Dmitrii Sadokov[3,4] and Olesia Sadokova[5]

[1] Chair of Ecology and Nature Protection, Yerevan State University, Armenia
[2] Scientific Center of Zoology and Hydroecology, Institute of Hydroecology and Ichthyology, National Academy of Sciences of the Republic of Armenia, Armenia
[3] Shenzhen MSU-BUT University, International University Park Road, 1, Dayun New Town, Longgang Dist., Shenzhen, Guangdong, Province, 518172, China
[4] Information Technologies, Mechanics and Optics (ITMO) University, Lomonosov Str., 9, St. Petersburg, 191002, Russia
[5] St. Petersburg State University", Universitetskaya Emb., 7/9, St. Petersburg, 199034 Russia

Abstract: For the first time, microplastics have been studied in the lakes of the Republic of Armenia. Samples of water and bottom sediments were taken from two mid-mountain lakes of the Lori plateau: the Clear Liman (area 10.7 ha) and the Horse Liman (area 2.4 ha). These lakes are designated as a Freshwater Key Biodiversity Area (FKBA). Sampling was carried out in the center of each lake. Bottom sediments were sampled using a cone sampler, surface water was sampled using a 75 μm polyamide mesh. The concentration of microplastics found in water samples is 100–153 MP/m^3. Among the detected particles, fibers predominated (87%), and the rest were identified as fragments (13%). Bottom sediments contained 60–80 MP/kg (wet weight) of microplastics shaped like fibers. The detected fibers ranged in length from 100 μm to 10 mm, with their thickness varied from 13–45 μm. Qualitative analysis of microplastics was performed using an ATR-FTIR spectrometer (Nicolet iS50, Thermo Scientific). Not all particles were analyzed for polymer type, due to the limitations of the ATR lower detection limit of 20 μm. The analyzed particles were identified as polypropylene (PP), polyethylene (PE), and polyethylene terephthalate (PET). The degradation of polymers in the environment is influenced by various factors:

*Corresponding author: lusinehambaryan@ysu.am

ultraviolet radiation, weathering, currents, waves, as well as the impact of biota. Alterations in microplastic properties are a matter of special interest in the way surficial biofilms over the microplastic particles can change their buoyancy, affect transport in the water column, and contribute to its degradation. Water and sediment samples were examined to establish algae populations in lake water. In the microalgae composition of the sediments, the diatom group was characterized by the greatest species diversity, where more than 26 species were recorded, with the dominance of *Fragilaria capucina*, *Diatoma hiemale*, and *Cocconeis placentula*. In the bottom communities of both lakes were dominated by diatom species (Clear Liman – 58%, Horse Liman – 49%). Subdominant groups were cyanobacteria and green algae, accounting for 41% and 31% in the Clear Liman and the Horse Liman, respectively. The data obtained will be the subject of future research on the microplastics and phytoplankton interaction.

1. Introduction

Research of non-chemical pollution and its impact on the environment has become highly relevant in different regions of the Earth and biomes and carries significant potential for scaling up (Barnes et al., 2009; Browne et al., 2011; Rochman, 2018; González-Pleiter et al., 2021; Allen et al., 2022). Microplastic (MP) particles lie within the size range between 1 μm (and supramolecular) and visually seen debris up to 5 mm. MP should be and is considered an emerging object of research with specific physical and chemical properties (Thompson et al., 2004). The latter can vary depending on the environment where particles are disposed of and also on their size, shape, polymer chemical composition, and concentration in solution (Kooi et al., 2018). Numerous studies of MP pollution were carried out on marine environments and noticeably fewer investigations concern freshwater reservoirs (Lambert & Wagner, 2018). Wide-scale MP research started in the 2000s, while the freshwater environment for the first time was investigated on this subject only in 2013 (Eriksen et al., 2013; Rochman, 2018).

Distribution patterns and sources of MP in the environment are surprisingly rarely discussed in the literature (Rochman, 2018), unlike issues of its biological and biochemical interactions, some of which are reviewed in our paper. However, it is evident that origins and ways of transport precondition many intrinsic properties of MP and are meaningful for spatial analysis or modeling of the MP behavior through the whole life cycle from its formation to ultimate sedimentation in natural sinks or biological utilization (Wagner et al., 2014; Rochman, 2018). After all, quantitative and spatial analysis of MP input in natural waters is a certain way to apply scientific investigation as a justified basis for waste management on the local or regional scale. Outputs and approaches of MP distribution analysis will naturally vary in accordance with national and territorial features (Talbot & Chang, 2022).

Armenia is a typical mountainous country, with elevation changes between 375 m above sea level (a.s.l.) (on the north-east, in River Debed valley) and 4,095 m a.s.l. (Mount Aragats). Most of the terrain of Armenia (90%) elevates more than 1,000 m above sea level. Such differences in altitude determine a variety of climatic zones and landscapes. The regional climate of Armenia is dry, due to elevated relief, remoteness, and isolation from seas by the mountains of Southern Caucasus. The mean temperature in mid-mountainous areas of Armenia is +22°C in July and –3°C in January, with extreme values of +35°C and –34°C, respectively (Berdnilov et al., 2022). Overall yearly precipitation in Armenia is close to 600 mm, with the lowest value recorded in the south (Ararat Valley and Zangezur Mountains).

Lake Sevan is the greatest reservoir of the Caucasus, which is important as a source of freshwater for the whole country. Apart from the Sevan, more than hundreds of small lakes and reservoirs with areas from several dozens of hectares to first square kilometers exist in Armenia, with Arpi Lake being the second largest one after the Sevan. Armenia is a small country where the hydrological network is poorly developed, which is why many lakes and rivers undergo heavy stress due to increased pollution from local industries or household wastewater. It is a matter of prior importance to provide a deep understanding of water quality levels and specific pollution rates, especially for the provinces where water resources are scarce (2012).

Since MP distribution is closely related to the larger plastic objects occurrence in the environment, it is necessary to introduce and evaluate regional waste management plans. National Law No3P-159 (2004) regulating waste management in the Republic of Armenia was signed in 2004 and introduces elements of responsibility, control, and regulation of waste disposal. However, the implementation of the law relies on many socio-economic issues that do not provide a sufficient basement for sustainable waste management practice, even for recycling facilities, which are poorly developed in Armenia. More importantly, communal waste disposal is not facilitated or certified, so waste is uncontrollably dumped in rural areas and within freshwater basins, which inevitably causes plastic pollution of natural waters.

2. Behavior Microplastics in Freshwater

2.1. Physical Qualities of Plastics

The distribution of MPs in the water column depends on many conditions: the physical properties of the plastic (density, size, shape), interaction with the microbiota, and environmental conditions (water density, wind, current, tides). In addition, abiotic and biotic factors can cause the degradation of plastic (Lambert & Wagner, 2018), but it is difficult to distinguish those contributing to this process most. First of all, in the natural environment, plastic undergoes degradation (aging) under the influence of physical and mechanical factors, heat, abrasion, weathering, and photodegradation induced by ultraviolet radiation. It is photooxidation that is the main factor in the aging of plastics, accompanied by the destruction of molecular bonds on the surface and the addition of oxygen (Luo et al., 2022). It was reported that plastic in the aquatic environment is less susceptible to photo- and thermal degradation, due to the limited penetration of ultraviolet radiation (UV) (Andrady, 2011; Kalogerakis et al., 2017; Mao et al., 2020). Water salinity also affects the rate of photooxidation, which was confirmed by the lower aging degree of polystyrene in freshwater than in salt water under the same UV conditions (Mao et al., 2020). In addition, MP aging is accompanied by a change in the physical properties of plastics: hydrophobicity, endurance, roughness, and porosity, which stimulate the adsorption of pollutants and microorganisms on the MP surface (Lin et al., 2020). The appearance of biofilms on plastic surfaces resulting from the activity of some species of microalgae also may cause biodegradation (Sapozhnikov et al., 2022), lead to embrittlement, and alter the conditions of sedimentation (Figure 1).

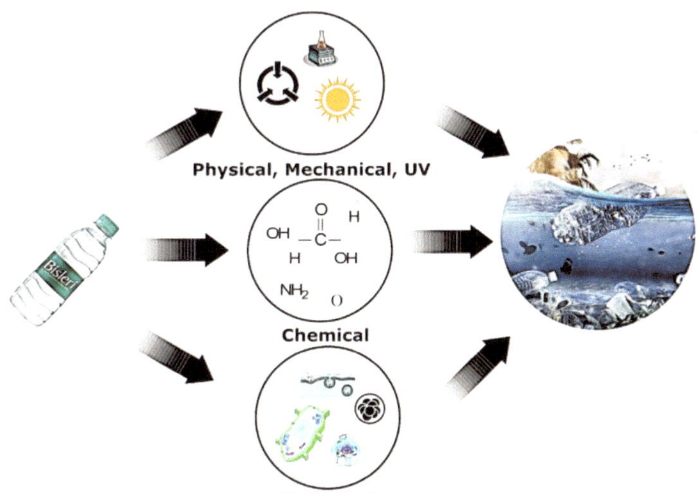

Figure 1: Plastic degradation processes in aquatic environments.

MPs are prone to the formation of biofilms on the surface, i.e., colonies of microorganisms that use the plastic surface as a substrate (Lobelle & Cunliffe, 2011; Kozlovskii & Blinovskaia, 2015; Pedrotti et al., 2022; Vaseashta et al., 2021; Vladimir et al., 2023). Biofilm-coated aggregates are called 'plastispheres' (Zettler et al.,, 2013). Plastispheres can be consumed by living organisms and transmitted along the food chain, at least in invertebrates (Cole et al., 2013; Lusher, 2015). Buoyancy and hydrophobicity of plastics can be changed due to biofilm formation, which forces them to eventually sink to the bottom or float up (Lobelle & Cunliffe, 2011; Kalogerakis et al., 2017). Thus, an individual set of factors is attributed to MP in the water in different regions individually. For example, the drainless lakes of Lori province are small freshwater reservoirs largely overgrown with water lilies. Shallow depths, poor mixing, and vigorous vegetation development within lake basins stimulate certain conditions that may favor MP accumulation and fouling.

2.2. Chemical Properties of Plastics

In addition to the physical properties of an MP, its chemical composition is of great importance, which in most cases turns out to be heterogeneous and unpredictable. Additives (plasticizers, antioxidants, flame retardants, pigments) are commonly found in plastic packages (Andrady & Neal, 2009). These additives are designed to increase the abrasion resistance and durability of goods, so they contribute to a longer presence of plastic in the environment and slow down the process of its degradation. Furthermore, such additives can cause harmful effects on organisms due to the enhanced toxicity of some of them. Abrasion, weathering, and embrittlement of MPs contribute to the release of these substances into the aquatic environment (Capolupo et al., 2020; Liu et al., 2020). In addition to their chemical composition, MPs can adsorb persistent organic pollutants and heavy metals on their surface. For example, Cd, Cu, Ni, and Zn demonstrate higher adsorption capacity on aged polystyrene particles (Mao et al., 2020). Enclosed (non-drainage) water bodies provide conditions for plastic toxicity to increase due to lack of water runoff and dilution.

MPs' impact on freshwater biota has not been sufficiently studied so far, while negative effects on marine mammals and birds caused by MP swallowing and accumulation in the digestive tract were supported by numerous studies (Lavers et al., 2014; Vaseashta et al., 2022). Evidence of MPs' toxic impact on the lymphatic system and reproductive functions was also reported for marine organisms (Lusher, 2015; Capolupo et al., 2020). Nevertheless, analogous effects on freshwater ecosystems and organisms are expected to be similar. Thus, it is necessary to understand the characteristics of the physicochemical, biological, and toxicological effects of MP particles (Figure 2).

2.3. Microplastics and Phytoplankton

Phytoplankton is an important link in aquatic ecosystems, and in addition to the photosynthetic activity, phytoplankton communities establish habitat conditions and participate in the biogeochemical cycle of nitrogen and phosphorus. As a rule, physiological parameters, species composition and qualitative characteristics of phytoplankton can alter under certain levels of pollution (Reynolds, 2006), which also to some extent can be used as biological indicator proxies of the effects induced by enhanced MPs concentration in aquatic systems (Hitchcock, 2022). Nevertheless, relevant data on the direct influence of MP on phytoplankton *in situ* is insufficient so far (for freshwater ecosystems), while a number of *ex situ* and *in vitro* studies with laboratory microalgae are known. One issue that

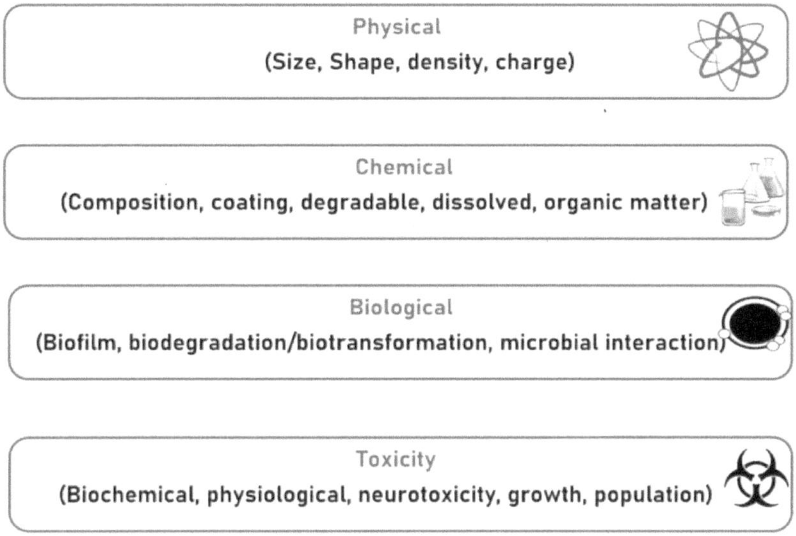

Figure 2: Different properties of microplastic particles.

remains to be resolved is the question of the dosage of MPs that are expected to cause any noticeable effect.

Algae growth inhibition tests were carried out using marine microalgae *Skeletonema costatum* in order to investigate possible toxic effects of MPs and the influence of various shading on phytoplankton (Zhang et al., 2017). It was discovered that MP of millimeter-scale and larger did not significantly affect the growth of the microalgae. An adverse effect of MPs on algae growth was demonstrated, but this effect declined with the increasing size of MP particles (Besseling et al., 2014; Bergami et al., 2017). In addition, a significant problem in comparing the results of laboratory experiments on the toxic effects of MPs with natural studies is the differences in the observed concentrations of MPs. In most studies aimed at investigating the impact of MPs on biochemical processes, solutions with high concentrations of MP particles (up to 250–1000 mg/l) were used, which many times exceed its content in natural water bodies (Besseling et al., 2014; Sjollema et al., 2016). Thus, application of the laboratory research to natural reservoirs with low (background) levels of MP pollution can be problematic.

Although some enhanced MP toxicity at 'background' concentration has been reported recently (Zhang et al., 2017; Bergami et al., 2017), these studies were mainly carried out when the phytoplankton reached its logarithmic phase of growth, when algae commonly show the highest capability to maintain activity under abiotic or biotic stressors (Tian et al., 2014). Since the growth of algae has four phases (lag, logarithmic, stationary, and death phases), it is very advisable that MP toxicity be assessed also for other phases except the logarithmic.

The growth of phytoplankton over MP surfaces can cause the formation of aggregates that may possess properties differing from those of communities free of microplastics. Such aggregates, in turn, can introduce toxic or invasive algal species into new environments via their mechanical transfer. Algae species considered toxic can often be found adsorbed on the plastic surface (Carson et al., 2013; Morohoshi et al., 2018). It was shown, for example, for the following groups as marine and freshwater cyanobacteria: *Chroococcales* (genus *Microcystis*), *Oscillatoriales* (genus *Rivularia*), *Nostocales* (genera *Calothrix* and *Scytonema*), *Pleurocapsales* (genus *Pleurocapsa*), and *Synechococcales* (genera *Synechococcus*, *Prochlorothrix,* and *Leptolyngbya*) (Bryant et al., 2016; Muthukrishnan et al., 2019). It is well known that cyanobacteria have high ecological plasticity, adaptability, and ability to allelopathically suppress competition of other species (by releasing toxic compounds). Also, cyanobacteria can quickly and effectively adapt to many adverse conditions and abrupt environmental changes (Leoni et al., 2014; Marti et al., 2016; Nava et al., 2017; Hambaryan et al., 2020), that is why it is important to consider this group in as part of the research of possible species transfer on MP aggregates.

Of primary interest is the accumulation of plastic in the marine environment, which clearly impacts both marine biota and humans, as microplastics can enter the food chain and cause a range of health effects in humans. It is known that algae can colonize artificial substrates such as polyethylene surfaces in wastewater (Table 1), and these colonizing algae have been found to be environmentally friendly and non-toxic (Sharma et al., 2014).

The adhesion of algae to the surface initiates biodegradation, and their production of ligninolytic and exopolysaccharide enzymes is key to the biodegradation of plastic by interacting with macromolecules present on the plastic surface and triggering biodegradation (Sarmah et al., 2018a; Chinaglia et al., 2018). The most common species found on the surface of plastic waste were from the blue-green algae (Cyanobacterium) division – *Anabaena spiroides* showed the highest percentage (8.18%), followed by the diatom *Navicula pupula* (4.44%) and the green alga *Scenedesmus dimorphus* (3.74%) (Kumar et al., 2017). A study by Sarmah and Rout (Sarmah et al., 2018b) also concluded that freshwater non-toxic cyanobacteria (*Phormidium lucidum* and *Oscillatoria subbrevis*) are able to rapidly grow and colonize the surface of various types of plastic and can effectively contribute to their use in the development of future biodegradation methods.

MPs and microalgae interaction can be in some way important also at the ecosystem level, especially if the productivity rate of an ecosystem has already been affected by other anthropogenic impacts (including plastic pollution) (Crooks, 2002; Zettler et al., 2013; Gevorgyan et al., 2016; Sakharova et al., 2020). Ultimately the ecosystem effects may be meaningful to cause deterioration of environmental services. Thus, it is necessary to bring laboratory experiments on the MP interaction with algae closer to real environmental conditions and to evaluate this interaction at all stages of microalgae growth.

Table 1: The most common species that form algal colonies (biofilms) on plastic surfaces (Attribution under Creative Commons license (Chia et al., 2020))

Algae Species	Plastic	Waterbody
Scenedesmus dimorphus (green alga), *Anabaena spiroides* (blue-green alga), and *Navicula pupula* (diatom)	Dumped waste polyethylene bags	Domestic wastewater in Chennai city, Tamil Nadu, India
Oscillatoria princeps, O. acuminate, O. subbrevis, O. willei, O. amoena, O. splendida, O. vizagapatensis, O. limnetica, O. earlei, O. peronata, O. formosa, O. okeni, O. geitleriana, O. limosa, O. chalybea, O. salina, O. rubescens, O. curviceps, O. tenuis, and *O. laete-virens*	Submerged polythene	Domestic sewage water bodies of Silchar town, Assam
O. limnetica, P. lucidum, Phormidium calcicola, O. earlei, Lyngbya cinerascens, Nostoc carneum, Nostoc linckia, Spirulina major, Hydrocoleum sp., *Chlorella* sp., *Pithophora* sp., *Scenedesmus quadricauda, Calothrix fusca, Stigeoclonium tenue, Calothrix marchica, Anomoeoneis* sp., *Oedogonium* sp., *Arthrospira platensis, Navicula minuta, Nitzschia* sp., *Navicula dicephala, Nitzschia intermedia, Spirogyra* sp., and *Synedra tabulata*	Polythene carry bags	Solid domestic sewage dumping sites
Phormidium lucidum, Oscillatoria subbrevis, Lyngbya diguetii, Nostoc carneum, and *Cylindrospermum muscicola*	Dumped waste polythene bags	Domestic sewage water drains of Silchar town, Assam
Phormidium tenue, Oscillatoria tenuis, Monoraphidium contortum, Microcystis aeruginosa, Closterium constatum, Chlorella vulgaris, and *Amphora ovalis*	Waste polythene materials	Various ponds, lakes, and water bodies of Kota city in the state of Rajasthan, India
Coleochaete scutata, Coleochaete soluta, Chaetophora, Aphanochaete, Gloeotaenium, Oedogonium, Oocystis, Oscillatoria, Phormidium, Chroococcus, Aphanothece, Fragilaria, Cocconeis, Navicula, and *Cymbella*	Polythene	Oligotrophic water bodies of Lucknow, Uttar Pradesh

3. Case Study of Armenian Lakes

The following investigations were conducted on samples taken from lakes in the northern part of the Republic of Armenia, specifically from Clear Liman and Horse Liman, as shown in Figure 3. Sample collection and analysis are described below.

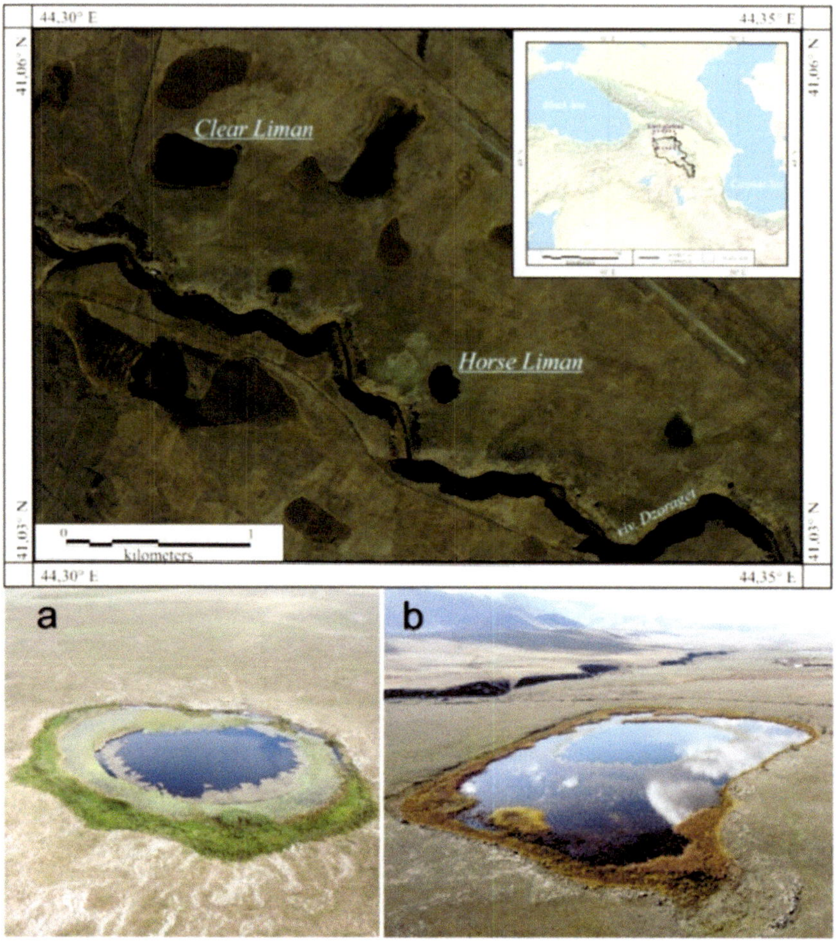

Figure 3: Sampling sites on the Lori plateau (top). (a) Horse Liman Lake, (b) Clear Liman Lake, with an inset showing location in Armenia.

3.1. Sample Selection

Samples were taken from two freshwater lakes of Lori province, northern Armenia, namely Clear Liman and Horse Liman. The lakes are located at 1470–1480 m above sea level, in the central part of the Lori plateau, between the canyons of Tashir and Dzoraget rivers (Figure 1) (Sadokov, 2022). Surface water was sampled from each lake from a depth of 0.5 m. A total of 150 liters of water was poured manually through a nylon net filter with a mesh size of 75 μm. Bottom sediments were sampled using a metal cone sampler, as shown in Figure 4.

Figure 4: Figure of a metal cone sampler.

3.2. Study of the Phytoplankton Population

The study of phytoplankton in water samples from both Clear Liman and Horse Liman lakes showed that cyanobacteria predominated in terms of abundance, accounting for 48% and 43%, respectively, and euglena algae were dominant in terms of biomass – 48% and 42%. Cyanobacteria (Cyanophyta), diatoms (Bacillariophyta), green algae (Chlorophyta), and euglenophytes (Euglenophyta) were noted as the main groups of the community in the water of both studied lakes. Dinophyta were found only in the Clear Liman plankton. Cyanobacteria were found in the phytoplankton of both lakes; in terms of abundance, the quantitative maximum in Horse Liman was 68.0 thousand cells/l, and in terms of biomass, the maximum was in Clear Liman (0.39 g/m^3). The main Cyanophyta species recorded were *Aphanothece clathrata, A. stagnina, Microcystis aeruginosa, Nostoc punctitorne, Spirulina platensis,* and *Merismopedia elegans.* In the water samples of the Clear Liman, diatoms were poorly represented (14.0–20.0 thousand cells/L), and species were noted: *Cyclotella comta, Synedra ulna, Cocconeis placentula, Navicula* sp., *Stephanodiscus astrae, Cymbella lanceolata.* In the plankton of the Horse Liman, green algae accounted for 54.0 thousand cells/L, 0.4 g/m^3, the main species were *Ankistrodesmus spirilliformis, A. stipitatus, A. angustus, Monoraphidium griffithii, Kirchneriella contorta,* and *Oocystis solitaria* (Figure 5).

In the bottom sediments of both lakes (Figure 6), diatoms were characterized by the highest species diversity (26 species were recorded), of which *Fragilaria*

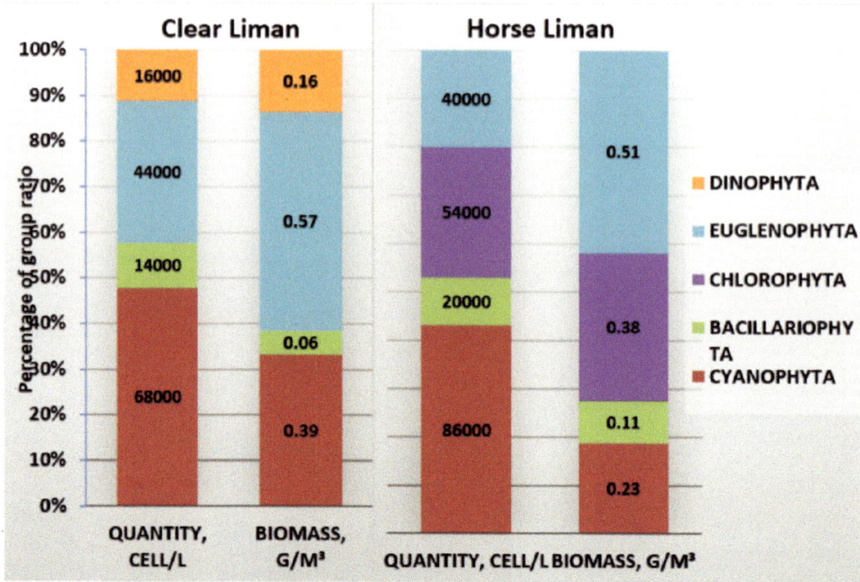

Figure 5: Percentage (%) of group share of phytoplankton in the water of Horse Liman and Clear Liman lakes by abundance (cell/L.) and biomass (g/m^3).

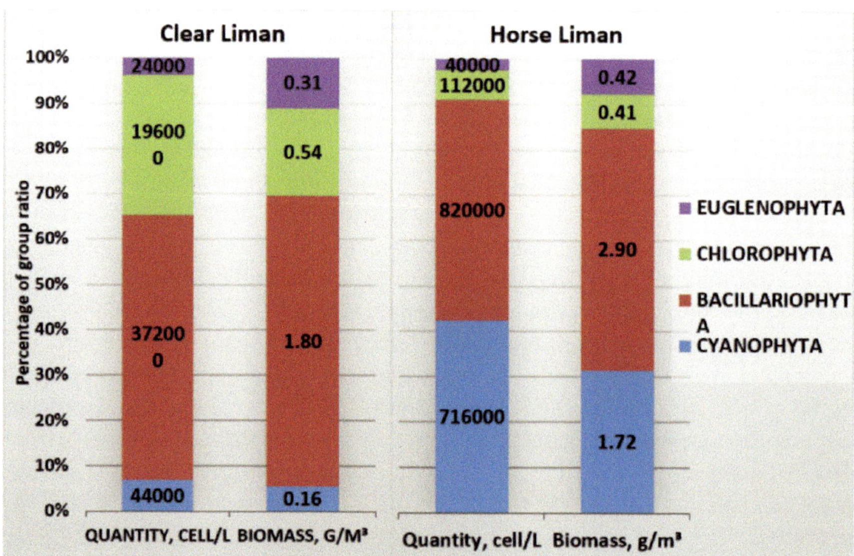

Figure 6: Percentage (%) of group share of microalgae in sediments of Horse Liman and Clear Liman lakes by abundance (cell/L.) and biomass (g/m^3).

capucina, Diatoma hiemale, and *Cocconeis placentula* were noted to be the most numerous. In second place in Horse Liman was a group of green algae of more than 6 species, and in Clear Liman – more than 4 species of cyanobacteria. In the bottom communities of both lakes, Clear Liman diatoms quantitatively dominated – 49%, blue-green 42% were subdominant, and in Horse Liman diatoms accounted for 58%, green algae were in second place, accounting for 31% of the total share of the community.

Thus, revealed phytoplankton indicator species characterized the lake category as slightly polluted, where the presence of cyanobacteria and their dominance carries risks that there may be toxic water blooms, which will be a manifestation of eutrophication. The presence of large-celled taxa of the Euglenophyta: *Trachelomonas volvocina* and *Euglena* sp., both in phytoplankton and in sediments, indicates the presence of persistent organic water pollution. The overall composition of microalgae populations indicates the possibility of MPs fouling. It is necessary to assess the way of interaction between MPs and algae communities *in situ* in future research.

3.3. Microplastics Identification and Research

Since there is still no accepted standard for sample preparation, the choice of methodology is based on the known published data sources. The main stages of laboratory work aimed at sample treatment and MP identification include removal of organic matter, density separation, filtering, visual identification, and determination of the polymer type (Saprykin & Samoylov, 2021).

Chemical and enzymatic methods are used to remove organic matter. Chemical approaches are based on the use of strong acids, alkalis, and oxidizing agents, and proteinase-K, protease, and trypsin are more often used as enzymes. The use of alkali or acid causes the deformation of plastic, while enzymes are not widely used because of their high cost and low efficiency. Fenton's reagent (H_2O_2 + Fe^{2+}) or 30% hydrogen peroxide is the main chemical for removing organics, most commonly used due to a reasonable cost-effectiveness ratio (Saprykin & Samoylov, 2021). As a result of organic substances oxidation, Fenton's reagent causes a violent exothermic reaction. The temperature regime should be controlled to avoid splashing and destruction or loss of studied MP in a sample (Liu et al., 2019).

Density separation is carried out with the use of heavy liquid to separate polymer particles from a denser precipitate. The most commonly used heavy liquids are sodium chloride, sodium metatungstate, potassium iodide, and zinc chloride (Coppock et al., 2017). Although sodium chloride is the cheapest and safest compound among other heavy liquids, its maximum density is limited by 1.2 g/m^3, which is insufficient to separate some of the polymers (PET, PVC) (Zobkov & Evsyukova, 2018). Zinc chloride is mentioned as a heavy liquid in many studies, with the only drawback being its toxicity, which imposes an obligation on a laboratory to be equipped so that dangerous compounds could be utilized properly (Coppock et al., 2017). We used a solution of zinc chloride with a density of 1.5 g/cm^3, which is sufficient to extract MPs abundantly found in the environment (Vermeiren et al., 2020).

Decantation or centrifugation methods can be used to separate heavy mineral sediment from floating plastic. Various vessels can be used as separating columns during decantation (glass beakers, cylinders with and without overflow, separating funnels, conical funnels, columns with a ball valve) (Julie et al., 2015; Coppock et al., 2017; Crichton et al., 2017; Vermeiren et al., 2020). Samples were processed based on the methodology recommended by the National Oceanic and Atmospheric Administration (NOAA) (Julie et al., 2015) and the laboratory protocol presented by Vermeiren P (Vermeiren et al., 2020) with some modifications. Since the bottom sediment samples contained a large amount of organic matter, the sample was not dried to avoid its sintering, so oxidation of organic matter was carried out before density separation (Vermeiren et al., 2020).

A 100 g of the wet-weight sediment was placed in a 500 cm^3 conical flask, added with 50 ml of 25% H_2O_2, stirred periodically, and the reaction was monitored. If the reaction ran very vigorously, the flask was cooled in water. After 18 hours, 20 ml of Fenton's reagent was added: 10 ml 50% H_2O_2 + 10 ml Fe^{2+} (2.5 g $FeSO_4$*$7H_2O$, 165 ml H_2O, and 1 ml strong H_2SO_4). A day later, we proceeded to the stage of density separation. For this, a glass funnel (Julie et al., 2015) was used, the outlet of which was blocked with a glass test tube with a rubber tip (Figure 7).

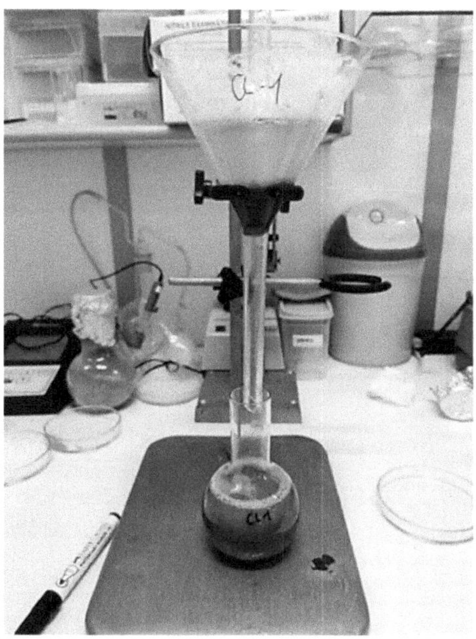

Figure 7: Apparatus using glass funnel for density separation.

The sediment was gradually drained from the solution through the funnel outlet, weakening the tightness of the rubber tip. The supernatant residual solution was transferred to the vacuum filter. A 38-micron polyamide mesh was used as a filter. After filtering and washing, the mesh was placed into a petri dish and dried for at least 2 hours. Next, the mesh was viewed under an AmScope stereo microscope with a magnification of 7-45X. MP particles were identified visually, transferred to a clean petri dish when possible, and photographed using an AmScope 10 MP camera. The type of polymer was determined using a Fourier Transform Infrared Spectrometer in Attenuated Total Internal Reflection (ATR-FTIR).

4. Results and Discussion

As a result of the Clear Liman samples research, 100 MP/m^3 was found in surface water and 60 MP/(kg of wet weight) was estimated for bottom sediments. In the Horse Liman, the concentration of MPs was assessed at 153 MP/m^3 and 80 MP/(kg of wet weight) in surface water and bottom sediments, respectively. In water samples, detected particles were represented by the following forms: fibers – 87%, fragments – 13%. In bottom sediments, all discovered MP particles were determined as fibers. Spheres and films were not found anywhere. The length of the fibers ranged from 100 μm to 10 mm, and the thickness varied from 13–45 μm. The fragments had a size from 380–450 μm. Three types of polymers have been identified: polyethylene terephthalate (PET), polypropylene (PP), and polyethylene (PE). Microscopic images of two investigated MP fragments are presented in Figure 8.

These preliminary observations are in good agreement with the known data that fibers are the dominant form of MP in freshwater bodies (Su et al., 2016; Wang et al., 2018; González-Pleiter et al., 2020; Alfonso et al., 2020). Since no household sewage is reported to inflow into the researched lakes, an atmospheric source of fiber pollution can be assumed for the Horse Liman and the Clear Liman (Zhang et al., 2020; González-Pleiter et al., 2020).

Visual pre-identification of MP can be the reason for many small particles to be underestimated or, conversely, overestimated because of the similarity of some natural pieces and plastic particles. Detection methods such as IR, Raman microscopy, and FTIR imaging based on the FPA focal plane enable the elimination of the human factor risks and determine polymer types and sizes more accurately and quickly. ATR-FTIR is a more profound technique for investigation of polymer type of black particles compared to Raman microscopy, which causes fluorescence of such objects, but the disadvantage of ATR-FTIR is a limited size of measured particles (no smaller than 20 μm) (Käppler et al., 2016). We also experienced difficulties in detecting small fibers with ATR-FTIR, likewise, it was mentioned in other studies (Stanton et al., 2019).

To account for potential background external contamination in the laboratory, 'ambient' MP particles were surveyed and calculated on a clean petri dish which

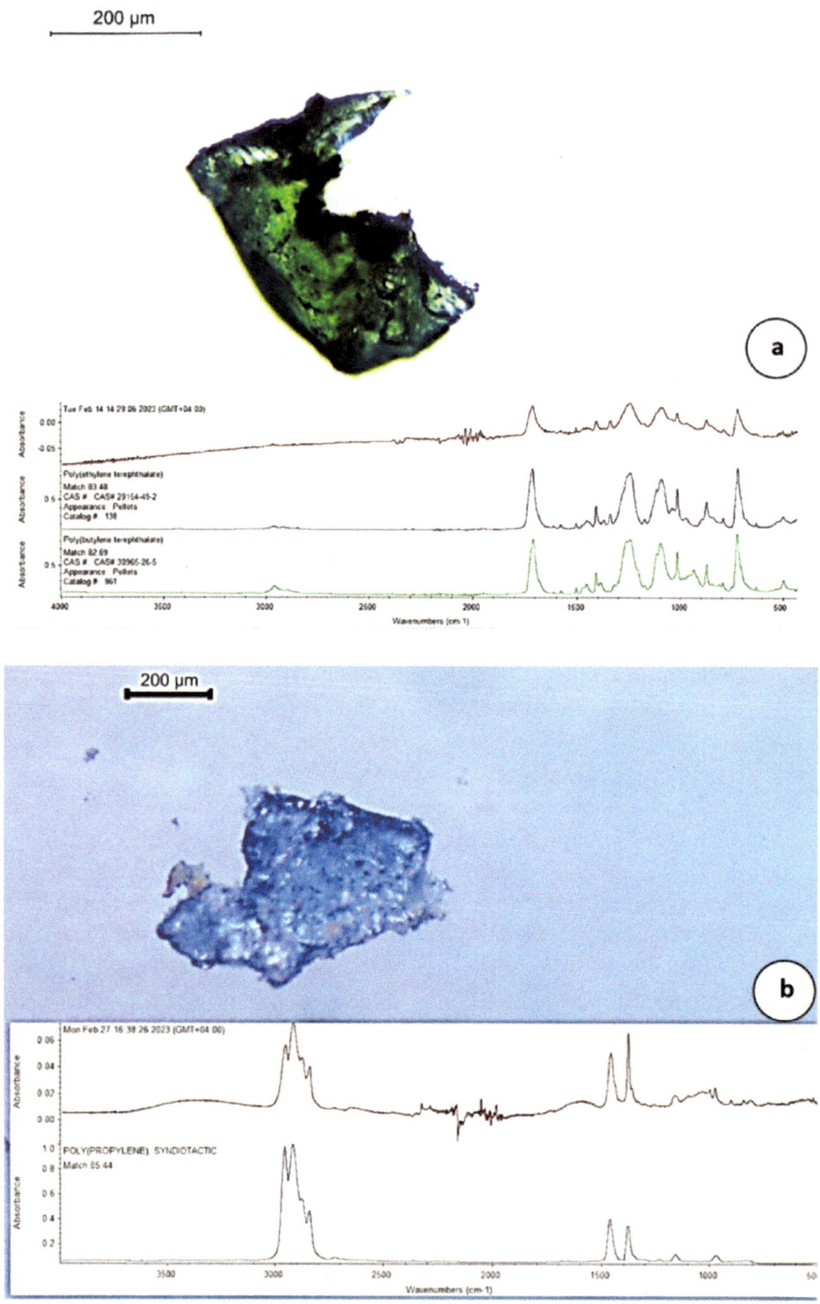

Figure 8: (a) image of a fragment of polyethylene terephthalate, (b) image of a fragment of polypropylene.

accumulated during the whole laboratory session. A total of 12 fiber particles were noticed to have ambient laboratory origins, with particularly many red particles among them (Figure 9 a,b,c), although laboratory staff wore cotton gowns and latex gloves, and no red clothes were worn during the work. White fibers were also found on the petri dish and identified by FTIR as cellulose (Figure 9 d,e), so similar fibers found in the samples from the lakes were not counted. Thus, it is not always possible to determine a particular source of MP contamination, and a more insightful investigation of this issue should be carried out, additionally, strict sterility should be secured during laboratory work and sampling.

On a global scale, microplastics have been found in all mountain lakes where research has been carried out. The degree of MP pollution in our studied lakes is estimated as low compared with the studies on similar water bodies (Imhof et al., 2013; Eriksen et al., 2013; Free et al., 2014; Su et al., 2016; Wang et al., 2018; Alfonso et al., 2020; Tsering et al., 2022). For example, the average MP concentration value in the surface waters of Patagonian lakes was estimated at 0.9 ± 0.6 MPs/m^3, and in bottom sediments of a high-altitude freshwater Anchar Lake (NW Himalaya) MP abundance was estimated at 606 ± 360 MP/kg. However, comparison of results from different case studies is often not reliable mainly due to differences in sampling, processing, and measurement units.

Considering our observations and most common laboratory practices of microplastic identification, it is essential to outline principal trajectories, problems, restrictions, and approaches of similar research cases. First of all, methodological standards should be agreed upon internationally and considering different types of aquatic systems (depending on salinity, size, hydromechanics, and human impact

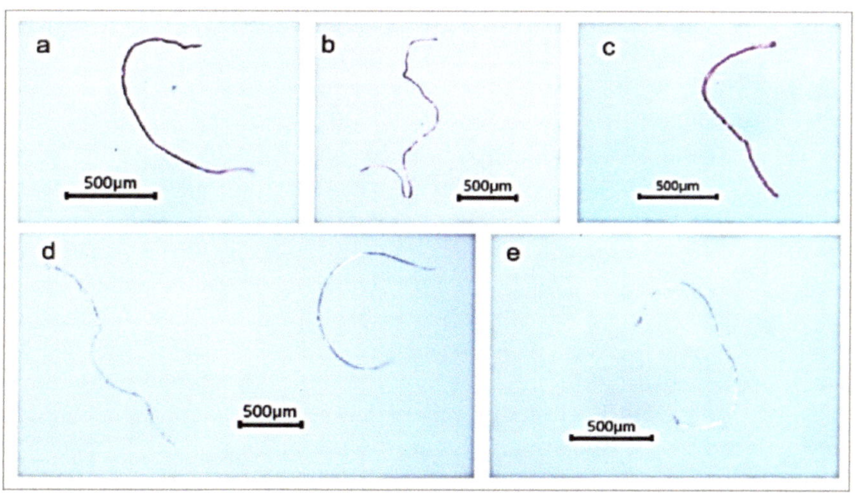

Figure 9: Image of filaments under a microscope (shown with enhanced contrast). (a, b) – red fibers found in a blank petri dish; (c) – red fibers found in the sediment sample; (d) – white fibers of cellulose on a blank petri dish; (e) – fibers found in the sample.

level). These standards are key tools to estimate local MPs' pollution degree and to make it comparable on the regional or even global scale. Particle number per volume is one of the widespread measurement units, which we suggest being used as a priority. It is also possible and reasonable to develop a bioindication gradation in order to practice certain species wellness and stability monitoring as a proxy of MPs contamination and effects.

Secondly, when elaborated, the standards need to be signed, unified, and ratified in different countries. Moreover, the methods presented as standards must take into consideration the analysis of waste origination and transportation to its ultimate sink in water bodies. In this way, interdisciplinary approaches should be employed to obtain a comprehensive picture within research, including studies of landscape ecology, geochemical and hydrochemical barriers, environmental toxicology, urban ecology, and, obviously, waste management legislation analysis. Citizen science, monitoring, and grassroots voluntary initiatives must not be ignored, since they can bring up many new issues to discuss and these are key points to sustain ecosystem services provided by lakes or seas with the involvement of indigenous people.

5. Conclusion

Once MP particles were discovered (for the first time) in the mountainous lakes Clear Liman and Horse Liman water and sediments, it proved the evidence of universal MP pollution of aquatic ecosystems.

MP abundance in water bodies likely depends on the remoteness of settlements (especially households), the vicinity of tourist places, recreational and industrial areas, and regional waste management specifics (Talbot & Chang, 2022). In scientific reports and studies, little attention is usually paid to the sources of MP entry into water bodies and to the factors of the spatial distribution of inorganic matter (i.e., geomorphological, hydrological, and climatic factors). A series of measurements is required to trace and validate the dynamics of MP pollution. Therefore, in our future research, we also aim to collect more relevant information about the individual features of the region at the sampling sites. The lack of standardized protocols or methodological guidelines for evaluating the concentrations of plastic MNPs, including their sampling and treatment creates significant difficulties, which make it impossible to compare the results obtained by different research groups. It is essential to create a database of known MP studies, which may be useful for assessing the status of ecosystems or the degree of their damage, as well as profile information networks that can significantly contribute to combating plastic pollution. Construction of accurate models of MPs compositional changes should be performed, and risks of transformation of freshwater biota into 'plastispheres' need to be evaluated. Results, presented in the chapter, have been the first step to initiate research on MP pollution in freshwater systems in Armenia.

Acknowledgments

The work was performed in frames of the grant project "Assessment of microplastic pollution in different water bodies of RA and analysis of the state of bioindicator microalgae as a result of this impact" supported by the internal grant program of Yerevan State University. The authors are grateful for the assistance and consultation of Daria Tikhonova (Institute of Limnology of the Russian Academy of Sciences, SPC RAS) and Alexander Esayan (Yerevan State University).

Declarations

Conflict of Interest Statement
The authors declare no competing interest, financial or otherwise.

Authors' Contributions
All authors have contributed to the preparation of this article

Use of Human Subjects or Animals in Research
Research Involving Animals: N.A.
Research Involving Human Subjects: N.A.

References

Alfonso, M.B., Scordo, F., Seitz, C., MavoManstretta, G.M., Ronda, A.C, Arias, A.H. et al. (2020). First evidence of microplastics in nine lakes across Patagonia (South America). *Science of The Total Environment*, 733. https://doi.org/10.1016/j.scitotenv.2020.139385.

Allen, S., Allen, D., Karbalaei, S., Maselli, V., & Walker, T.R. (2022). Micro(nano) plastics sources, fate, and effects: What we know after ten years of research. *Journal of Hazardous Materials Advances*, 6,100057. https://doi.org/10.1016/j.hazadv.2022.100057.

Andrady, A.L., & Neal, M.A. (2009). Applications and societal benefits of plastics. *Philosophical Transactions of the Royal Society B: Biological Sciences*, 364, 1977–1984. https://doi.org/10.1098/rstb.2008.0304.

Andrady, A.L. (2011). Microplastics in the marine environment. *Mar. Pollut. Bull.*, 62, 1596–1605.

Barnes, D.K.A., Galgani, F., Thompson, R.C., & Barlaz, M. (2009). Accumulation and fragmentation of plastic debris in global environments. *Philosophical Transactions of the Royal Society B: Biological Sciences*, 364, 1985–1998. https://doi.org/10.1098/rstb.2008.0205.

Berdnilov, S.V., Seliutin, V.V., Shestak, S.A., Iakovleva, A.A., & Mikhaliuka, R.G. (2022). *Scientific Basis for the Conservation and Rehabilitation of Lake Sevan Natural Resources*. Federal Research Centre Southern Scientific Centre of the Russian Academy of Sciences National Academy of Sciences of the Republic of Armenia.

Bergami, E., Pugnalini, S., Vannuccini, M.L., Manfra, L., Faleri, C., Savorelli, F. et al. (2017). Long-term toxicity of surface-charged polystyrene nanoplastics to marine planktonic species *Dunaliella tertiolecta* and *Artemia franciscana*. *Aquatic Toxicology*, 189, 159–169. https://doi.org/10.1016/J.AQUATOX.2017.06.008.

Besseling, E., Wang, B., Lürling, M., & Koelmans, A.A. (2014). Nanoplastic affects the growth of *S. obliquus* and reproduction of *D. magna*. *Environ. Sci. Technol.*, 48. https://doi.org/10.1021/es503001d.

Browne, M.A., Crump, P., Niven, S.J., Teuten, E., Tonkin, A., Galloway, T., & Thompson, R. (2011). Accumulation of microplastic on shorelines worldwide: Sources and sinks. *Environ. Sci. Technol.*, 45, 9175–9179. https://doi.org/10.1021/es201811s.

Bryant, J.A., Clemente, T.M., Viviani, D.A., Fong, A.A., Thomas, K.A., Kemp, P. et al. (2016). Diversity and activity of communities inhabiting plastic debris in the North Pacific Gyre. *M. Systems*, 1. https://doi.org/10.1128/msystems.00024-16.

Capolupo, M., Sørensen, L., Jayasena, K.D.R., Booth, A.M., & Fabbri, E. (2020). Chemical composition and ecotoxicity of plastic and car tire rubber leachates to aquatic organisms. *Water Res.*, 169. https://doi.org/10.1016/j.watres.2019.115270.

Carson, H.S., Nerheim, M.S., Carroll, K.A., & Eriksen, M. (2013). The plastic-associated microorganisms of the North Pacific Gyre. *Mar. Pollut. Bull.*, 75, 126–132. https://doi.org/10.1016/J.MARPOLBUL.2013.07.054.

Chia, W.Y., Tang, D.Y., Shoo, K.S., Lup, A.N.K., & Chew, K.W. (2020). Nature's fight against plastic pollution: Algae for plastic biodegradation and bioplastics production. *Env. Sci. & Tech.*, 4. DOI: 10.1016/j.ese.2020.100065.

Chinaglia, S., Tosin, M., & Degli-Innocenti, F. (2018). Biodegradation rate of biodegradable plastics at the molecular level. *Polym. Degrad. Stabil.*, 147, 237–244.

Cole, M., Lindeque, P., Fileman, E., Halsband, C., Goodhead, R., Moger, J., & Galloway, T.S. (2013). Microplastic ingestion by zooplankton. *Environ. Sci. Technol.*, 47, 6646–6655. https://doi.org/10.1021/es400663f.

Coppock, R.L., Cole, M., Lindeque, P.K., Queirós, A.M., & Galloway, T.S. (2017). A small-scale, portable method for extracting microplastics from marine sediments. *Environmental Pollution*, 230, 829–837. https://doi.org/10.1016/j.envpol.2017.07.017.

Crichton, E.M., Noël, M., Gies, E.A., & Ross, P.S. (2017). A novel, density-independent and FTIR-compatible approach for the rapid extraction of microplastics from aquatic sediments. *Analytical Methods*, 9, 1419–1428. https://doi.org/10.1039/c6ay02733d.

Crooks, J.A. (2002). Characterizing ecosystem-level consequences of biological invasions: The role of ecosystem engineers. *Oikos*, 97.

Eriksen, M., Mason, S., Wilson, S., Box, C., Zellers, A., Edwards, W. et al. (2013). Microplastic pollution in the surface waters of the Laurentian Great Lakes. *Mar. Pollut. Bull.*, 77, 177–182. https://doi.org/10.1016/J.MARPOLBUL.2013.10.007.

Free, C.M., Jensen, O.P., Mason, S.A., Eriksen, M., Williamson, N.J., & Boldgiv, B. (2014). High-levels of microplastic pollution in a large, remote, mountain lake. *Mar. Pollut. Bull.*, 85, 156–163. https://doi.org/10.1016/J.MARPOLBUL.2014.06.001.

Gevorgyan, G.A., Mamyan, A.S., Hambaryan, L.R., Khudaverdyan, S.K., & Vaseashta, A. (2016). Environmental risk assessment of heavy metal pollution in Armenian River ecosystems: A case study of Lake Sevan and Debed river catchment basins. *Pol. J. Environ. Stud.*, 25. https://doi.org/10.15244/pjoes/63734.

González-Pleiter, M., Velázquez, D., Edo, C., Carretero, O., Gago, J., Barón-Sola, Á. et al. (2020). Fibers spreading worldwide: Microplastics and other anthropogenic litter in an Arctic freshwater lake. *Science of The Total Environment*, 722. https://doi.org/10.1016/j.scitotenv.2020.137904.

González-Pleiter, M., Edo, C., Aguilera, Á., Viúdez-Moreiras, D., Pulido-Reyes, G., González-Toril, E. et al. (2021). Occurrence and transport of microplastics sampled within and above the planetary boundary layer. *Science of The Total Environment*, 761. https://doi.org/10.1016/j.scitotenv.2020.143213.

Hambaryan, L.R., Stepanyan, L.G., Mikaelyan, M.V., & Gyurjyan, Q.G. (2020). The bloom and toxicity of Cyanobacteria in Lake Sevan. *Proceedings of the YSU B: Chemical and Biological Sciences*, 54. https://doi.org/10.46991/pysu:b/2020.54.2.168.

Hitchcock, J.N. (2022). Microplastics can alter phytoplankton community composition. *Science of The Total Environment*, 819, 153074. https://doi.org/10.1016/j.scitotenv.2022.153074.

Imhof, H.K., Ivleva, N.P., Schmid, J., Niessner, R., & Laforsch, C. (2013). Contamination of beach sediments of a subalpine lake with microplastic particles. *Current Biology*, 23, R867–R868. https://doi.org/10.1016/J.CUB.2013.09.001.

Julie, M., Joel, B., Gregory, F., & Courtney, Arthur. (2015). *Laboratory Methods for the Analysis of Microplastics in the Marine Environment: Recommendations for Quantifying Synthetic Particles in Waters and Sediments*. NOAA Technical Memorandum NOS-OR&R-48.

Kalogerakis, N., Karkanorachaki, K., Kalogerakis, G.C., Triantafyllidi, E.I., Gotsis, A.D., Partsinevelos, P., & Fava, F. (2017). Microplastics generation: Onset of fragmentation of polyethylene films in marine environment mesocosms. *Front. Mar. Sci.*, 4. https://doi.org/10.3389/fmars.2017.00084.

Käppler, A., Fischer, D., Oberbeckmann, S., Schernewski, G., Labrenz, M., Eichhorn, K.J., & Voit, B. (2016). Analysis of environmental microplastics by vibrational microspectroscopy: FTIR, Raman or both? *Anal. Bioanal. Chem.*, 408, 8377–8391. https://doi.org/10.1007/s00216-016-9956-3.

Kocharyan, R., (2004). *The Law of the Republic of Armenia on Waste*. 9 pp. https://procurement- notices.undp.org/view_file.cfm?doc_id=160529.

Kooi, M., Besseling, E., Kroeze, C., Van Wezel, A.P., & Koelmans, A.A. (2018). Modeling the fate and transport of plastic debris in freshwaters: Review and guidance. *In: Handbook of Environmental Chemistry* (pp. 125–152). Springer Verlag.

Kozlovskii, N.V., & Blinovskaia, J.Y. (2015). Microplastic is a world ocean macro problem. *Ecology and Population Health*, 10, 159–162.

Kumar, R.V., Kanna, G., & Elumalai, S. (2017). Biodegradation of polyethylene by green photosynthetic microalgae. *J. Biorem. Biodegrad.*, *8*, 2.

Lambert, S., & Wagner, M. (2018). Microplastics are contaminants of emerging concern in freshwater environments: An overview. *In: The Handbook of Environmental Chemistry*. Springer Verlag.

Lavers, J.L., Bond, A.L., & Hutton, I. (2014). Plastic ingestion by flesh-footed shearwaters (Puffinus carneipes): Implications for fledgling body condition and the accumulation of plastic-derived chemicals. *Environmental Pollution*, 187, 124–129. https://doi.org/10.1016/J.ENVPOL.2013.12.020.

Leoni, B., Marti, C.L., Imberger, J., & Garibaldi, L. (2014). Summer spatial variations in phytoplankton composition and biomass in surface waters of a warm-temperate, deep, oligo-holomictic lake: Lake Iseo, Italy. *Inland Waters*, 4. https://doi.org/10.5268/IW-4.3.569.

Lin, J., Yan, D., Fu, J., Chen, Y., & Ou, H. (2020). Ultraviolet-C and vacuum ultraviolet inducing surface degradation of microplastics. *Water Res.*, 186. https://doi.org/10.1016/j.watres.2020.116360.

Liu, F., Olesen, K.B., Borregaard, A.R., & Vollertsen, J. (2019). Microplastics in urban and highway stormwater retention ponds. *Science of The Total Environment*, 671, 992–1000. https://doi.org/10.1016/j.scitotenv.2019.03.416.

Liu, P., Zhan, X., Wu, X., Li, J., Wang, H., & Gao, S. (2020). Effect of weathering on the environmental behavior of microplastics: Properties, sorption, and potential risks. *Chemosphere*,242,125193.https://doi.org/10.1016/J.CHEMOSPHERE.2019.125193.

Lobelle, D., & Cunliffe, M. (2011). Early microbial biofilm formation on marine plastic debris. *Mar. Pollut. Bull.*, 62, 197–200. https://doi.org/10.1016/J.MARPOLBUL.2010.10.013.

Luo, H., Liu, C., He, D., Xu, J., Sun, J., Li, J., & Pan, X. (2022). Environmental behaviors of microplastics in aquatic systems: A systematic review on degradation, adsorption, toxicity and biofilm under aging conditions. *J. Hazard. Mater.*, 423.

Lusher, A. (2015). Microplastics in the marine environment: Distribution, interactions, and effects. *In: Marine Anthropogenic Litter* (pp. 245–307). Springer International Publishing.

Mao, R., Lang, M., Yu, X., Wu, R., Yang, X., & Guo, X. (2020). Aging mechanism of microplastics with UV irradiation and its effects on the adsorption of heavy metals. *J. Hazard. Mater.*, 393. https://doi.org/10.1016/j.jhazmat.2020.122515.

Marti, C.L., Imberger, J., Garibaldi, L., & Leoni, B. (2016). Using time scales to characterize phytoplankton assemblages in a deep subalpine lake during the thermal stratification period: Lake Iseo, Italy. *Water Resour. Res.*, 52. https://doi.org/10.1002/2015WR017555.

Morohoshi, T., Oi, T., Aiso, H., Suzuki, T., Okura, T., & Sato, S. (2018). Biofilm formation and degradation of commercially available biodegradable plastic films by bacterial consortiums in freshwater environments. *Microbes Environ.*, 33. https://doi.org/10.1264/jsme2.ME18033.

Muthukrishnan, T., Al Khaburi, M., & Abed, R.M.M. (2019). Fouling microbial communities on plastics compared with wood and steel: Are they substrate- or location-specific? *Microbiol.*, 78. https://doi.org/10.1007/s00248-018-1303-0.

Nava, V., Patelli, M., Soler, V., & Leoni, B. (2017). Interspecific relationship and ecological requirements of two potentially harmful cyanobacteria in a deep South-Alpine Lake (L. Iseo, Italy). *Water (Switzerland)*, 9. https://doi.org/10.3390/w9120993.

Pedrotti, M.L., de Figueiredo Lacerda, A.L., Petit, S., Ghiglione, J.F., & Gorsky, G. (2022). Vibrio spp. and other potential pathogenic bacteria associated to microfibers in the North-Western Mediterranean Sea. *PLoS One*, 17, e0275284. https://doi.org/10.1371/journal.pone.0275284.

Reynolds, C.S. (2006). *Ecology of Phytoplankton: Ecology, Biodiversity and Conservation*. Cambridge University Press, Cambridge. http://dx.doi.org/10.1017/CBO9780511542145.

Rochman, C.M. (2018). Microplastics research – from sink to source. *Science* (1979), 360, 28–29.

Sadokov, D. (2022). *Environmental Value, Landscape, and Biological Diversity of the Lori Plateau Lakes and Watershed* (Lori Region, Republic of Armenia). Final Report. Yerevan.

Sakharova, E.G., Krylov, A.V., Sabitova, R.Z., Tsvetkov, A.I., Gambaryan, L.R., Mamyan, A.S. et al. (2020). Horizontal and vertical distribution of phytoplankton in the Alpine Lake Sevan (Armenia) during the summer Cyanoprokaryota Bloom. *Contemp. Probl. Ecol.*, 13. https://doi.org/10.1134/S1995425520010072.

Sarmah, P., & Rout J. (2018a). Algal colonization on polythene carry bags in a domestic solid waste dumping site of Silchar town in Assam. *Phykos.*, 48, 67–77,

Sarmah, P., & Rout, J. (2018b). Efficient biodegradation of low-density polyethylene by cyanobacteria isolated from submerged polyethylene surface in domestic sewage water. *Environ. Sci. Pollut. Res.*, 25, 33508–33520. https:// doi: 10.1007/s11356-018-3079-7.

Sapozhnikov, P., Salimon, A., Korsunsky, A.M., Kalinina, O., Ilyina, O., Statnik, E., & Snigirova, A. (2022). Plastic in the aquatic environment: Interactions with microorganisms. *In: Handbook of Environmental Chemistry* (pp. 197–254). Springer Science and Business Media Deutschland GmbH.

Saprykin, A., & Samoylov, P. (2021). Micro- and nanoplastics in the environment, Analytics, sources, distribution, and environmental problems. *Ekol., Ser. Anal. Obz. Mirovoi Lit.,* 110, 1. https://www.spcl.nsc.ru/wp-content/uploads/2022/04/V_110.pdf.

Sharma, M.., Dubey, A., & Pareek, (2014). A. Algal flora on degrading polythene waste. *CIBTech. J. Microbiol.*, 3, 43–47.

Sjollema, S.B., Redondo-Hasselerharm, P., Leslie, H.A., Kraak, M.H.S., & Vethaak, A.D. (2016). Do plastic particles affect microalgal photosynthesis and growth? *Aquatic Toxicology*, 170, 259–261. https://doi.org/10.1016/j.aquatox.2015.12.002.

Stanton, T., Johnson, M., Nathanail, P., MacNaughtan, W., & Gomes, R.L. (2019). Freshwater and airborne textile fiber populations are dominated by 'natural', not microplastic, fibers. *Science of The Total Environment*, 666, 377–389. https://doi.org/10.1016/j.scitotenv.2019.02.278.

Su, L., Xue, Y., Li, L., Yang, D., Kolandhasamy, P., Li, D., & Shi, H. (2016). Microplastics in Taihu Lake, China. *Environmental Pollution*, 216, 711–719. https://doi.org/10.1016/J.ENVPOL.2016.06.036.

Talbot, R., & Chang, H. (2022). Microplastics in freshwater: A global review of factors affecting spatial and temporal variations. *Environmental Pollution*, 292, 118393. https://doi.org/10.1016/J.ENVPOL.2021.118393.

Thompson, R.C., Olsen, Y., Mitchell, R.P., Davis, A., Rowland, S.J., John, A.W.G. et al. (2004). Lost at sea: Where is all the plastic? *Science* (1979), 304, 838–838. https://doi.org/10.1126/science.1094559.

Tian, F., Zhou, J., Sun, Z., Cai, Z., Xu, N., An, M., & Duan, S. (2014). Inhibitory effects of Chinese traditional herbs and herb-modified clays on the growth of harmful algae, Phaeocystisglobosa and Prorocentrum Donghai Tense. *Harmful Algae*, 37. https://doi.org/10.1016/j.hal.2014.05.015.

Tsering, T., Sillanpää, M., Viitala, M., & Reinikainen, S.P. (2022). Variation of microplastics in the shore sediment of high-altitude lakes of the Indian Himalaya using different pretreatment methods. *Science of The Total Environment*, 849, 157870. https://doi.org/10.1016/J.SCITOTENV.2022.157870.

Vaseashta, A., Ivanov, V., Stabnikov, V., & Marinin, A. (2021). Environmental safety and security investigations of neustonic microplastic aggregates nearwater-air interphase. *Polish Journal of Environmental Studies*, 30(4), 3457–3469. https://doi.org/10.15244/pjoes/131947.

Vaseashta, A., Duca, G., & Travin, S.O. (Eds.). (2022). *Handbook of Research on Water Sciences and Society*. Engineering Science Reference. DOI: 10.4018/978-1-7998-7356-3.

Vermeiren, P., Muñoz, C., & Ikejima, K. (2020). Microplastic identification and quantification from organic rich sediments: A validated laboratory protocol. *Environmental Pollutio*n, 262. https://doi.org/10.1016/j.envpol.2020.114298.

Vladimir, M., Tatiana, R., Evgeniy, S., Veerasingam, S., & Bagaev, A. (2023). Vertical and seasonal variations in biofilm formation on plastic substrates in coastal waters of the Black Sea. *Chemosphere*, 317, 137843. https://doi.org/10.1016/J.CHEMOSPHERE.2023.137843.

Wagner, M., Scherer, C., Alvarez-Muñoz, D., Brennholt, N., Bourrain, X., Buchinger, S. et al. (2014). Microplastics in freshwater ecosystems: What we know and what we need to know. *Environ. Sci. Eur.*, 26, 1–9. https://doi.org/10.1186/s12302-014-0012-7.

Wang, W., Yuan, W., Chen, Y., & Wang, J. (2018). Microplastics in surface waters of Dongting Lake and Hong Lake, China. *Science of The Total Environmen*t, 633, 539–545. https://doi.org/10.1016/j.scitotenv.2018.03.211.

Zettler, E.R., Mincer, T.J., & Amaral-Zettler, L.A. (2013). Life in the "plastisphere": Microbial communities on plastic marine debris. *Environ. Sci. Technol.*, 47, 7137–7146. https://doi.org/10.1021/es401288x.

Zhang, C., Chen, X., Wang, J., & Tan, L. (2017). Toxic effects of microplastic on marine microalgae *Skeleton emacostatum*: Interactions between microplastic and algae. *Environmental Pollution*, 220, 1282–1288. https://doi.org/10.1016/J.ENVPOL.2016.11.005.

Zhang, Y., Kang, S., Allen, S., Allen, D., Gao, T., & Sillanpää, M. (2020). Atmospheric microplastics: A review on current status and perspectives. *Earth Sci. Rev.*, 203, 103118. https://doi.org/10.1016/J.EARSCIREV.2020.103118.

Zobkov, M., & Evsyukova, E. (2018). Microplastic v morskoy rede: obzor metodo vobora, podgotovkiianalizaprobvody, donnyhotlozheniyiberegovyhnanosov. *Okeanologiya*, 58, 149–157. https://doi.org/10.7868/s0030157418010148.

10 | Hazards Associated with Micro/Nano Plastics in Agricultural Soils

Surya Singh[1,2]

[1] Division of Environmental Monitoring and Exposure Assessment (Water & Soil), ICMR – National Institute for Research in Environmental Health, Bhopal – 462 030, India

[2] Academy of Scientific and Innovative Research (AcSIR), Ghaziabad – 201 002, India

Abstract: Microplastics (MPs) have been identified as one of the emerging contaminants posing risks to the environment as well as human health. However, research in this direction has progressed in the recent decade only. Being small in size and light in weight, these particles are ubiquitous in various environmental matrices; viz., water, air, and soil; and thus pose ecological as well as human health risks. Increasing plastic usage in agricultural practices in the form of packaging, mulching, etc. has introduced this contaminant in agricultural soil as well. Agriculture is one of the fundamental requirements for the sustenance of human lives. Considering the role of agriculture in human lives and the economy, it is of utmost importance to keep the quality of soil intact and free from any kind of pollutants. Nevertheless, the United Nations' sustainable development goals (SDGs) also emphasize ending hunger, achieving food security, improving nutrition, and promoting sustainable agriculture. The microplastic particles, upon mixing with the soil, disturb the natural soil composition and properties. As plastic particles are carbon-rich polymers, these also have the potential to influence the carbon and nitrogen ratio of the soil affecting crop growth. Effects on crop growth and/or crop's nutritional properties are expected to negatively affect human health as well. Therefore, it is of utmost importance to comprehend the presence of micro/nanoplastics (MNPs) in agricultural soil and its associated hazards.

*Corresponding author: suryasingh.nireh@icmr.gov.in

1. Introduction

Soil is the loose material covering the surface of most of the land. Fundamental components of the soil include organic matter and inorganic particles. These fundamental components of the soil are structural support for the plants (agriculture) as well as the water source and nutrients. Contamination of this organic matter and/or disturbances in the inorganic composition of the soil, therefore, affects the soil quality which in turn deteriorates the plant growth. Therefore, conservation and preservation of the soil is not only important for the environment but for the sustained growth of agriculture as well.

In the present scenario, several inorganic and organic contaminants are found in the soil, such as metals/metalloids, non-metals, halogenated, non-halogenated compounds, and so on. Examples of these categories include cadmium, lead, arsenic, cyanides, chloroform, Dichlorodiphenyltrichloroethane (DDT), polychlorobiphenyls, ethane, benzene, toluene, polyaromatic hydrocarbons etc.

Table 1: Emerging contaminants in soil and their principal sources (Adapted from Maddela et al. 2022)

Emerging Contaminants	Potential Source	Reference
Per- and poly-fluoroalkyl substances (PFAS)	Food packaging material, stain- and water-repellent fabrics, non-stick products (e.g., Teflon), polishes, waxes, paints, cleaning products, fire-fighting foams, industrial facilities, landfill wastewater treatment plants, and living organisms due to accumulation and persistence over time	USEPA, 2018
Micro/nanoplastics (MNPs)	Unscientifically disposed/mismanaged plastic waste	Singh et al., 2024a,b
Additives – phthalates	Plastic additives	Benjamin et al., 2015
Additives – biphenyls	Dielectric fluids, additives in lubricant fluids, adhesives, building sealants, plasticizers, fireproofing agents, paints, and ink products	Breivik et al., 2016
Polybrominated diphenyl ethers (PBDEs)	Electrical and electronic devices, furniture, textiles, plastics, baby products	Maddela et al., 2020a
Organophosphorus flame retardants (OPFRs)	Polymers used in polyurethane (PU) foam, furniture, baby products, etc.	Maddela et al., 2020b
Metals/metalloids	Fertilizers, geogenic weathering, etc.	Ashraf et al., 2014
Nanoparticles	Cosmetics, pharmaceuticals, coatings, paints, and catalytic additives	Ahmed et al., 2021

(IUPAC, 2009). Apart from these conventional contaminants, numerous emerging contaminants are also on the rise affecting soil quality, such as per- and polyfluoroalkyl substances (PFAS), micro/nanoplastics (MNPs), phthalates, biphenyls, polybrominated diphenyl ethers (PBDEs), organophosphorus flame retardants (OPFRs), nanoparticles, etc. (Table 1) (Maddela et al. 2022). Among these, MNPs are of special concern owing to their ubiquity and non-biodegradability in the environment.

MNPs are small plastic particles below the size range of 5 mm (Frias & Nash 2019). These particles have been reported from nearly all the environmental matrices, including soil. Increasing plastic usage in agricultural practices in the form of packaging, mulching, etc., has introduced this contaminant in agricultural soil as well. Agriculture, which is one of the essential necessities for the survival of human lives, is negatively impacted by the presence of such contaminants (Figure 1). Considering the importance of agriculture, the quality of soil needs to be maintained, and the presence of such contaminants has to be minimized. The Sustainable Development Goals (SDGs) of the United Nations also focus on ending hunger, achieving food security, improving nutrition, and promoting sustainable agriculture (Goal nos. 3 and 6) (SDGs, 2015).

Figure 1: Micro/nano plastics in agricultural soil and associated hazards.

2. Micro/nano Plastics in Soil

The presence of MNPs has been reported globally in agricultural soil by various researchers. Although there are variations in sampling methods and analytical procedures, the reported concentration of microplastics (MPs) indicates the severity of soil contamination. Yang et al. (2021a) reported the concentration of MNPs in Chile (22001–34000 items per kg), Australia (1271–4191 items per kg), USA (0–1270 items per kg), Spain (1271–4191 items per kg), and China (9801–22000 items per kg) (Yang et al., 2021a). Wide variation in the concentration of MNPs depends upon the type of agricultural soil from which the samples were collected. Another study reported the agricultural MNPs concentration in Europe (~63000–430000 tons), America (~44000–300000 tons), and Australia (~2800–19000 tons) (Maddela et al., 2022). Huang et al. (2020) studied the mulched

Table 2: Micro/nano plastic pollution reported in agricultural soil (Adapted from Yu et al., 2022 and Sa'adu and Farsang, 2023)

Site	Land Use	Main Sources	Sampling Depth (cm)	Concentration (items/kg)	Size (mm)	Shape	Polymer Type*	References
United Kingdom	Agricultural soil	Plastic crop cover	0–5	3680 ± 129.1	–	–	–	Cusworth et al., 2024
Germany	Long-term soil monitoring sites	–	0–25 to 0–30	29.04 ± 18.85	Mean: 1.17 ± 1.23 Median: 0.65	–	PE, Synthetic rubber polymer	Weber et al., 2024
Germany	Arable land	Sludge	0–30; 30–60; 60–90	14.6	–	–	PES, PA, PVC, PAN	Tagg et al., 2022
Schleswig-Holstein, Germany	Winter rapeseed – winter wheat – winter barley	Wastewater irrigation, organic fertilizer	0–10, 10–20	0–217.8	1–5	Fragment, fiber, film	PE, PP, PA	Harms et al., 2021
Southeast Germany	–	–	0–5	0.34 ± 0.36	1–5	Film, fragment, fiber	PE, PS, PP	Piehl et al., 2018
Switzerland	Vegetables	Mulching	Drainage water sample from agricultural soil	10.5 ± 9.5 number per liter	>0.1	–	PA, PE	Bigalke et al., 2022

Switzerland	Arable land	Organic compost	–	22.4 ± 3.3 tons per year	–	–	–	Kawecki et al., 2021
Bhopal, India	Agricultural soil	Littering, atmospheric deposition	0–5	30.7 ± 9.19 (area 1) 6.9 ± 4.95 (area 2)	1–5	Fiber, fragment	PE, PP	Singh et al., 2024a
Indore, India	Agricultural soil	Littering, atmospheric deposition	0–5	11	2–5	Fiber, particles, fragment	PET, PP, PA, PES, CA	Singh et al., 2024b
Maharashtra, India	Agricultural soil	Mulching	Un-mulched: 0 and 15 Mulched: 0, 15, and 30	40.46 (mulched; 15 cm); 20.54 (un-mulched; 15 cm)	–	Fragment, fibers, foam	Polyacrylamide, PE, PP, CA, PET	Mahesh et al., 2022
Karnataka, India	Agricultural soil	Mulching	Un-mulched: 0 and 15 Mulched: 0, 15, and 30	8.45 (mulched; 30 cm) 2.83 (un-mulched; 15 cm)	–	Fragment, fibers, foam	Polyacrylamide, PE, PP, CA, PET	Mahesh et al., 2022
India	Arable land (tomatoes)	Mulching	0–10; 11–20; 21–30	37.97%, 35.07%, 36.99% plastic residue	–	Film, fiber	–	Kumar & Sheela, 2021
Hungary	Arable land (tomatoes)	Greenhouse	0–20; 20–40	225 ± 61.69	2–3	–	PE, PVC, PP	Sa'adu and Farsang, 2022

Contd...

Contd...

Site	Land Use	Main Sources	Sampling Depth (cm)	Concentration (items/kg)	Size (mm)	Shape	Polymer Type*	References
Yeoju, Korea	Orchard field, greenhouse field, paddy field, upland	Mulching film, car tires, atmospheric deposition	0–5	664–3440	<5	Fragment, film, fiber, pellet	PE, PP, PS, PVC	Choi et al., 2021
Yong-In, Korea	Paddy land, mulching land, greenhouse land	Mulching film, littering	0–5	20–325, 10–265 75–7630	<5	Fragment, fiber, sheet	PE, PP, PET	Kim et al., 2021
Korea	Rice and vegetables	Mix	0–5	664	1–5	Fragment, film, fiber, sphere	PE, PP	Choi et al., 2021
Fars, Iran	–	Wind erosion	0–10	156 ± 153	0.040–0.74	–	Light density MPs	Rezaei et al., 2019
Japan	Paddy land	Coated fertilizer	15 cm	144 mg per kg	–	–	PE	Katsumi et al., 2021
Thailand	Horticultural land	Mix	0–5 cm; 20 cm	12–117 items per m^2	1–3	Fragment, fiber	PE, LDPE, PP, PS	Fakour et al., 2021
Mellipilla, Chile	Corn land	Sludge application	0–25	600–10400	<5	Fiber	–	Corradini et al., 2019
Hainan Island, China	Agricultural soil (Paddy, vegetables,	Plastic mulching	0–30	15461.52	0.02–0.2	Fragment	PE, PP	Khan et al., 2023

	maize, melon, banana etc.)							
China	Agricultural land	–	0–20	314	0.5–1	Fragment, foam, fiber, film	PP, PES	Zhou et al., 2023
China	Arable land (maize)	Mulching	0–100	8885 (at 0–10 cm); 2899 (at 80–100 cm)	–	Film, fiber, and granules	PE, PP, PET	Li et al., 2022
Qinghai – Tibet Plateatu, China	Farmland, mulching land, greenhouse land	Mulching film, wastewater irrigation, littering	0–3, 3–6	53.2 ± 29.7, 43.9 ± 22.3	<0.5 <0.5	Film, fragments, fibers, foams, spherules,	PP, PE, PS, PA	Feng et al., 2021
Hebei, Shandong, China	Wheat land	Sludge, wastewater irrigation, mulching	–	3910 ± 1031	<5	Fiber, fragment, film, sphere	PVC, PA, PP, PS, PE, PES	Wang et al., 2021 a
Enshi, China	Tobacco field	Mulching film, sludge application, organic fertilizer	0–20	646.67–2840	≤1	Fragment, film, fiber, foam, pellet	–	Zhang et al., 2021
Shouguang, China	Vegetable land	Mulching film, littering, organic fertilizer, runoff	0–5, 5–10, 10–25	275–5411, 179–7175, 307–4507	<0.5	Fiber, fragment, film, pellet, foam	PP, PE, PS, PES, PU, PMMA	Yu et al., 2021a

Contd...

Contd...

Site	Land Use	Main Sources	Sampling Depth (cm)	Concentration (items/kg)	Size (mm)	Shape	Polymer Type*	References
China	Arable land	Plastic gauze	–	1629.68 tons per year	–	–	–	Mo et al., 2021
Shaanxi, China	–	Sludge, wastewater irrigation, mulching	0–10	1430–3410	0–0.49	Fiber, fragment, film	PS, PE, PP, PVC, PET	Ding et al., 2020
Shihezi, China	Cotton fields	Mulching film	0–40	80.3 ± 49.3, 308 ± 138.1	<5	Fragment	PE	Huang et al., 2020
Harbin, China	Vegetable land, cornfield	Mulching film	0–20, 20–30	50 ± 87, 400 ± 692	0.05–5	–	PE	Zhang et al., 2020b
Wuhan, China	Vegetable land	Sludge application, mulching film	0–5	320–12560	<0.2	Fiber, pellet, fragment, foam	PA, PP, PS, PE, PVC	Chen et al., 2020a
Hangzhou Bay, China	Vegetable land	Sludge application, mulching film, wastewater irrigation, littering	0–10	571 (mulching), 263 (non-mulching)	0.06–5, 1–3	Film, fiber, fragment	PE, PP, PA, PES, Nylon, Acrylamide	Zhou et al., 2020
Wuhan, China	Vegetable land	Sludge application, mulching film, wastewater	0–5	4.3×10^4–6.2×10^5	0.01–0.1	Fragment, fiber, pellet, film, foam	PE, PA, PP, PS, PVC	Zhou et al., 2019

		irrigation, littering						
Kunming, China	Vegetable land	Sludge application, mulching film, wastewater irrigation	0–5, 5–10	13470–42960	0.05–0.25	Film, fragment	–	Zhang & Liu, 2018
Valencia	Cereal field, olive field	Sludge application	0–10	Without sludge: 930 ± 740 (LD), 1100 ± 570 (HD), With sludge: 2130 ± 950 (LD), 3060 ± 1680 (HD)	>0.05	Fragment, fiber, film	Low-density or high-density MPs	Berg et al., 2020
Ontario, Canada	–	Sludge application	0–15	4–541	>0.3	Fiber, fragment	PS, PE, PP, PES, PU, PA, PMMA	Crossman et al., 2020
Greece	Watermelons and tomatoes	Greenhouse	0–30	301 ± 140 and 69 ± 38	–	–	PE	Isari et al., 2021
Spain and Netherlands	Broccoli, celery, watermelon	Mix	0–10; 10–30	2242 ± 984 (Spanish soil) 888 ± 500 (Dutch soil)	–	–	PE, PP	Schothorst et al., 2021

Contd...

Contd...

Site	Land Use	Main Sources	Sampling Depth (cm)	Concentration (items/kg)	Size (mm)	Shape	Polymer Type*	References
Mauritius	Vegetables	Mix	0–10; 10–20	320 ± 112.2 (0–10); 420 ± 244 (10–20)	0.25–4.9	Fiber, fragment, flake	PP, PA	Ragoobur et al., 2021
Tunisia	Mix	–	–	13.21 ± 0.89 to 852.24 ± 124.2	0.22–1.22	–	PEVA, PE, PP	Boughattas et al. ,2021

*Abbreviations: PE: Polyethylene; PP: Polypropylene; PVC: Polyvinyl chloride; PET: Polyethylene terephthalate; PES: Polyester; PA: Polyamide; PU: Polyurethane; PS: Polystyrene; CA: Cellulose acetate; PAN: Polyacrylonitrile; LDPE: Low-density polyethylene; PEVA: Polyethylene vinyl acetate; PMMA: Polymethyl methacrylate agricultural soils in China and reported that soil under the influence of mulching had a total number of 1075.6 ± 346.8 items per kg (Huang et al., 2020). Effect of soil depth was also taken under consideration for evaluating the concentration of MPs. The amount of microplastics was found more in shallow depth (0–3 cm), viz., 53.2 ± 29.7 items per kg as compared to the quantity, such as 43.9 ± 22.3 items per kg in deep soil (3–6 cm) (Feng et al., 2021). In a similar study conducted by Liu et al. (2018) in vegetable farmlands, MPs were found more by the numbers in the shallow depth (0–3 cm) as compared to deep soil (3–6 cm), i.e., 78 ± 12.91 items per kg vs. 62.50 ± 12.97 items/kg (Liu et al., 2018). MPs' abundance of 40.35 mg per kg has also been reported in arid agricultural soil under the influence of mulching (Li et al., 2020 a). Zhou et al. (2020) studied the abundance of MPs in agricultural soil of the coastal plain and found 571 items per kg in mulched soil and 263 items per kg in non-mulched soil (Zhou et al., 2020). Use of sewage sludge as fertilizer is also used in many agricultural lands which is an important source of MPs in soil. Chinese soil, fertilized with sewage sludge, was reported to have 545.9 and 87.6 MP items per kg upon application of 30 tons per hectare and 15 tons per hectare of sludge as fertilizer, respectively (Zhang et al., 2020a). Detailed information about the global abundance of MPs in agricultural soil is summarized in Table 2.

3. Sources of Micro/Nano Plastics in Soil

MPs reach up to the soil environment through multiple direct and indirect pathways. Some of the common routes are the application of sewage sludge, use of compost, irrigation through wastewater, surface run-off, atmospheric fall-out, wind-blown deposition, etc.

3.1. Application of Sewage Sludge

Utilization of sewage sludge as the soil fertilizer leads to the dispersing of MNPs into the agricultural soil. Sewage receives MPs from domestic sources, such as personal care products (PCPs), textile washing, etc. (Singh et al., 2021). Particularly, large size MPs settle during various processes in sewage treatment plants and are finally collected as sewage sludge (Gao et al., 2020). Being rich in organic matter, trace elements, and microbes, this sludge is applied as fertilizer in the agricultural soil in order to enhance productivity. Apart from the sludge directly applied into the agricultural soil, some of the sludge is also improperly dumped in soil for long periods, which ultimately becomes a source of MNPs contamination in soil. Through various indirect pathways, these MPs may also reach up to the nearby agricultural fields. According to an estimate, globally approximately half of the sewage sludge is applied in farmlands. Moreover, Blasing and Amelung (2018) reported that up to 72% of the sludge is applied in agricultural soil in Finland and Ireland (Blasing & Amelung, 2018). Various studies have reported the difference in MP contamination of exposed and unexposed soil to sewage sludge (Table 2). A study showed that the application of approximately 23 tons of sludge per hectare resulted in 7–43 particles per g of MPs in China (Zhang & Liu, 2018).

3.2. Application of Water in Irrigation

The presence of micro/nanoplastics in water used for irrigation is one of the common sources of soil contamination. The presence of MNPs in surface water (rivers, lakes, reservoirs, etc.) and groundwater is already established (Mintenig et al., 2019; Singh et al., 2022; Singh & Bhagwat, 2022). Further, the considerable presence of MNPs in treated wastewater/sewage is also present because current wastewater/sewage treatment facilities are not equipped to remove MPs completely (Singh et al., 2021). Therefore, the utilization of treated wastewater/sewage in resource-scarce settings results in the transfer of MNPs from water to soil (Blasing & Amelung, 2018). Moreover, the application of surface/groundwater sources is another potential source of MNPs in agricultural soil.

3.3. Application of Compost

In order to enhance fertility and productivity, compost is generally applied in agricultural soil. The compost resulting from biological waste is reported to contain a significant quantity of plastics due to improper waste segregation and disposal (Blasing & Amelung, 2018). In a study, Weithmann et al. (2018) reported the presence of MPs in fertilizer produced from biological waste fermentation and composting, with many of the particles greater than 1 mm in size (Weithmann et al., 2018). In China, agricultural soil added with sludge compost was reported to have 545.9 and 87.6 MP items per kg (Zhang et al., 2020a).

3.4. Surface Run-off and Flooding Episodes

Surface run-off from plastic-contaminated areas and flooding situations are also responsible for accumulating plastic particles in agricultural areas. Improperly disposed of plastic waste on roads and particles generated from tire abrasion are some of the sources, which result in MP contamination of soil. The fields which are adjacent to the roads are the most affected due to this source of MPs. In a study conducted by Baensch-Baltruschat et al. (2021), MP transport in the road banks and adjacent soils was found to be 57300–65400 tons per annum (Baensch-Baltruschat et al., 2021), which shows that farm soils nearby the roads have ample chances to get contaminated through surface run-off and flooding.

3.5. Atmospheric Fallout and Wind-blown Deposition

Since MPs are also present in the aerial environment, atmospheric fallout is another mechanism through which they may get deposited in the soil. Further,

MPs, being light-weighed, are also blown away through wind movement resulting in their deposition in far-off areas (Singh et al., 2023). The occurrence of MPs in Antarctica, the Arctic, the Tibetan plateau, etc., supports the mechanism of atmospheric fallout and windblown deposition (Dong et al., 2021a; Cincinelli et al., 2017; Lusher et al., 2015; Aves et al., 2022; Jiang et al., 2019; Ambrosini et al., 2019; Long et al., 2022).

4. Migration of Micro/Nano Plastics in Soil

MNPs in soil are in constant movement owing to various natural and anthropogenic factors, such as soil characteristics (porous structure and cracks in soil), soil biotic organisms (bacteria, fungi, worms, plants, rodents, and other animals, etc.), soil management activities (plowing, sowing, cultivation, harvesting, etc.), and climatic variations (hot and dry seasons, wind/air movements, etc.) (Perez-Reveron et al., 2022; Yu et al., 2022). This movement of MPs may be horizontal as well as vertical among different layers of the soil (Yu et al., 2019; Al-Jaibachi et al., 2019).

Atmospheric transport or air movement is one of the important reasons for MPs' migration, especially in arid/semi-arid regions, where small plastic particles can easily re-suspend and re-enter the soil environment (Scheurer & Bigalke, 2018; Allen et al., 2019; Rezaei et al., 2019, 2022). Natural plant growth can also facilitate MP movement in the soil, such as the expansion of roots, movement of roots, root–water interaction, etc. (Yang et al., 2021a). Growth of roots in some plants results in extensive soil pores which makes the horizontal and vertical movement of MPs easier (Li et al., 2021). Researchers have found that plants having a higher number of primary and secondary root systems result in upward (vertical) movement of MPs through floatation of less dense plastic particles in the water-filled pore spaces created through increased soil porosity. On the other hand, the tertiary root network keeps the MPs aggregated with soil particles (O'Connor et al., 2019; Li et al., 2021; Lwanga et al., 2022; Li et al., 2022). Yu et al. (2022) reported that drying and wet cycles also assist in the vertical movement of MPs (Yu et al., 2022). The development of cracked surfaces in the soil owing to dry weather can make it possible for MPs to reach up to deeper layers of soil (Li et al., 2020b). It has been reported that coarse-textured soils having higher porosity, and relatively less dense soil types result in a higher rate of transfer of MPs toward deeper layers (Schell et al., 2022). Nevertheless, the presence of dense vegetation restricts MP movement during surface runoff in the wet season as plants act as physical barriers (Rehm et al., 2021).

In agricultural soil, where various manual and machine activities are carried out throughout the year, disturbances among various layers of the soil are usual, which result in the vertical movement of MPs. Moreover, tillage, ridging, and other agricultural practices turn the soil upside down which results in the direct transfer of MPs from one place to another (Piehl et al., 2018). Further, burrowing organisms also result in transferring microplastics up to deeper layers of soil. Earthworms, ingesting MPs along with soil, may also result in horizontal movement of the same (Lwanga et al., 2016; Rillig et al., 2017a). A study reported the presence of MPs in sheep feces and estimated that approximately 10^6 particles per hectare per year may be transported by a herd of 1,000 sheep (Beriot et al., 2021). Similarly, microorganisms such as bacteria and fungi also assist in MPs' transport since bacterial cells and fungal hyphae act as vectors of MPs (Guo et al., 2020; Gkoutselis et al., 2021; Singh & Bhagwat, 2022; Zhao et al., 2022).

Apart from natural and anthropogenic factors, MPs' migration is affected by the inherent properties of MPs as well (Yu et al., 2022). Among these properties, chemical composition, hydrophobicity, weathering pattern of MP particles, size, and shape are significant. O'Connor et al. (2019) have found that polyethylene particles penetrate up to deeper layers of soil if compared with polypropylene particles (O'Connor et al. 2019). Likewise, polystyrene particles having a hydrophilic nature are more mobile as compared to the particles having hydrophobic properties (Dong et al., 2019). It has also been seen that MP particles having carboxylic functional groups (–COOH) migrate relatively easily in the column test (Dong et al., 2019). The size of the MP particles also plays an important role in the movement with smaller size particles having higher chances to migrate farther (O'Connor et al. 2019). The shape of the particles also determines the course of movement as spherical-shaped particles get easily transported, while fibers generally get entangled with soil aggregates restricting the easy movement (Zhang et al., 2019). In addition to the various layers of soil containing MPs present in the terrestrial environment, it is also able to make entry into the groundwater and/or nearby surface water resources.

5. Hazards Associated with Micro/Nano Plastics in Soil

MPs, having carbon molecules as the main framework, affect soil and subsequently crops in various ways. Moreover, since these particles do not biodegrade easily, their continuous presence in soil influences the physicochemical parameters which ultimately affect the entire agricultural system. Another important aspect here is to address the vector potential of MPs. These particles act as efficient carriers of several contaminants, such as metals, metalloids, polyaromatic hydrocarbons, polychlorinated biphenyls, hexachlorocyclohexanes, pharmaceuticals and personal care products, per- and polyfluorinated substances, bacteria, viruses, fungus, etc. (Hartmann et al., 2017; Huffer et al., 2019; Upadhyay et al., 2022). Another concern is the presence of various additives and plasticizers which are added during plastic manufacturing processes to enhance the utility and performance of plastic products for various applications. These additives are released slowly into the soil environment, which has the potential to affect soil flora and fauna (Hahladakis et al., 2018). Thus, MPs have the potential to affect the soil quality, its microorganisms, and crop growth patterns. Depending upon the MPs' chemical composition, size, shape, concentration, additives, etc., these consequences may be positive, negative, or neutral.

5.1. Effects on Physicochemical Quality of Soil and Nutrient Cycling

Once MP particles enter the soil, these get agglomerated with the soil organic matter and are also acted upon by the microbes and thus become a part of the soil structure (Rillig et al., 2017b). Soil structure remains in a dynamic process and hence its aggregates form and disintegrate continuously. The entrapped MP particles in soil aggregates affect the soil porosity, water-holding capacity, and bulk density, and therefore influence the natural structural composition of the soil (Machado et al., 2018, 2019; Wan et al., 2019; Zhang & Zhang, 2020). Apart from structural imbalance, MP particles may also alter the pH of the soil. Researchers have reported that polyamide and high-density polyethylene particles increase the soil pH, while polystyrene and polytetrafluoroethylene have been found to reduce the pH (Yang et al., 2021b; Dong et al., 2021b). Moreover, this effect is also dependent on the type of crop grown; some crops are resistant to any change in soil properties (Lian et al., 2021).

Nutrient cycling is also reported to be affected by the presence of MPs in soil. MPs are made up of carbon framework, thus their presence creates an imbalance in the soil's carbon-nitrogen ratio (Liu et al., 2017; Zhang et al., 2020c). Increased concentration of polypropylene MPs has been seen to enhance the release of dissolved organic carbon, dissolved organic nitrogen, as well as dissolved organic phosphorous (Liu et al., 2017). MPs have also been reported to decrease the amount of soil organic matter, available phosphorous, nitrogen, and potassium which are the macronutrients in soil (Zhang et al., 2020c; Dong et al., 2021b; Wang et al., 2022). A decrease in the amount of macronutrients is certainly a risk for soil health.

5.2. Effects on Soil Enzymes

Soil enzymes are one of the important components of the soil system, which influence the rate of various reactions, such as the decomposition of organic materials, the release of nutrients, etc. Thus, enzymes affect the soil microbial metabolism and biochemical cycling processes, which ultimately regulate nutrient cycling (Burns et al., 2013). Soil enzyme activities may be enhanced, decreased, or remain unaffected owing to the presence of MPs. It has been reported that the activities of dehydrogenase, urease, and fluorescein diacetate hydrolase get reduced (Yi et al., 2020; Yu et al., 2020). The concentration, shape, and chemical composition of MPs play an important role in posing effects on enzymatic activities. Where low concentrations of polypropylene MPs pose no significant effect on phenoloxidase and fluorescein diacetate hydrolase, high concentration (~28% w/w) activates these enzymes (Liu et al., 2017). Feng et al. (2022) reported that MPs made up of polylactic acid, polybutylene succinate, and polyhydroxybutyrate at a concentration of 2% w/w enhanced the activities of phosphatase, urease, and catalase enzymes, while one-tenth concentration of the same reduced the activities of phosphatase (Feng et al., 2022). Similarly, the composition of MPs also matters, as it has been seen that polyethylene MPs affect the activities of urease and catalase while posing no effect on sucrase (Huang et al., 2019). Similarly, the type of shape of MP particles and exposure duration in the soil also matter (Yi et al., 2020; Chen et al., 2020b). Yi et al. (2020) investigated that fiber-shaped polypropylene exerts more effect on urease and alkaline phosphatase compared to sphere-shaped particles, while the effect is vice-versa for dehydrogenase (Yi et al. 2020).

5.3. Effects on Soil Microorganisms

Depending upon the chemical characteristics, different types of MPs affect soil microorganisms in different ways. Some microbes are attracted by the plastic particles and populate around them, while others resist the same. It has been seen that polyethylene and polyvinyl chloride tend to decrease the abundance and diversity of microbes (Fei et al., 2020). However, some other studies have reported that there is no significant effect on the population diversity and activities of microorganisms (Judy et al., 2019; Blöcker et al., 2020). The concentration and exposure time of MPs also affect the behavior of microbes. Kumar et al. (2020) reported that high concentrations of MPs generally increase soil microbial activities, however, the opposite effect was observed at low concentrations of MPs (Kumar et al., 2020). Similarly, studies on exposure time variation found that microbial species richness and diversity generally increase upon initial exposure to MPs, while the same is reduced upon prolonged exposure (Ren et al., 2020). These variations in impacts of MPs may be either due to the differences in chemical composition, size, and shape of MPs; or due to variations in soil types.

Another interesting aspect of microbial interactions with MP particles is the development of the plastisphere, where microbes colonize the plastic surfaces creating a unique ecological habitat (Yu et al., 2021b; Ghosh et al., 2022; Vaseashta et al., 2021). Studies reported that the presence of bacterial communities on the surface of MPs was higher as compared to the neighboring environment (Zettler et al., 2013). These communities have been reported to play an important role in the biogeochemical cycling of carbon and sulfur (Xie et al., 2021). Development of such kinds of ecological niches may have unusual consequences on the ecological structure as well as functions in the long run.

5.4. Effects on Plants

As of current date, few studies are showing the effects of MPs on plants (Qi et al., 2018), however, impacts on various physiological and biochemical processes of plant communities have been seen. The presence of 1% polyethylene

MP particles in the soil samples has been reported to result in reduced number and weight of the wheat grain biomass (Kumar et al., 2020). MP particles may affect the plants directly or indirectly. Under the direct effects, nano-sized plastic particles may enter the plant roots through physiological processes and thus enter into the systemic circulation, which may result in cell damage, oxidative stress, genotoxicity, etc. (Qi et al., 2018; Boots et al., 2019; Hernandez-Arenas et al., 2021). Entry of polystyrene nanoplastic particles (less than 100 nm) through the process of endocytosis has been found to take place in tobacco cells (Bandmann et al., 2012). Similarly, a plant culture study conducted by Li et al. (2019) reported the incorporation of polystyrene MP particles (0.2 μm) in edible plants (Li et al., 2019).

Moreover, additives, plasticizers, and other chemicals added to the plastics during manufacturing processes also affect the crops (Rillig et al., 2019). Indirect effects may be caused due to changes in the soil physicochemical properties and soil microbial communities (Yu et al., 2021c). Organic contamination of MP particles in the agricultural soil is also a possibility through the application of livestock manure or deposition from open defecation practices of humans (Quilliam et al., 2023). Pathogens present in these organic waste materials might get an opportunity to survive better through binding with the MPs as plastic surfaces may offer protection against grazing, interspecies competition, and other environmental stresses (Amaral-Zettler et al., 2020). Moreover, pathogens present on such surfaces remain in continuous association with the rhizosphere. Thus, the chances of pathogens' entry into the plant root systems increase. Adverse effects on plants due to MPs may pose repercussions further in the food chain.

5.5. Effects on Soil Fauna

Soil faunal species play an important role in maintaining the soil quality, thus indirectly affecting the crop growth pattern. However, there are very limited studies on the effects of MPs on soil faunal species. Earthworms, mites, burrowing animals, springtails, etc., are some of the important organisms that maintain the agroecosystem functions through their physicochemical and biochemical activities (George et al., 2017). Tiny plastic particles may be ingested by these organisms, which may affect their survival by affecting the growth pattern, immune disorders, neurotoxicity, reproductive abnormalities, and mortality patterns (Sarker et al., 2020; Wang et al., 2020; Wang et al., 2021b). A study was conducted on soil-inhabiting nematodes, viz., *Caenorhabditis elegans*, in which the organism was exposed to 1 mg per liter of polystyrene particles of different sizes (0.1, 0.5, 1, 2, and 5 μm) for the duration of 72 hours. The organism exposed to 1 μm size of plastic particles showed a shortening of body length, and increased mortality with a shorter life span owing to uptake and accumulation of MP particles (Lei et al., 2018). Zhu et al. (2018) conducted a similar type of study in the oligochaete, *Enchytraeus crypticus*, through exposure to polystyrene MP particles in varied concentrations. It was found that low concentration (0.025 weight % in oatmeal) resulted in gaining body weight, while high concentration (10 weight %) exerted a negative impact. Nevertheless, a concentration of 0.5 weight % did not result in any significant change (Zhu et al., 2018). Lwanga et al. (2016) also supported these findings by observing the increased mortality of *Lumbricus terrestris* after exposure to polyethylene MP particles in the concentration of 0.8–1.7 wt. %; however, no significant effect was observed at low concentrations (0.2 wt. %) (Lwanga et al., 2016). Effects might also be on the gut microbiota of animals, such as earthworms, thus affecting feeding behavior and development (Eltemsah & Bohn, 2019). Apart from experiencing changes in the growth pattern, physiological processes, and reproductive abnormalities; these soil faunal species may also play a significant role in passing over the microplastic particles in the food chain (Lwanga et al., 2016), ultimately leading to adverse health effects in humans.

6. Conclusion and Future Recommendations

Agricultural soil is one of the important components of the ecosystem, which attracts special consideration as it is responsible for maintaining the quality of crop growth and food production. Various emerging contaminants, including MNPs, have invaded the agricultural soil, threatening the soil quality. Although research on the contamination potential of micro/nanoplastics has emerged recently; it presents some important aspects to take into account. As per the limited database available to date, micro/nanoplastics certainly pose significant effects on the soil ecosystem upon long-term exposure. At present, there is no standard methodology for the detection and quantification of these plastic particles which often creates inconsistencies among the reported studies. However, with the development of standard sampling and analytical protocols in micro/nanoplastics' research, it would be easier to correlate various studies and draw appropriate conclusions. Hence, there is a critical need for standardization of procedures. The cytotoxic and genotoxic effects of MNPs also needed to be studied extensively for terrestrial floral and faunal species. Further, the vector potential of MNPs needs to be explored in the context of agroecosystem. Another crucial aspect to focus on is the leaching of these plastic particles through soil to the underground water resources. In this, the type of soil, flow pattern, and geology of the area also need to be considered. As far as the management of MNP pollution is concerned, concerted efforts of the scientific community, various stakeholders, community, policymakers, and regulators are required. Effective segregation of waste at source and proper recycling/disposal of plastic waste need to be strengthened. Behavioral change and community participatory behavior are some of the key issues that must be looked into in this particular aspect.

References

Ahmed, B., Rizvi, A., Ali, K., Lee, J., Zaidi, A., Khan, M.S., & Musarrat, J. (2021). Nanoparticles in the soil-plant system: A review. *Environmental Chemistry Letters*, 19, 1545–1609. DOI: 10.1007/s10311-020-01138-y.

Al-Jaibachi, R., Cuthbert, R.N., & Callaghan, A. (2019). Examining effects of ontogenic microplastic transference on Culex mosquito mortality and adult weight. *Science of The Total Environment*, 651, 871–876. DOI: 10.1016/j.scitotenv.2018.09.236.

Allen, S., Allen, D., Phoenix, V.R., Roux, G.L., Jimenez, P.D., Simonneau, A. et al. (2019). Atmospheric transport and deposition of microplastics in a remote mountain catchment. *Nature Geoscience*, 12, 339–344. DOI:10.1038/s41561-019-0335-5.

Amaral-Zettler, L.A., Zettler, E.R., & Mincer, T.J. (2020). Ecology of the plastisphere. *Nature Reviews Microbiology*, 18, 139–151. DOI: 10.1038/s41579-019-0308-0.

Ambrosini, R., Azzoni, R.S., Pittino, F., Diolaiuti, G., Franzetti, A., & Parolini, M. (2019). First evidence of microplastic contamination in the supraglacial debris of an alpine glacier. *Environmental Pollution*, 253, 297–301. DOI: 10.1016/j.envpol.2019.07.005.

Ashraf, M.A., Maah, M.J., & Yusoff, I. (2014). Soil contamination, risk assessment, and remediation. *In:* Hernandez-Soriano, M.C. (Ed.), *Environmental Risk Assessment of Soil Contamination*. IntechOpen Publishing, United Kingdom. DOI: 10.5772/57287.

Aves, A.R., Revell, L.E., Gaw, S., Ruffell, H., Schuddeboom, A., Wotherspoon, N.E. et al. (2022). First evidence of microplastics in Antarctic snow. *The Cryosphere*, 16, 2127–2145. DOI: 10.5194/tc-16-2127-2022.

Baensch-Baltruschat, B., Kocher, B., Kochleus, C., Stock, F., & Reifferscheid, G. (2021). Tyre and road wear particles: A calculation of generation, transport, and release to water and soil with special regard to German roads. *Science of The Total Environment*, 752, 141939. DOI: 10.1016/j.scitotenv.2020.141939.

Bandmann, V., Muller, J.D., Kohler, T., & Homann, U. (2012). Uptake of fluorescent nano beads into BY2-cells involves clathrin-dependent and clathrin-independent endocytosis. *FEBS Letters*, 586, 3626–3632. DOI: 10.1016/j.febslet.2012.08.008.

Benjamin, S., Pradeep, S., Josh, M.S., Kumar, S., & Masai, E. (2015). A monograph on the remediation of hazardous phthalates. *Journal of Hazardous Materials*, 298, 58–72. DOI: 10.1016/j.jhazmat.2015.05.004.

Berg, P.V., Lwanga, E.H., Corradini, F., & Geissen, V. (2020), Sewage sludge application as a vehicle for microplastics in eastern Spanish agricultural soils. *Environmental Pollution*, 261, 114198. DOI: 10.1016/j.envpol.2020.114198.

Beriot, N., Peek, J., Zornoza, R., Geissen, V., & Lwanga, E.H. (2021). Low-density microplastics detected in sheep feces and soil: A case study from the intensive vegetable farming in Southeast Spain. *Science of The Total Environment*, 755, 142653. DOI: 10.1016/j.scitotenv.2020.142653

Bigalke, M., Fieber, M., Foetisch, A., Reynes, J., & Tollan, P. (2022). Microplastics in agricultural drainage water: A link between terrestrial and aquatic microplastic pollution. *Science of The Total Environment*, 806, 150709. DOI: 10.1016/j.scitotenv.2021.150709.

Blasing, M., & Amelung, W. (2018). Plastics in soil: Analytical methods and possible sources. *Science of The Total Environment*, 612, 422–435. DOI: 10.1016/j.scitotenv.2017.08.086.

Blöcker, L., Watson, C., & Wichern, F. (2020). Living in the plastic age: Different short-term microbial response to microplastics addition to arable soils with contrasting soil organic matter content and farm management legacy. *Environmental Pollution*, 267, 115468. DOI: 10.1016/j.envpol.2020.115468.

Boots, B., Russell, C.W., & Green, D.S. (2019). Effects of microplastics in soil ecosystems: Above and below ground. *Environmental Science & Technology*, 53, 11496–11506. DOI: 10.1021/acs.est.9b03304.

Boughattas, I., Hattab, S., Zitouni, N., Mkhinini, M., Missawi, O., Bousserrhine, N., & Banni, M. (2021). Assessing the presence of microplastic particles in Tunisian agriculture soils and their potential toxicity effects using *Eisenia andrei* as bioindicator. *Science of The Total Environment*, 796, 148959. DOI: 10.1016/j.scitotenv.2021.148959

Breivik, K., Armitage, J.M., Wania, F., Sweetman, A.J., & Jones, K.C. (2016). Tracking the global distribution of persistent organic pollutants accounting for e-waste exports to developing regions, *Environmental Science & Technology*, 50, 798–805. DOI: 10.1021/acs.est.5b04226.

Burns, R.G., DeForest, J.L., Marxsen, J., Sinsabaugh, R.L., Stromberger, M.E., Wallenstein, M.D. et al. (2013). Soil enzymes in a changing environment: Current knowledge and future directions. *Soil Biology and Biochemistry*, 58, 216–234. DOI: 10.1016/j.soilbio.2012.11.009.

Chen, Y., Leng, Y., Liu, X., & Wang, J. (2020a). Microplastic pollution in vegetable farmlands of suburb Wuhan, Central China. *Environmental Pollution*, 257, 113449. DOI: 10.1016/j.envpol.2019.113449

Chen, H., Wang, Y., Sun, X., Peng, Y., & Xiao, L. (2020b). Mixing effect of polylactic acid microplastic and straw residue on soil property and ecological function. *Chemosphere*, 243, 125271. DOI: 10.1016/j.chemosphere.2019.125271.

Choi, Y.R., Kim, Y.-N., Yoon, J.H., Dickinson, N., & Kim, K.-H. (2021). Plastic contamination of forest, urban, and agricultural soils: A case study of Yeoju city in the Republic of Korea. *Journal of Soils and Sediments*, 21, 1962–1973. DOI: 10.1007/s11368-020-02759-0.

Cincinelli, A., Scopetani, C., Chelazzi, D., Lombardini, E., Martellini, T., Katsoyiannis, A. et al. (2017). Microplastic in the surface waters of the Ross Sea (Antarctica): Occurrence, distribution and characterization by FTIR. *Chemosphere*, 175, 391–400. DOI: 10.1016/j.chemosphere.2017.02.024.

Corradini, F., Meza, P., Eguiluz, R., Casado, F., Huerta-Lwanga, E., & Geissen, V. (2019). Evidence of microplastic accumulation in agricultural soils from sewage sludge disposal. *Science of the Total Environment*, 671, 411–420. DOI: 10.1016/j.scitotenv.2019.03.368.

Crossman, J., Hurley, R.R., Futter, M., & Nizzetto, L. (2020). Transfer and transport of microplastics from biosolids to agricultural soils and the wider environment. *Science of The Total Environment*, 724, 138334. DOI: 10.1016/j.scitotenv.2020.138334.

Cusworth, S.J., Davies, W.J., McAinsh, M.R., & Stevens, C.J. (2024). A nationwide assessment of microplastic abundance in agricultural soils: The influence of plastic crop covers within the United Kingdom. *Plants, People, Planet*, 6, 304–314. DOI: 10.1002/ppp3.10430.

Ding, L., Zhang, S., Wang, X., Yang, X., Zhang, C., Qi, Y., & Guo, X. (2020). The occurrence and distribution characteristics of microplastics in the agricultural soils of Shaanxi Province, in north-western China. *Science of the Total Environment*, 720, 137525. DOI: 10.1016/j.scitotenv.2020.137525.

Dong, H., Wang, L., Wang, X., Xu, L., Chen, M., Gong, P., & Wang, C. (2021a). Microplastics in a remote lake basin of the Tibetan Plateau: Impacts of atmospheric transport and glacial melting. *Environmental Science & Technology*, 55, 12951–12960. DOI: 10.1021/acs.est.1c03227.

Dong, Y., Gao, M., Qiu, W., & Song, Z. (2021b). Effect of microplastics and arsenic on nutrients and microorganisms in rice rhizosphere soil. *Ecotoxicology and Environmental Safety*, 211, 111899. DOI:10.1016/j.ecoenv.2021.111899.

Dong, Z., Zhu, L., Zhang, W., Huang, R., Lv, X.W., Jing, X. et al. (2019). Role of surface functionalities of nanoplastics on their transport in seawater-saturated sea sand. *Environmental Pollution*, 255, 113177. DOI: 10.1016/j.envpol.2019.113177.

Eltemsah, Y.S., & Bohn, T. (2019). Acute and chronic effects of polystyrene microplastics on juvenile and adult *Daphnia Magna. Environmental Pollution*, 254, 112919. DOI: 10.1016/j.envpol.2019.07.087

Fakour, H., Lo, S.-L., Yoashi, N.T., Massao, A.M., Lema, N.N., Mkhontfo, F.B. et al. (2021). Quantification and analysis of microplastics in farmland soils: Characterization, sources, and pathways. *Agriculture*, 11, 330. DOI: 10.3390/agriculture11040330.

Fei, Y., Huang, S., Zhang, H., Tong, Y., Wen, D., Xia, X. et al. (2020), Response of soil enzyme activities and bacterial communities to the accumulation of microplastics in an acid cropped soil. *Science of The Total Environment*, 707, 135634. DOI: 10.1016/j.scitotenv.2019.135634.

Feng, S., Lu, H., & Liu, Y. (2021). The occurrence of microplastics in farmland and grassland soils in the Qinghai-Tibet plateau: Different land use and mulching time in facility agriculture. *Environmental Pollution*, 279, 116939. DOI: 10.1016/j.envpol.2021.116939.

Feng, X., Wang, Q., Sun, Y., Zhang, S., & Wang, F. (2022). Microplastics change soil properties, heavy metals availability and bacterial community in a Pb-Zn contaminated soil. *Journal of Hazardous Materials*, 424, 127364. DOI: 10.1016/j.jhazmat.2021.127364.

Frias, J.P.G.L., & Nash, R. (2019). Microplastics: Finding a consensus on the definition. *Marine Pollution Bulletin*, 138, 145–147. DOI: 10.1016/j.marpolbul.2018.11.022.

Gao, D., Li, X-Y., & Liu, H-T. (2020). Source, occurrence, migration, and potential environmental risk of microplastics in sewage sludge and during sludge amendment to soil. *Science of The Total Environment*, 742, 140355. DOI:10.1016/j.scitotenv.2020.140355.

George, P.B.L., Keith, A.M., Creer, S., Barrett, G.L., Lebron, I., Emmett, B.A. et al. (2017). Evaluation of mesofauna communities as soil quality indicators in a national-level monitoring programme. *Soil Biology and Biochemistry*, 115, 537–546. DOI: 10.1016/j.soilbio.2017.09.022.

Ghosh, A., Woodward, E.M.S., Saha, R., Nelson, C.E., & Bhadury, P. (2022). Nitrogen driven niche differentiation in bacterioplankton communities of northeast coastal Bay of Bengal. *Environmental Research Communications*, 4, 035006. DOI: 10.1088/2515-7620/ac5a69.

Gkoutselis, G., Rohrbach, S., Harjes, J., Obst, M., Brachmann, A., Horn, M.A., & Rambold, G. (2021). Microplastics accumulate fungal pathogens in terrestrial ecosystems. *Scientific Reports*, 11, 13214. DOI: 10.1038/s41598-021-92405-7.

Guo, J.-J., Huang, X.-P., Xiang, L., Wang, Y.-Z., Li, Y.-W., Li, H. et al. (2020), Source, migration and toxicology of microplastics in soil. *Environment International*, 137, 105263. DOI: 10.1016/j.envint.2019.105263

Hahladakis, J.N., Velis, C.A., Weber, R., Iacovidou, E., & Purnell, P. (2018). An overview of chemical additives present in plastics: Migration, release, fate, and environmental impact during their use, disposal, and recycling. *Journal of Hazardous Materials*, 344, 179–199. DOI: 10.1016/j.jhazmat.2017.10.014.

Harms, I.K., Diekotter, T., Troegel, S., & Lenz, M. (2021). Amount, distribution, and composition of large microplastics in typical agricultural soils in northern Germany. *Science of the Total Environment*, 758, 143615. DOI: 10.1016/j.scitotenv.2020.143615.

Hartmann, N.B., Rist, S., Bodin, J., Jensen, L.H.S., Schmidt, S.N., Mayer, P. et al. (2017). Microplastics as vectors for environmental contaminants: Exploring sorption, desorption, and transfer to biota. *Integrated Environmental Assessment and Management*, 13, 488–493. DOI: 10.1002/ieam.1904.

Hernandez-Arenas, R., Beltran-Sanahuja, A., Navarro-Quirant, P., & Sanz-Lazaro, C. (2021). The effect of sewage sludge containing microplastics on growth and fruit development of tomato plants. *Environmental Pollution*, 268, 115779. DOI: 10.1016/j.envpol.2020.115779.

Huang, Y., Zhao, Y., Wang, J., Zhang, M., Jia, W., & Qin, X. (2019). LDPE microplastic films alter microbial community composition and enzymatic activities in soil. *Environmental Pollution*, 254, 112983. DOI: 10.1016/j.envpol.2019.112983

Huang, Y., Liu, Q., Jia, W., Yan, C., & Wang, J. (2020). Agricultural plastic mulching as a source of microplastics in the terrestrial environment. *Environmental Pollution*, 260, 114096. DOI: 10.1016/j.envpol.2020.114096.

Huffer, T., Metzelder, F., Sigmund, G., Slawek, S., Schmidt, T.C., & Hofmann, T. (2019). Polyethylene microplastics influence the transport of organic contaminants in soil. *Science of the Total Environment*, 657, 242–247. DOI: 10.1016/j.scitotenv.2018.12.047.

Isari, E.A., Papaioannou, D., Kalavrouziotis, I.K., & Karapanagioti, H.K. (2021). Microplastics in agricultural soils: A case study in cultivation of watermelons and canning tomatoes. *Water*, 13, 2168. DOI: 10.3390/w13162168.

IUPAC. (2009). *Systematic Classification of the Major Pollutants Found in Soils According to IUPAC Guidelines*. Available online at https://rune.une.edu.au/web/bitstream/1959.11/52965/5/openpublished/MainWilson2021BookChapter.pdf, accessed on Mar. 6th, 2024.

Jiang, C., Yin, L., Li, Z., Wen, X., Luo, X., Hu, S. et al. (2019). Microplastic pollution in the rivers of the Tibet Plateau. *Environmental Pollution*, 249, 91–98. DOI: 10.1016/j.envpol.2019.03.022.

Judy, J.D., Williams, M., Gregg, A., Oliver, D., Kumar, A., Kookana, R., & Kirby, J.K. (2019). Microplastics in municipal mixed-waste organic outputs induce minimal short- to long-term toxicity in key terrestrial biota. *Environmental Pollution*, 252, 522–531. DOI: 10.1016/j.envpol.2019.05.027

Katsumi, N., Kusube, T., Nagao, S., & Okochi, H. (2021), Accumulation of microcapsules derived from coated fertilizer in paddy fields. *Chemosphere*, 267, 129185. DOI: 10.1016/j.chemosphere.2020.129185.

Kawecki, D., Goldberg, L., & Nowack, B. (2021). Material flow analysis of plastic in organic waste in Switzerland. *Soil Use and Management*, 37, 277–288. DOI: 10.1111/sum.12634.

Khan, M.A., Huang, Q., Khan, S., Wang, Q., Huang, J., Fahad, S. et al. (2023). Abundance, spatial distribution, and characteristics of microplastics in agricultural soils and their relationship with contributing factors. *Journal of Environmental Management*, 328, 117006. DOI: 10.1016/j.jenvman.2022.117006.

Kim, S.-K., Kim, J.-S., Lee, H., & Lee, H.-J. (2021). Abundance and characteristics of microplastics in soils with different agricultural practices: Importance of sources with internal origin and environmental fate. *Journal of Hazardous Materials*, 403, 123997. DOI: 10.1016/j.jhazmat.2020.123997.

Kumar, M., Xiang, X., He, M., Tsang, D.C.W., Gupta, J., Khan, E. et al. (2020). Microplastics as pollutants in agricultural soils. *Environmental Pollution*, 265, 114980. DOI: 10.1016/j.envpol.2020.114980.

Kumar, V.M., & Sheela, M.A. (2021). Effect of plastic film mulching on the distribution of plastic residues in agricultural fields. *Chemosphere*, 273, 128590. DOI: 10.1016/j.chemosphere.2020.128590.

Lei, L., Liu, M., Song, Y., Lu, S., Hu, J., Cao, C. et al. (2018). Polystyrene (nano) microplastics cause size-dependent neurotoxicity, oxidative damage, and other adverse effects in *Caenorhabditis elegans*. *Environmental Science: Nano*, 5, 2009–2020. DOI: 10.1039/C8EN00412A.

Li, H., Lu, X., Wang, S., Zheng, B., & Xu, Y. (2021). Vertical migration of microplastics along soil profile under different crop root systems. *Environmental Pollution*, 278, 116833. DOI: 10.1016/j.envpol.2021.116833.

Li, J., Song, Y., & Cai, Y. (2020b). Focus topics on microplastics in soil: Analytical methods, occurrence, transport, and ecological risks. *Environmental Pollution*, 257, 113570. DOI: 10.1016/j.envpol.2019.113570.

Li, L., Zhou, Q., Yin, N., Tu, C., & Luo, Y. (2019). Uptake and accumulation of microplastics in an edible plant. *Chinese Science Bulletin*, 64, 928–934. DOI: 10.1360/N972018-00845.

Li, S., Ding, F., Flury, M., Wang, Z., Xu, L., Li, S. et al. (2022). Macro- and microplastic accumulation in soil after 32 years of plastic film mulching. *Environmental Pollution*, 300, 118945. DOI: 10.1016/j.envpol.2022.118945.

Li, W., Wufuer, R., Duo, J., Wang, S., Luo, Y., Zhang, D., & Pan, X. (2020a). Microplastics in agricultural soils: Extraction and characterization after different periods of polythene film mulching in an arid region. *Science of The Total Environment*, 749, 141420. DOI: 10.1016/j.scitotenv.2020.141420.

Lian, J., Liu, W., Meng, L., Wu, J., Zeb, A., Cheng, L. et al. (2021). Effects of microplastics derived from polymer-coated fertilizer on maize growth, rhizosphere, and soil properties. *Journal of Cleaner Production*, 318, 128571. DOI: 10.1016/j.jclepro.2021.128571.

Liu, H., Yang, X., Liu, G., Liang, C., Xue, S., Chen, H. et al. (2017). Response of soil dissolved organic matter to microplastic addition in Chinese loess soil. *Chemosphere*, 185, 907–917. DOI: 10.1016/j.chemosphere.2017.07.064.

Liu, M., Lu, S., Song, Y., Lei, L., Hu, J., Lv, W. et al. (2018). Microplastic and mesoplastic pollution in farmland soils in suburbs of Shanghai, China. *Environmental Pollution*, 242, 855–862. DOI: 10.1016/j.envpol.2018.07.051.

Long, X., Fu, T.-M., Yang, X., Tang, Y., Zheng, Y., Zhu, L. et al. (2022). Efficient atmospheric transport of microplastics over Asia and adjacent ocean. *Environmental Science & Technology*, 56, 6243–6252. DOI: 10.1021/acs.est.1c07825.

Lusher, A.L., Tirelli, V., O'Connor, I., & Officer, R. (2015). Microplastics in Arctic polar waters: The first reported values of particles in surface and sub-surface samples. *Scientific Reports*, 5, 14947. DOI: 10.1038/srep14947.

Lwanga, E.H., Gertsen, H., Gooren, H., Peters, P., Salanki, T., Ploeg, M.V. et al. (2016). Microplastics in the terrestrial ecosystem: Implications for *Lumbricus terrestris* (*Oligochaeta*, Lumbricidae). *Environmental Science & Technology*, 50, 2685–2691. DOI: 10.1021/acs.est.5b05478.

Lwanga, E.H., Beriot, N., Corradini, F., Silva, V., Yang, X., Baartman, J. et al. (2022). Review of microplastic sources, transport pathways, and correlations with other soil stressors: A journey from agricultural sites into the environment. *Chemical and Biological Technologies in Agriculture*, 9, 20. DOI: 10.1186/s40538-021-00278-9.

Machado, A.A.S., Lau, C.W., Till, J., Kloas, W., Lehmann, A., Becker, R., & Rillig, M.C. (2018). Impacts of microplastics on the soil biophysical environment. *Environmental Science & Technology*, 52, 9656–9665. DOI: 10.1021/acs.est.8b02212.

Machado, A.A.S., Lau, C.W., Kloas, W., Bergmann, J., Bachelier, J.B., Faltin, E. et al. (2019). Microplastics can change soil properties and affect plant performance. *Environmental Science & Technology*, 53, 6044–6052. DOI: 10.1021/acs.est.9b01339.

Maddela, N.R., Venkateswarlu, K., Kakarla, D., & Megharaj, M. (2020a). Inevitable human exposure to emissions of polybrominated diphenyl ethers: A perspective on potential health risks. *Environmental Pollution*, 266, 115240. DOI: 10.1016/j.envpol.2020.115240.

Maddela, N.R., Venkateswarlu, K., & Megharaj, M. (2020b). Tris(2-chloroethyl) phosphate, a pervasive flame retardant: Critical perspective on its emissions into the environment and human toxicity. *Environmental Science: Processes & Impacts*, 22, 1809–1827. DOI: 10.1039/D0EM00222D.

Maddela, N.R., Ramakrishnan, B., Kakarla, D., Venkateswarlu, K., & Megharaj, M. (2022). Major contaminants of emerging concern in soils: A perspective on potential health risks. *RSC Advances*, 12, 12396–12415. DOI: 10.1039/D1RA09072K.

Mahesh, P., Chauhan, A., & Warrier, A.K. (2022). *Plastic Mulching: Microplastics in Agricultural Soils*. DOI: 10.13140/RG.2.2.29673.16480. Available online at https://www.researchgate.net/publication/360773890_Plastic_Mulching_Microplastics_in_Agricultural_Soil#fullTextFileContent.

Mintenig, S.M., Loder, M.G.J., Primpke, S., & Gerdts, G. (2019). Low numbers of microplastics detected in drinking water from groundwater sources. *Science of The Total Environment*, 648, 631–635. DOI: 10.1016/j.scitotenv.2018.08.178.

Mo, X., Li, H., Lian, Y., Zheng, B., Dong, J., & Lu, X. (2021). Estimation of soil microplastic input derived from plastic gauze using a simplified model. *Science of The Total Environment*, 793, 148577. DOI: 10.1016/j.scitotenv.2021.148577.

O'Connor, D., Pan, S., Shen, Z., Song, Y., Jin, Y., Wu, W-M., & Hou, D. (2019). Microplastics undergo accelerated vertical migration in sand soil due to small size and wet-dry cycles. *Environmental Pollution*, 249, 527–534. DOI: 10.1016/j.envpol.2019.03.092.

Perez-Reveron, R., Alvarez-Mendez, S., Kropp, R.M., Perdomo-Gonzalez, A., Hernandez-Borges, J., & Diaz-Pena, F.J. (2022). Microplastics in agricultural systems: Analytical methodologies and effects on soil quality and crop yield. *Agriculture*, 12, 1162. DOI: 10.3390/agriculture12081162.

Piehl, S., Leibner, A., Loder, M.G.J., Dris, R., Bogner, C., & Laforsch, C. (2018). Identification and quantification of macro- and microplastics on an agricultural farmland. *Scientific Reports*, 8, 17950. DOI: 10.1038/s41598-018-36172-y.

Qi, Y., Yang, X., Pelaez, A.M., Lwanga, E.H., Beriot, N., Gertsen, H. et al. (2018). Macro- and micro-plastics in soil-plant system: Effects of plastic mulch film residues on wheat (*Triticum aestivum*) growth. *Science of The Total Environment*, 645, 1048–1056. DOI: 10.1016/j.scitotenv.2018.07.229.

Quilliam, R.S., Pow, C.J., Shilla, D.J., Mwesiga, J.J., Shilla, D.A., & Woodford, L. (2023). Microplastics in agriculture – A potential novel mechanism for the delivery of human pathogens on to crops. *Frontiers in Plant Science*, 14. DOI: 10.3389/fpls.2023.1152419.

Ragoobur, D., Lwanga, E.H., & Somaroo, G.D. (2021). Microplastics in agricultural soils, wastewater effluents, and sewage sludge in Mauritius. *Science of The Total Environment*, 798, 149326. DOI: 10.1016/j.scitotenv.2021.149326.

Rehm, R., Zeyer, T., Schmidt, A., & Fiener, P. (2021). Soil erosion as transport pathway of microplastic from agriculture soils to aquatic ecosystems. *Science of The Total Environment*, 795, 148774. DOI: 10.1016/j.scitotenv.2021.148774.

Ren, X., Tang, J., Liu, X., & Liu, Q. (2020). Effects of microplastics on greenhouse gas emissions and the microbial community in fertilized soil. *Environmental Pollution*, 256, 113347. DOI: 10.1016/j.envpol.2019.113347.

Rezaei, M., Riksen, M.J.P.M., Sirjani, E., Sameni, A., & Geissen, V. (2019). Wind erosion as a driver for transport of light density microplastics. *Science of The Total Environment*, 669. 273–281. DOI: 10.1016/j.scitotenv.2019.02.382.

Rezaei, M., Abbasi, S., Pour Mahmood, H., Oleszczuk, P., Ritsema, C., & Turner, A. (2022). Microplastics in agricultural soils from a semi-arid region and their transport by wind erosion. *Environmental Research*, 212, 113213. DOI: 10.1016/j.envres.2022.113213.

Rillig, M.C., Ziersch, L., & Hempel, S. (2017a). Microplastic transport in soil by earthworms. *Scientific Reports*, 7, 1362. DOI: 10.1038/s41598-017-01594-7.

Rillig, M.C., Ingraffia, R., & Machado, A.A.S. (2017b). Microplastic incorporation into soil in agroecosystems. *Frontiers in Plant Science*, 8, 01805. DOI: 10.3389/fpls.2017.01805.

Rillig, M.C., Lehmann, A., Machado, A.A.S., & Yang, G. (2019). Microplastic effects on plants. *New Phytologist*, 223, 1066–1070. DOI: 10.1111/nph.15794.

Sa'adu, I., & Farsang, A. (2022). Greenhouse farming as a source of macroplastic and microplastics contamination in agricultural soils: A case study from Southeast-Hungary. *Agrochemistry and Soil Science*, 71, 43–57. DOI: 10.1556/0088.2022.00120.

Sa'adu, I., & Farsang, A. (2023). Plastic contamination in agricultural soils: A review. *Environmental Sciences Europe*, 35, 13. DOI: 10.1186/s12302-023-00720-9.

Sarker, A., Deepo, D.M., Nnadi, R., Rana, J., Islam, S., Rahman, S. et al. (2020). A review of microplastics pollution in the soil and terrestrial ecosystems: A global and Bangladesh perspective. *Science of The Total Environment*, 733, 139296. DOI: 10.1016/j.scitotenv.2020.139296.

Schell, T., Hurley, R., Buenaventura, N.T., Mauri, P.V., Nizzetto, L., Rico, A., & Vighi, M. (2022). Fate of microplastics in agricultural soils amended with sewage sludge: Is surface water runoff a relevant environmental pathway? *Environmental Pollution*, 293, 118520. DOI: 10.1016/j.envpol.2021.118520.

Scheurer, M., & Bigalke, M. (2018). Microplastics in Swiss floodplain soils. *Environmental Science & Technology*, 52, 3591–3598. DOI: 10.1021/acs.est.7b06003.

Schothorst, B.V., Beriot, N., Lwanga, E.H., & Geissen, V. (2021). Sources of light density microplastic related to two agricultural practices: The use of compost and plastic mulch. *Environments*, 8, 36. DOI: 10.3390/environments8040036.

SDGs. (2015). *United Nations Sustainable Development Goals*. Available online at https://sdgs.un.org/goals. Accessed on Jan. 21st, 2023.

Singh, S., Kalyanasundaram, M., & Diwan, V. (2021). Removal of microplastics from wastewater: Available techniques and way forward. *Water Science & Technology*, 84, 3689–3704. DOI:10.2166/wst.2021.472.

Singh, S., Trushna, T., Kalyanasundaram, M., Tamhankar, A.J., & Diwan, V. (2022). Microplastics in drinking water: A macro issue. *Water Supply*, 22, 5650–5674. DOI: 10.2166/ws.2022.189.

Singh, S., & Bhagwat, A. (2022). Microplastics: A potential threat to groundwater resources. *Groundwater for Sustainable Development*, 19, 100852. DOI: 10.1016/j.gsd.2022.100852.

Singh, S., Chakma, S., Alawa, B., Kalyanasundaram, M., & Diwan, V. (2023). Identification, characterization, and implications of microplastics in soil: A case study of Bhopal, Central India. *Journal of Hazardous Materials Advances*, 9, 100225. DOI: 10.1016/j.hazadv.2022.100225

Singh, S., Chakma, S., Alawa, B., Kalyanasundaram, M., & Diwan, V. (2024a). Assessment of microplastic pollution in agricultural soil of Bhopal, Central India. *Journal of Material Cycles and Waste Management*, 26, 708–722. DOI: 10.1007/s10163-023-01805-6.

Singh, S., Chakma, S., Alawa, B., Kalyanasundaram, M., & Diwan, V. (2024b). Microplastic pollution in terrestrial environment: Identification, characterization, and risk assessment in Indore, Central India. *Soil Use and Management*, 40, e13053. DOI: 10.1111/sum.13053.

Tagg, A.S., Brandes, E., Fischer, F., Fischer, D., Brandt, J., & Labrenz, M. (2022). Agricultural application of microplastic-rich sewage sludge leads to further uncontrolled contamination. *Science of The Total Environment*, 806, 150611. DOI: 10.1016/j.scitotenv.2021.150611.

Upadhyay, R., Singh, S., & Kaur, G. (2022). Sorption of pharmaceuticals over microplastics' surfaces: Interaction mechanisms and governing factors. *Environmental Monitoring and Assessment*, 194, 803. DOI: 10.1007/s10661-022-10475-0.

USEPA. (2018). *United States Environmental Protection Agency. PFOA, PFOS, and other PFASs*. Available online at https://www.epa.gov/pfas/pfas-explained, accessed on Aug. 30th, 2024.

Vaseashta, A., Ivanov, V., Stabnikov, V., & Marinin, A. (2021). Environmental safety and security investigations of neustonic microplastic aggregates near water-air interphase. *Polish Journal of Environmental Studies*, 30(4), 3457–3469. https://doi.org/10.15244/pjoes/131947.

Wan, Y., Wu, C., Xue, Q., & Hui, X. (2019). Effects of plastic contamination on water evaporation and desiccation cracking in soil. *Science of The Total Environment*, 654, 576–582. DOI: 10.1016/j.scitotenv.2018.11.123.

Wang, F., Wang, Q., Adams, C.A., Sun, Y., & Zhang, S. (2022). Effects of microplastics on soil properties: Current knowledge and future perspectives. *Journal of Hazardous Materials*, 424, 127531. DOI: 10.1016/j.jhazmat.2021.127531.

Wang, J., Li, J., Liu, S., Li, H., Chen, X., Peng, C. et al. (2021a). Distinct microplastic distributions in soils of different land-use types: A case study of Chinese Farmlands. *Environmental Pollution*, 269, 116199. DOI: 10.1016/j.envpol.2020.116199.

Wang, Q., Adams, C.A., Wang, F., Sun, Y., & Zhang, S. (2021b). Interactions between microplastics and soil fauna: A critical review. *Critical Reviews in Environmental Science and Technology*, 52, 3211–3243. DOI: 10.1080/10643389.2021.1915035.

Wang, W., Ge, J., Yu, X., & Li, H. (2020). Environmental fate and impacts of microplastics in soil ecosystems: Progress and perspective. *Science of The Total Environment*, 708, 134841. DOI: 10.1016/j.scitotenv.2019.134841.

Weber, C.J., Bastijans, J.-E., Lugger, K., & Heller, C. (2024). Microplastic in long-term soil monitoring: First spatial and temporal data on plastics in agricultural topsoils. *Journal of Soils and Sediments*, 24, 692–706. DOI: 10.1007/s11368-023-03663-z.

Weithmann, N., Moller, J.A., Loder, M.G.J., & Piehl, S. (2018). Organic fertilizer as a vehicle for the entry of microplastic into the environment. *Science Advances*, 4, DOI: 10.1126/sciadv.aap8060.

Xie, H., Chen, J., Feng, L., He, L., Zhou, C., Hong, P. et al. (2021). Chemotaxis-selective colonization of mangrove rhizosphere microbes on nine different microplastics. *Science of The Total Environment*, 752, 142223. DOI: 10.1016/j.scitotenv.2020.142223.

Yang, L., Zhang, Y., Kang, S., Wang, Z., & Wu, C. (2021a). Microplastics in soil: A review on methods, occurrence, sources, and potential risk. *Science of the Total Environment*, 780, 146546. DOI: 10.1016/j.scitotenv.2021.146546.

Yang, W., Cheng, P., Adams, C.A., Zhang, S., Sun, Y., Yu, H., & Wang, F. (2021b). Effects of microplastics on plant growth and arbuscular mycorrhizal fungal communities in a soil spiked with ZnO nanoparticles. *Soil Biology and Biochemistry*, 155, 108179. DOI: 10.1016/j.soilbio.2021.108179.

Yi, M., Zhou, S., Zhang, L., & Ding, S. (2020). The effects of three different microplastics on enzyme activities and microbial communities in soil. *Water Environment Research*, 93, 24–32. DOI: 10.1002/wer.1327.

Yu, M., Ploeg, M., Lwanga, E.H., Yang, X., Zhang, S., Ma, X. et al. (2019). Leaching of microplastics by preferential flow in earthworm (*Lumbricus terrestris*) burrows. *Environmental Chemistry*, 16, 31–40. DOI: 10.1071/EN18161.

Yu, H., Fan, P., Hou, J., Dang, Q., Cui, D., Xi, B., & Tan, W. (2020). Inhibitory effect of microplastics on soil extracellular enzymatic activities by changing soil properties and direct adsorption: An investigation at the aggregate-fraction level. *Environmental Pollution*, 267, 115544. DOI: 10.1016/j.envpol.2020.115544.

Yu, L., Zhang, J., Liu, Y., Chen, L., Tao, S., & Liu, W. (2021a). Distribution characteristics of microplastics in agricultural soils from the largest vegetable production base in China. *Science of The Total Environment*, 756, 143860. DOI: 10.1016/j.scitotenv.2020.143860.

Yu, H., Zhang, Y., & Tan, W. (2021b). The "Neighbor Avoidance Effect" of microplastics on bacterial and fungal diversity and communities in different soil horizons. *Environmental Science and Ecotechnology*, 8, 100121. DOI: 10.1016/j.ese.2021.100121.

Yu, H., Qi, W., Cao, X., Hu, J., Li, Y., Peng, J. et al. (2021c). Microplastic residues in wetland ecosystems: Do they truly threaten the plant-microbe-soil system? *Environment International*, 156, 106708. DOI: 10.1016/j.envint.2021.106708.

Yu, H., Zhang, Y., Tan, W., & Zhang, Z. (2022). Microplastics as an emerging environmental pollutant in agricultural soils: Effects on ecosystems and human health. *Frontiers in Environmental Science*, 10, 855292. DOI: 10.3389/fenvs.2022.855292.

Zettler, E.R., Mincer, T.J., & Amaral-Zettler, L.A. (2013). Life in the "Plastisphere": Microbial communities on plastic marine debris. *Environmental Science & Technology*, 47, 7137–7146. DOI: 10.1021/es401288x.

Zhang, D., Ng, E.L., Hu, W., Wang, H., Galaviz, P., Yang, H. et al. (2020c). Plastic pollution in croplands threatens long-term food security. *Global Change Biology*, 26, 3356–3367. DOI: 10.1111/gcb.15043.

Zhang, G.S., & Liu, Y.F. (2018). The distribution of microplastics in soil aggregate fractions in southwestern China. *Science of The Total Environment*, 642, 12–20. DOI: 10.1016/j.scitotenv.2018.06.004.

Zhang, G.S., Zhang, F.X., & Li, X.T. (2019). Effects of polyester microfibers on soil physical properties: Perception from a field and a pot experiment. *Science of The Total Environment*, 670, 1–7. DOI: 10.1016/j.scitotenv.2019.03.149.

Zhang, G.S., & Zhang, F.S. (2020). Variations in aggregate-associated organic carbon and polyester microfibers resulting from polyester microfibers addition in a clayey soil. *Environmental Pollution*, 258, 113716. DOI: 10.1016/j.envpol.2019.113716.

Zhang, L., Xie, Y., Liu, J., Zhing, S., Qian, Y., & Gao, P. (2020a). An overlooked entry pathway of microplastics into agricultural soils from application of sludge-based fertilizers. *Environmental Science & Technology*, 54, 4248–4255. DOI: 10.1021/acs.est.9b07905.

Zhang, S., Liu, X., Hao, X., Wang, J., & Zhang, Y. (2020b). Distribution of low-density microplastics in the mollisol farmlands of northeast China. *Science of The Total Environment*, 708, 135091. DOI: 10.1016/j.scitotenv.2019.135091.

Zhang, Z., Peng, W., Duan, C., Zhu, X., Wu, H., Zhang, X., & Fang, L. (2022). Microplastics pollution from different plastic mulching years accentuate soil microbial nutrient limitations. *Gondwana Research*, 108, 91–101. DOI: 10.1016/j.gr.2021.07.028.

Zhao, S., Zhang, Z., Chen, L., Cui, Q., Cui, Y., Song, D., & Fang, L. (2022), Review on migration, transformation, and ecological impacts of microplastics in soil. *Applied Soil Ecology*, 176, 104486. DOI: 10.1016/j.apsoil.2022.104486.

Zhou, B., Wang, J., Zhang, H., Shi, H., Fei, Y., Huang, S. et al. (2020). Microplastics in agricultural soils on the coastal plain of Hangzhou Bay, east China: Multiple sources other than plastic mulching film. *Journal of Hazardous Materials*, 388, 121814. DOI: 10.1016/j.jhazmat.2019.121814.

Zhou, Y., Liu, X., & Wang, J. (2019). Characterization of microplastics and the association of heavy metals with microplastics in suburban soil of Central China. *Science of The Total Environment*, 694, 133798. DOI: 10.1016/j.scitotenv.2019.133798.

Zhou, Y., Jia, Z., Zheng, G., Chen, L., Zhang, Q., Su, B., & Zhou, S. (2023). Microplastics in agricultural soils on the coastal plain: Spatial characteristics, influencing factors, and sources. *Science of The Total Environment*, 901, 165948. DOI: 10.1016/j.scitotenv.2023.165948.

Zhu, B.-K., Fang, Y.-M., Zhu, D., Christie, P., Ke, X., & Zhu, Y.-G. (2018). Exposure to nanoplastics disturbs the gut microbiome in the soil oligochaete *Enchytraeus crypticus*. *Environmental Pollution*, 239, 408–415. DOI: 10.1016/j.envpol.2018.04.017.

11 | Micro-Nano Plastics in Aquatic Environments: Associated Health Impacts and Mitigation Strategies

Ludmila Sidorenko[1], Tatiana Gutul[2], Inna Rastimesina[3], Shushanik Kazaryan[4] Ashkhen Hovhannisyan[4], Susanna Tiratsuyan[4], Anatolie Sidorenko[2] and Ashok Vaseashta[2,5,6*(0000-0002-5649-0067)]

[1] Department of Molecular Biology and Human Genetics, State Medical and Pharmaceutical University "Nicolae Testemitanu", bd. Ştefan cel Mare şi Sfânt, 165 Chişinău, MD-2004, Republica Moldova

[2] Technical University of Moldova, Ghitu Institute of Electronic Engineering and Nanotechnology, Academiei str.3/3, Chişinău, Moldova, MD-2028, Republic of Moldova

[3] Technical University of Moldova, Ghitu Institute of Electronic Engineering and Nanotechnology, Academiei str.3/3, Chişinău, Moldova, MD-2028, Republic of Moldova

[4] Department of Medical Biochemistry and Biotechnology, Russian-Armenian University 123 Hovsep Emin street, Yerevan, 0051 Armenia

[5] Office of Strategic Research, International Clean Water Institute, Applied Research Division, Manassas, VA 20108-0258, USA

[6] Research Institute of the University of Bucharest, 90 Panduri Street, 050663, Bucharest Romania

Abstract: Micro-nanoplastics (MNPs) are small plastic particles (<5 mm and <100 nm in size, respectively) that have become prevalent in aquatic environments due to pollution. These particles pose significant risks to human and environmental health due to their persistence, ability to adsorb other pollutants, and potential to enter the food chain cycle, as they accumulate and are biologically transferred into the aquatic ecosystem. The biological transfer of MNPs is a complex process influenced by various factors. The movement of aquatic organisms, both vertically and horizontally, through different

*Corresponding author: prof.vaseashta@ieee.org

habitats plays a significant role in the dispersion of plastics within the aquatic environment. This process involves multiple interacting elements, and further research is essential to gain a deeper understanding of the mechanisms behind accumulation and the pathways of migration MNPs. Such insights are crucial for developing effective strategies to address and mitigate plastic pollution in the aquatic environment. Plastics degrade over a certain period of time and leach harmful chemicals in aquatic environments leading to harmful health impacts to gut microbiota and cardiovascular, neurological, and reproductive disorders, among others. Further research is essential to achieve a comprehensive understanding of the hazards posed by MNP pollution. Detailed clinical investigations would serve as a foundation for the development of effective strategies for pollution management and control. In this chapter, we present a review of health impacts associated with MNPs in aquatic environments. We have introduced several MNP contamination mitigation strategies, viz., microbial degradation, photocatalytic dissociation of MNPs, and capture through nanomaterial-based membranes. We discuss the suitability of these strategies for large-scale *in-situ* MNP mitigation.

1. Introduction

The widespread use of plastic materials has brought convenience and versatility to modern life but has also led to serious environmental problems. This has resulted in environmental pollution being increasingly recognized as a pervasive pollutant in aquatic environments. Improper disposal and the subsequent fragmentation or degradation of plastics have significantly increased the prevalence of micro- and nano-plastics (MNPs) in the environment. These MNPs can be classified into two categories: primary MNPs, which are intentionally manufactured at micro- or nanoscale sizes for specific applications, and secondary MNPs, which result from the breakdown of larger plastic items over time (Habumugisha et al., 2022). They pose significant health and ecological risks due to their small size and bioaccumulation of plastic particles in living organisms through various food chains and food webs in ecosystems (Kandpal & Srivastava, 2024; Nelis et al., 2023). The aquatic ecosystem, a vital habitat for diverse organisms, is profoundly impacted by disruptions caused by MNPs. A comprehensive evaluation of the toxicological effects of MNPs is critical, encompassing their exposure, toxicity, and transfer through the food web. This includes examining the physical, chemical, and biological interactions between MNPs and aquatic organisms, with a focus on their effects on growth, reproduction, behavior, immune function, and overall health across species—from microscopic plankton to larger organisms such as fish and marine mammals (Benson et al., 2022; Singh et al., 2022).

Most marine organisms tend to consume plastic debris that can easily accumulate into vertebrates, invertebrates, and planktonic entities (Gangadoo et al., 2020; Zaki & Aris, 2022). Primary producers, such as phytoplankton, can absorb and accumulate MNPs. These plastics can then be ingested by smaller organisms like zooplankton, facilitating their transfer through the food web (Gholamhosseini et al., 2023). This process raises concerns about the bioaccumulation and biomagnification of MNPs, which increase exposure for top predators, including humans (Benson et al., 2022). The risks associated with MNPs in food webs are a critical focus of toxicological studies (Feng et al., 2023; Gomes et al., 2022; Revel et al., 2018). Several research studies have explored how these plastics disrupt the balance and functioning of food webs, potentially leading to cascading effects such as altered population dynamics, shifts in species composition, and biodiversity loss. Additionally, the transfer of MNPs to humans through seafood consumption highlights significant concerns about human health, underscoring the need for in-depth research on their impacts.

There is a close relationship between the size of plastic particles and their toxicological effects on the body. MNPs can bind pollutants and heavy metals, pass through biological barriers, or are adsorbed on various microbial surfaces (Kokilathasan & Dittrich, 2022) and physically obstruct light transmission, and disrupt photosynthesis or respiration, thereby leading to oxidative stress (Dissanayake et al., 2022; Gao et al., 2021), causes the generation of reactive oxygen species (ROS) (Das, 2023; Shi et al., 2023; Singh et al., 2022), damaging DNA, lipids, and other biomacromolecules (You et al., 2022). In addition, various plasticizers present in plastics can also lead to ROS development. Furthermore, MNPs can interact with the biomolecules of the cell wall, which leads to the formation of an eco-crown. The cumulative effects of MNPs are amplified up the food chain in organisms living at higher trophic levels through a process known as biomagnification. Experiments have shown that in vertebrates, MNPs can cause early death, inflammation, metabolic disturbances, hormonal imbalances, and other adverse effects due to accumulation in the liver and kidneys (Elizalde-Velázquez et al., 2020). MNPs can enter the human body through various pathways, as reported in several studies (Rubio et al., 2020; Yuan et al., 2022). These particles have been detected in multiple organs and biological fluids, highlighting their widespread distribution in the human system (Mofijur et al., 2021). Human biomonitoring, widely regarded as the gold standard for assessing environmental exposure to MNP pollutants and their metabolites, utilizes biological samples such as blood, saliva, breast milk, and urine. This methodology has been extensively reviewed (Cook & Halden, 2020; Smolders et al., 2009).

Thermoplastics consist of various types of plastics, such as polyethylene (PE), polypropylene (PP), polystyrene (PS), and polyvinyl chloride (PVC). Among PE variants, high-density polyethylene (HDPE) is widely used in products like detergent bottles, milk jugs, and molded plastic crates, while low-density polyethylene (LDPE) is commonly found in outdoor furniture, siding, floor tiles, shower curtains, and flexible packaging. PE and PS are the most commonly used plastics in consumer products, characterized by their shorter lifespan compared to other plastics. Notably, it is estimated that 82% of PVC waste is disposed of in landfills, 15% is incinerated, and only 3% is recycled (Suresh et al., 2017).

The combination of high production volumes, short life cycles, and significant environmental release makes these materials a key focus of ongoing scientific research. LDPE is one of the most widely utilized plastics, forming the backbone of packaging industries and daily consumer products, including disposable bags, containers, and films. Its desirable properties include flexibility, chemical resistance, and low production cost, which has rendered LDPE indispensable (Tuteja et al., 2024). However, these same properties contribute to its persistence in the environment, making it one of the most challenging plastics to degrade.

Globally, plastic production has surged, reaching over 400 MT annually as of 2022, with LDPE constituting a significant share. Despite advancements in recycling technologies, only a small fraction of LDPE waste, around 5%, is effectively recycled, as reported by Plastics Europe, a knowledge hub that tracks activities related to plastics in Europe, for circular economy and sustainable development. The rest accumulate in landfills and find their way to the aquatic ecosystems, where they present long-term environmental and ecological risks. The Republic of Moldova, for instance, imports nearly 100,000 tons of plastic annually, with an average recycling rate of just 3%, underscoring the urgent need for alternative waste management solutions (Geyer, 2023). LDPE's environmental persistence arises from its chemical composition. The polymer comprises a hydrophobic backbone of long hydrocarbon chains, which are highly resistant to biological, chemical, and physical degradation. This durability results in a decomposition timeline spanning hundreds of years. Under the influence of environmental factors such as weathering, solar radiation, mechanical forces, and microorganisms, plastic materials constantly decompose, forming microplastics and nanoplastics (Dhali et al., 2024). Over time, fragmented LDPE contributes to nano-microplastic pollution, affecting soil, human health, aquatic life, food chains, and drinking water (Bergmann et al., 2015; Brachner et al., 2020; Ramsperger et al., 2023; Vitali et al., 2023).

MNPs can be classified into primary and secondary types based on their formation process (Avio et al., 2017; Vaseashta, Ivanov, et al., 2021). Primary MNPs are pre-manufactured plastic particles generally incorporated into personal care products (PCPs) (Anderson et al., 2016; Godoy et al., 2019), which can easily enter the environment following use (Adawi et al., 2018; Cheung & Fok, 2016). Plastics can leach harmful chemicals, plasticizers such as bisphenol A (endocrine 'disruptor' by mimicking estrogen)—linked to hormonal imbalances, reproductive disorders, and developmental issues, associated with an increased risk of cardiovascular diseases (CVD), diabetes, and certain cancers. Phthalates and other substances known as endocrine disruptors, affecting testosterone levels and fertility, are associated with neurological disorders such as attention-deficit/hyperactivity disorder (ADHD) in children, and metabolic disorders (Prattichizzo et al., 2024; Zhang et al., 2024; Zhu et al., 2023). Plastic products may contain lead, cadmium, or mercury, which are neurotoxic. MNPs can accumulate in cells, and tissues, which leads to mechanical damage, cytotoxicity, genotoxicity, immunotoxicity, neurotoxicity, disruption of homeostasis and development, metabolic disorders, and oxidation-reduction reactions potentially triggering "immune responses" and inflammation. In contrast, secondary MNPs are generated through the breakdown of larger plastic items through processes such as UV radiation exposure, physical degradation (photo, thermal), thermo-oxidative, hydrolysis, and environmental biodegradation (Au et al., 2017; Auta et al., 2017; Hebner & Maurer-Jones, 2020; Lambert & Wagner, 2016; Weinstein et al., 2016). Qualitative methods are used to identify the presence of MNPs, while quantitative methods determine their concentration and size in various environments, including soil, water, air, and sediment, using a scanning electron microscope (SEM), energy-dispersive X-ray spectroscopy (EDX), Fourier transform infrared spectroscopy (FTIR), micro-FTIR, micro-Raman spectroscopy, particle size analyzer, SEM-EDX, transmission electron microscopy (TEM), scanning transmission X-ray microscope (STXM), attenuated total internal reflection (ATR) FTIR (ATR-FTIR), dissecting microscope, stereomicroscope, focal plane array (FPA), dynamic light scattering (DLS), and principal component analysis (PCA) (Anderson et al., 2016; Bigalke et al., 2018; Dehghani et al., 2017; Gao et al., 2007; Gray et al., 2018; Karthik et al., 2018; Mintenig et al., 2019; Peeken et al., 2018; Peng et al., 2017, 2018; Pivokonsky et al., 2018; Schymanski et al., 2018; Wang et al., 2018; Yang et al., 2015; Yu et al., 2018), to name a few. While MNPs have been detected across marine, terrestrial, and atmospheric ecosystems, as well as in food chains, no standardized qualitative or quantitative methods currently exist to identify or measure them accurately. This lack of uniformity poses significant challenges in determining the true abundance of MNPs in the environment, thus necessitating the establishment of qualitative and quantitative indicators for testing both MP and NP in the environment.

The direct impacts of MNPs on human health remain poorly understood. Additionally, MNPs often contain plasticizers and can adsorb harmful chemicals and heavy metals, many of which are known to cause endocrine disruption and other toxic effects. However, the extent to which MNPs influence the toxicity of these additives and adsorbed substances remains largely unexplored (Jiang et al., 2020; Vaseashta, Ivanov, et al., 2021). Continuous MNP pollution significantly impacts ecosystems, leading to reduced biodiversity, weakened fisheries sustainability, disruption of phytoplankton and zooplankton life cycles, and deterioration of human and animal health, while also threatening economic stability, are posing an ever-increasing ecological threat (Wang et al., 2023; Wang et al., 2021; Zhao et al., 2022). Considering these environmental aspects, strategic solutions include regulations and mitigation. Since regulations take time for approval and implementation, the quickest and most effective way to reduce the environmental burden is by mitigation strategies of MNPs. These strategies conjoin enforcing restrictions, bans on fossil-based plastic production, a ban on the use of toxic additives in products, and establishing standards for plastic production that prioritize environmentally friendly and biodegradable materials. The strategy should also include improved waste management strategy, including circularity of plastic waste to reach close to zero return to landfills.

These processes should be complemented with advanced nanoscale filtration mechanisms in treatment plants to capture MNPs before they contaminate aquatic ecosystems. Yet another mechanism to reduce the environmental burden by effective recycling and upcycling of plastic products. This chapter focuses on the mitigation strategies, more specifically on the microbial aspects of MNP degradation, either as a standalone or in conjunction with other techniques for a proof-of-concept system, with the ability of a large-scale unit that can remove/ degrade MNPs to approx. 100 L/min. We further discuss the suitability of these strategies for commercial-scale units.

2. Effects of MNPs on Human Health

Recent research has highlighted the critical and often overlooked relationship between environmental chemical exposures and CVD. One emerging area of concern is the role of MNPs, which are small plastic particles derived either from the breakdown of larger plastics or from various products, such as PCPs, textiles, paints, etc. MNPs, defined as particles smaller than 5 μm (microplastics) or 1 μm (nano plastics), have been linked to various biological disruptions, such as oxidative stress, inflammation, cellular aging, and immune system activation (Kazaryan et al., 2022; Madni et al., 2021; Stabnikova et al., 2022; Tumoyan et al., 2024; You et al., 2012). These processes contribute to cardiovascular and metabolic disturbances that could ultimately lead to diseases such as atherosclerosis, hypertension, and even premature death (Zheng et al., 2024). Recent studies have identified MNPs in a variety of cardiovascular tissues, including the heart, blood vessels, and atherosclerotic plaques (Prattichizzo et al., 2024). While plastic pollution's impact on the environment has been well-documented, emerging concerns about its effects on human health—particularly cardiovascular health—are becoming more prominent and need thorough investigation.

Traditionally, plastic-related health risks were linked to additives like plasticizers—Bisphenol A (BPA) and phthalates, known for their endocrine-disrupting properties (Kahn et al., 2020). However, in recent years, MNPs—tiny plastic particles that are either manufactured for use in products like cosmetics or created through the degradation of larger plastic items—have gained attention due to their potential to cause harm. These particles are present in the air, water, food, and even drinking water, and humans are exposed to millions of MNPs every year. Groundwater is also under threat from MNP pollution, as it poses a risk to drinking water supplies (Vaseashta et al., 2022; Vaseashta, Gevorgyan, et al., 2021). MNPs can enter the soil through a variety of routes, such as mulching with plastic, compost, and sewage sludge, as well as precipitation, and have negative impacts on the soil, food safety, and living organisms. These particles can adsorb harmful chemicals like pesticides, polycyclic aromatic hydrocarbons, and polychlorinated biphenyls on their surface (Kandpal & Srivastava, 2024). Accumulated MNPs in water detrimentally affect the structural and functional traits of aquatic biota and the related ecosystem functions (Adegoke et al., 2023; Di Pippo et al., 2023). Aquatic organisms may serve as a potential route of human exposure through seafood consumption (Khalid et al., 2021; Le et al., 2024; Zhang et al., 2022). MNPs directly affect living beings, either through ingestion or toxicity, and could act as a vehicle for invasive species by adsorption on their surface, or via other synthetic chemical pollutants, thus potentiating or synergizing their toxic capability due to their ingredients (Vethaak & Legler, 2021).

The accumulation of these particles in human tissues has raised concerns about their impact on health, particularly concerning CVD. Early evidence suggests that MNPs might contribute to the development or progression of CVD by triggering various pathophysiological processes as a novel cardiovascular risk factor (Zhu et al., 2023). Their small size allows them to penetrate various tissues, with smaller particles exhibiting greater tissue distribution. Additionally, negatively charged particles are more likely to accumulate in vascularized organs due to their ability to cross cellular membranes. Studies in animal models have shown that exposure to MNPs—such as polystyrene and polyethylene particles—leads to the accumulation of these particles in the blood and heart. In particular, research in rodents has shown that MNPs cause several cardiovascular problems, including cardiac fibrosis, capillary congestion, and inflammation in heart tissues. These changes are believed to result from the activation of inflammatory pathways such as NLRP3/caspase-1 and disruptions in cellular communication, leading to myocardial damage. At the microvascular level, MNP exposure has been shown to increase thrombotic activity, triggering inflammatory responses that damage the endothelium (Shiwakoti et al., 2022). Inflammation in the vasculature has been linked to the development of atherosclerosis, a condition where plaque builds up in the arteries, leading to restricted blood flow and increased risk of heart attack or stroke. MNPs have been shown to exacerbate atherosclerosis in experimental models by altering lipid metabolism and promoting foam cell formation in arterial walls (Wang et al., 2023; Zhong et al., 2024).

Several studies have reported the presence of MNPs in the pericardium, epicardial adipose tissue, myocardium, and arterial walls (Liu et al., 2024). A study analyzed tissues from patients undergoing heart surgery and found that MNPs were present in a variety of cardiovascular tissues, including the heart muscle and surrounding fat (Yang et al., 2023). Another study, focusing on thrombi from patients with cardiovascular disease, found that MNPs were present in 24 out of 30 samples. Most of these particles were PE, PVC, and PA, suggesting that common plastic materials contribute to this accumulation. Interestingly, the presence of MNPs in these thrombi was associated with more severe cardiovascular disease, supporting the hypothesis that MNPs may exacerbate or contribute to the progression of cardiovascular conditions. A new effective method of a quantitative investigation of the cardiovascular system state—cardiorhythmogram (CRG) analyses was elaborated (Sidorenko et al., 2018) based on the Independent Component Analysis (ICA) technique, as shown in Figure 1.

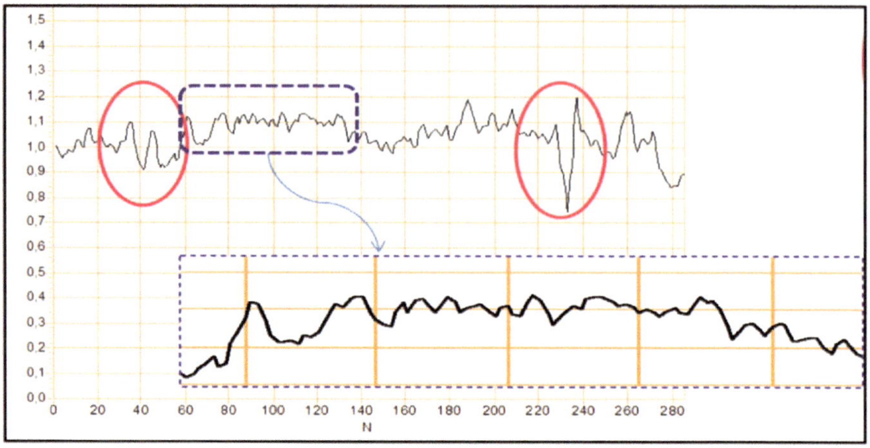

Figure 1: An advanced method of cardiorhythmogram analysis to assess the pathophysiologic state of regulation of the heart rhythm (Sidorenko et al., 2018).

Even low doses of MNPs over extended periods have been found to cause arterial thickening, suggesting that chronic exposure could lead to long-term cardiovascular damage. These findings underscore the need for further research to define the exposure thresholds at which MNPs begin to affect cardiovascular health. The presence of MNPs in human tissues has been increasingly documented and studies have shown that these particles are detectable in a variety of biological samples, including blood, urine, stool, and even semen. This widespread detection suggests that MNPs are absorbed by the body and distributed across different organ systems (Feng et al., 2023; Le et al., 2024). MPs accumulate in the gastrointestinal tract, disrupting the gut microbiome, and causing dysbiosis, which can be linked to various health issues, including gastrointestinal disorders, systemic inflammation, and chronic diseases, as well as disorders related to the Gut-Brain Axis (Bora et al., 2024). MNPs disrupt the blood-brain barrier, due to disturbances in oxidative and inflammatory balance, changes in neurotransmitters, and key enzymes associated with neurotransmission, leading to brain damage and affecting cognitive function, disrupting the functioning of neurons, and even changing the memory and behavior of organisms, having further potential negative effects on the central nervous system and contribute to the development of other diseases (Zheng et al., 2024). They can specifically affect protein folding, inducing the formation of aberrant proteins and, therefore, potentially triggering the development of systemic and local amyloidosis (Windheim et al., 2022). Furthermore, by penetrating the blood-brain barrier, MNPs increase the incidence of neurodegenerative diseases such as Alzheimer's disease (AD). They have been shown to significantly promote the formation of more Aβ oligomers and lead to apparent neurotoxicity (Gou et al., 2024; Prüst et al., 2020). Nano plastics are also a risk for the development of

Parkinson's disease (Liang et al., 2022). MNPs cause interaction between the gut and the nervous system along the gut-brain axis points, as an additional pathway of MNP-induced brain damage, the mechanisms of which remain unclear. Gut microbiota is being considered as a new therapeutic target for alleviating MNP-induced brain conditions (Shi et al., 2023). More concerning, however, is the accumulation of MNPs in key organs, such as the liver, lungs, and kidneys (Ali et al., 2024; Danso et al., 2022; Horvatits et al., 2022; Huang et al., 2020). It has been shown that in nano-biological interactions and pulmonary toxicity, MNPs likely play a predominant role due to the formation of hetero aggregates at the alveolar-capillary interface (Xu et al., 2023). MNPs accumulating in the urinary system can promote oxidative stress, inflammation, apoptosis, autophagy, and the gut-kidney axis, which leads to impaired metabolic function of the kidneys, and tissue fibrosis, thereby potentially increasing the risk of diseases of the urinary system (Huang et al., 2024; La Porta et al., 2023). In a study, higher concentrations of MNPs have been detected in the livers of individuals with cirrhosis and the stools of patients suffering from inflammatory bowel disease, indicating that MNPs may have pathological effects on specific organs (Horvatits et al., 2022). MNPs can cause pancreatic damage, glycogen accumulation, and hepatocellular edema as its exposure has been shown to aggravate Type II diabetes through partial inhibition of AKT/GSK3β phosphorylation and P-AKT/P-FoxO1 (Hsiao et al., 2025; Wang et al., 2023; Zhang et al., 2024). Longitudinal studies examining the absorption, distribution, and accumulation of MNPs, as well as their long-term health impacts, are essential. Small sizes of MNP limit the efficiency of detection of these particles and thus require urgent, sustainable solutions to mitigate pollution (Kandpal & Srivastava, 2024). To mitigate the health risks associated with MNPs, global efforts to reduce plastic production and pollution should be prioritized. In addition, public health initiatives aimed at reducing plastic exposure through changes in consumer behavior and policy could help protect human health and reduce the burden of CVD associated with plastic pollution. However, more urgently, it is critical to have an elaboration of novel, highly efficient methods of MNP degradation and elimination, as described in the following section.

3. MNP Mitigation Strategies

Traditional methods of plastic waste management—recycling, upcycling, incineration, and landfilling—have been used with limited success. In general, recycling LDPE, a commonly used monomer, is technically complex and economically burdensome, given the polymer's resistance to processing and contamination issues. Incineration, on the other hand, releases harmful gases. while landfilling merely delays environmental pollution without resolving the basic issue (Verma et al., 2016) of mitigating plastic pollution. Policy recommendations, such as limiting the use of plastic bags, and single-use plastic bottles, while replacing fossil-based plastics with bio-based plastics also have confronted significant limitations. Hence the strategic way to reduce the presence of MNPs in the environment is by its degradation to environmentally friendly components. While some preliminary investigations are in progress to degrade plastics, concerns still remain about capturing the additives that are used to modify plastic characteristics for some specific applications.

3.1. Photocatalytic Dissociation of MNPs

Photocatalytic degradation is a promising and environmentally friendly method to reduce microplastic pollution. The method uses UV radiation, from a light source or solar radiation to decompose plastic to a large extent and with minimal costs (Du et al., 2021). In general, IR rays (>700 nm) promote thermal oxidation, while UV rays (<400 nm) promote direct photodegradation (Thomas & Sandhyarani, 2013). When exposed to UV radiation, most plastic particles are partially degraded. The active oxygen species produced when exposed to visible light are incapable of triggering chain cleavage and subsequent oxidation processes. For photocatalytic degradation (PCD) of organic contaminants, semiconductor photocatalyst materials with a strong redox potential are used, such as TiO_2, ZnO, CdS, SnO_2, and ZnS, in reduced dimensions (Ariza-Tarazona et al., 2019). Highly reactive oxygen species (ROS) are generated, which ultimately degrade organic polymers, causing polymer chain breakage and even complete mineralization (Du et al., 2021). Photocatalysis, under optimum conditions, usually shows significantly higher degradation efficiency compared to biodegradation and is carried out under milder conditions as compared to chemical processes (Li et al., 2023; Pan et al., 2023). With an appropriate light source and catalytic agent, the process of photocatalysis is suitable for the degradation of a wide range of plastics. Using self-propelled microrobots to capture and decompose MPs, it was shown that photocatalysis can also be used in the collection of plastic waste (Beladi-Mousavi et al., 2021; Feng et al., 2022; Urso & Pumera, 2022; Wang et al., 2023). While there are many benefits of PCD of plastics, there are still many challenges, such as limitations in the specificity of photocatalysts, i.e., photocatalysts are often only suitable for degrading a certain type of plastic. Moreover, the process also cannot selectively affect the reactive centers of plastic (Ge et al., 2022). Additionally, TiO_2 photocatalysts perform better in the UV range but not in visible light conditions (He et al., 2023). Direct photocatalysis can occur only when the chemical compounds absorb incident light efficiently, as shown in Figure 2. Using semiconducting nanomaterial facilitates UV radiation absorption, causing ROS to be generated when in contact with water. In photogenerated catalysis, photocatalytic activity depends on the ability of the catalyst to create electron-hole pairs, which generate free radicals that can undergo secondary reactions. It was observed that the MNPs in the environment contain associated pollutants. Since photocatalysis is a surface reaction, photodegradation with accompanying contaminants may be difficult if opaque contaminants prevent light irradiation covering MNPs (Hamd et al., 2022; Sundar & Kanmani, 2020).

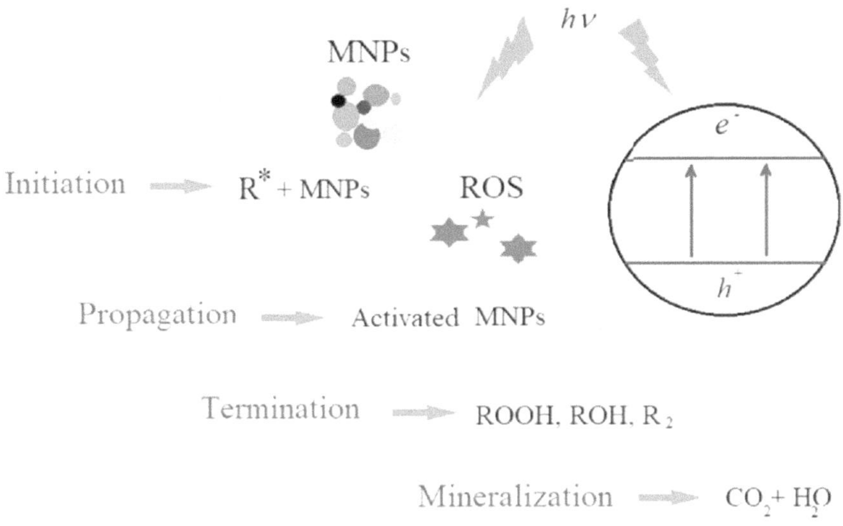

Figure 2: Photocatalytic dissociation of MNPS in biodegradable products.

3.2. Microbial Degradation of MNP

Biodegradation has emerged as a sustainable and viable alternative for mitigating plastic pollution. This process leverages the natural metabolic activities of microorganisms to break down complex polymeric structures into simpler, eco-friendly byproducts. Microbial biodegradation generally occurs in two phases. The initial phase involves the colonization of plastic surfaces by microorganisms and the formation of biofilms (Chen et al., 2024). Microorganisms attach to the surface of any material and over time form biofilms. The formation of biofilms is a multi-stage process that includes the formation of extracellular polysaccharides, proteins (many of which can be fibrous or form other higher-order forms), extracellular DNA and RNA, lipids, and other biomolecules that collectively form a protective and adhesive layer around the community (Bamford et al., 2023; Sharma et al., 2023). Biofilms are a form of existence of microorganisms common in water and soil environments (Roohi et al., 2017). These biofilms produce enzymes that cleave polymer chains into oligomers and monomers. Microorganisms such as bacteria, algae, yeasts, and fungi have demonstrated the ability to utilize polymers (Dhali et al., 2024; Zeenat et al., 2021) as a carbon source, facilitated by the secretion of specific plastic-degrading enzymes (PDE) like protease (Hu et al., 2018), esterase, hydrolase (Gricajeva et al., 2022), oxidase (Zampolli et al., 2023), cutinase (Dimarogona et al., 2015; Liu et al., 2019) as well as lipase (Andler et al., 2022; Caruso, 2015; Parida et al., 2023; Yang et al., 2015). These monomers are more easily metabolized by bacteria, leading to increased biodegradation. This process, in turn, supports bacterial growth, as the microorganisms derive both carbon and energy exclusively from plastic breakdown. The second phase sees these simpler compounds metabolized into carbon dioxide, water, or biomass (Mohanan et al., 2020). Since biomass growth serves as a direct indicator of biodegradation, the observed increase in biomass concentration and microbial activity highlights a significant enhancement in the degradation process (Mohammadi et al., 2009). Factors such as temperature, humidity, nutrient availability, and the microbial species involved significantly influence the efficiency of biodegradation (Montazer et al., 2020).

In addition, several processes, as described below, are used either in parallel or as a standalone process to dissociate long carbon chains, making the degradation of plastics somewhat easier. One such process is photooxidation accomplished by UV irradiation which induces oxidative changes on the polymer surface. This process breaks carbon chains in the polymer, forming carbonyl and hydroxyl groups that increase hydrophilicity and facilitate microbial attachment. Chemical treatment with acids, alkalis, or oxidizing agents (e.g., hydrogen peroxide) alters the chemical structure of the surface, creating functional groups, which renders microbial degradation easier, which is described below, by making host sites for microorganisms. Yet another process is mechanical treatment, such as grinding or the introduction of microcracks increases surface area and creates defects, enabling easier penetration by microorganisms and their enzymes. Applying nanocomposite coatings as host catalytic materials, such as $MgFe_2O_4$/PVP or other metal oxide nanoparticles, facilitates polymer bond activation under light or chemical reactions, making the polymer more susceptible to enzymatic degradation. In addition, low-pressure plasma treatment alters the chemical and physical structure of the polymer surface, making it more hydrophilic, which promotes interactions with microorganisms. Pre-incubation in mineral solutions by immersing the film in nutrient solutions with microorganisms or enzymes can initiate the formation of bioactive layers on the polymer surface, enhancing its biodegradability. Each of these methods aims to facilitate microbial attachment, activate enzymes, and accelerate biodegradation processes. The choice of method depends on the polymer type, research objectives, and available resources. For example, a combined method for LDPE0151—a commonly used plastic film involves a three-step process: first, the PE film undergoes photooxidation through UV irradiation; second, a layer of $MgFe_2O_4$/PVP nanocomposite is applied to the film surface by immersing it in an aqueous solution; and finally, the treated film is incubated in a mineral solution (Corcimaru et al., 2024).

Inducing oxidative stress in microorganisms using an oxidant stimulates the production of oxidative enzymes, thereby enhancing biodegradation. However, the effectiveness of this process relies on the appropriate choice of stimulatory agents. Bacterial oxidative enzymes, such as peroxidase and catalase, have demonstrated the capability to catalyze the oxidative cleavage of carbon-carbon bonds in various organic compounds, accelerating the biodegradation process (Mohammadi et al., 2022). Simultaneously, the increasing recognition of biodegradation as a viable solution has spurred research into novel microbial strains, consortia, and enzyme engineering (Anush et al., 2019; Morales Ramos et al., 2024). The other aspect of microorganisms used is a whole-cell biosensor based on synthetic biology for on-site detection of food contaminants (Chen et al., 2024) as well as for use in biomedical, environmental, and agricultural monitoring (Moraskie et al., 2021). Recent advances in molecular biology and bioprocessing technologies hold the promise of accelerating LDPE and degradation of other similar monomers by integrating biodegradation into existing waste management practices. Furthermore, studies on the ecological impacts of biodegradable products can pave the way for safer and more effective applications.

3.2.1. Microbial Consortia vs. Individual Strains

A microbial consortium or community consists of two or more bacterial or microbial groups working symbiotically. Such consortiums can be endosymbiotic or ectosymbiotic and may occasionally be both. The objective is to explore the biodegradation potential of these microbial consortia in transforming plastics into biodegradable byproducts, as compared to individual strains. Specific focus is placed on the role of microbial consortia, as their synergistic interactions offer a scalable solution for addressing persistent plastic waste in diverse environmental conditions. While individual microorganisms such as *Pseudomonas fluorescens*, *Penicillium verrucosum*, and *Aspergillus flavus* have shown promising biodegradation capabilities, microbial consortia often exhibit superior efficiency. For example, bacteria in the consortia may oxidize polymer surfaces, while fungi degrade the resulting fragments into simpler compounds. These interactions allow consortia to adapt to varied environmental conditions and degrade plastics more comprehensively than individual strains. The subject of metabolic engineering is still in its infancy, although recent advances in microbial consortia present a valuable approach for expanding the scope of this subject, more specifically for synthetic metabolic engineering. The study faces several unknowns and challenges, especially in intercellular communication and the construction of stable and controllable microbial consortia systems.

Recent studies have demonstrated that microbial consortia isolated from landfill soils contaminated with plastic waste can significantly alter plastic (Cao et al., 2022; Park & Kim, 2019; Postolachi et al., 2022). Under aerobic and anaerobic conditions, these consortia not only reduced tensile strength and elongation of plastics at the break but also caused visible surface erosion and chemical modifications (Rastimesina, 2023). In a series of experiments conducted at the laboratories of the authors, LDPE films with a thickness of 35 µm, produced by Kraus Folie Sp., Poland, were used for experiments. The films are typically sterilized with 70% ethanol, followed by UV sterilization. These treatments aimed to eliminate potential contaminants without altering the polymer's surface properties. Microbial consortia were isolated by transferring LDPE strips from the soil to a liquid mineral salt medium (MSM). Enrichment cultivation is performed under continuous stirring conditions at 28°C for 30 days, followed by serial dilution and plating on solid media such as Czapek-Dox agar, nutrient agar, and MSM agar. Microbial isolates were purified and identified based on colony morphology, gram staining, and microscopic examination. The study employed various MSM formulations to assess the growth and biodegradation efficiency of microorganisms. These media differed in their nutrient content, pH, and nitrogen-to-phosphorus ratios (Postolaschi et al., 2022). A 0.1% glucose solution was added as a growth inducer in the initial stages to facilitate microbial adaptation to the medium. The pH was adjusted to values ranging from 5.5–6.5, depending on the microbial strains and experimental objectives. The samples are then cultivated for representative microbial strains in both solid and liquid MSM. Cultivation was normally carried out at 28°C for 30–40 days. Microbial consortia were cultivated on LDPE samples in liquid MSM under continuous stirring for up to 270 days. Consortia were obtained from aerobic and anaerobic conditions and included mixed populations of fungi, bacteria, and yeasts. Both individual strains and consortia were tested for their ability to degrade LDPE and alter its physical properties. The samples are evaluated for plastic degradation using gravimetric analysis by measuring the weight loss of the films before and after incubation. Washed LDPE films were air-dried and weighed. The percentage of weight loss was calculated using the formula:

$$\text{Weight loss (\%)} = (\text{Initial Weight} - \text{Final Weight})/\text{Initial Weight} \cdot 100.$$

The Tensile Testing included changes in the mechanical properties of LDPE and was evaluated using a CQ-508B tensile testing machine, following ISO 527-1:2019 standards. Parameters such as tensile strength (MPa) and elongation at break (%) were recorded. Five replicates were tested for each condition, and the average values were calculated. The surface of LDPE films was examined under an Optica® B-510 PH microscope to identify microbial colonization and surface alterations, such as biofilm formation and erosion. Additionally, Fourier-transform infrared (FTIR) spectroscopy was performed to detect chemical changes in LDPE after microbial treatment. Absorption spectra were analyzed for the emergence of new bands or splitting of existing bands, indicating oxidation or cleavage of polymer chains. Data from all experiments were analyzed using Microsoft Excel. Confidence intervals were calculated, and differences were considered statistically significant at $P < 0.05$. Such findings highlight the potential of consortia in scalable biodegradation applications.

Despite promising results, several challenges remain in harnessing biodegradation for practical applications. LDPE's hydrophobic nature hinders microbial attachment, while its crystalline regions resist enzymatic attack. Moreover, variations in polymer properties, environmental conditions, and microbial strains create inconsistencies in degradation rates. Addressing these issues requires a multifaceted approach, including the optimization of microbial strains, the development of culture media (Postolaschi et al., 2022), and pretreatment methods to increase polymer susceptibility. Modifying the surface of polymer films aims to increase their accessibility to microorganisms and enhance biodegradation processes.

The ability of individual strains to degrade LDPE varied significantly based on microbial type and cultivation conditions. Among bacterial strains, *Pseudomonas fluorescens* CNM-PFB-01 achieved a biodegradation efficiency of up to 0.86% after 40 days in MSM 4 medium (N:P ratio 4.3:1). This strain showed notable biomass accumulation and biofilm formation on LDPE surfaces, as observed under microscopy. Its activity suggests a potential role in initial polymer oxidation due to its metabolic versatility and enzyme production. Fungal strains demonstrated higher degradation rates than bacteria, as shown in Figure 3. *Penicillium verrucosum* CNM-FP-02, for instance, achieved an average LDPE weight loss of 6.06% on solid media after 30 days. This efficiency is attributed to the extracellular secretion of hydrolytic enzymes like esterases and cutinases, which break down the hydrophobic polymer matrix into smaller, more accessible fragments. Although *Pseudomonas fluorescens* exhibited substantial degradation capabilities, fungi such as *Penicillium* spp. and *Aspergillus flavus* proved to be more effective overall. The average degradation rate for *Aspergillus flavus* was 4.98%, with significant biofilm development and visible surface erosion on LDPE samples.

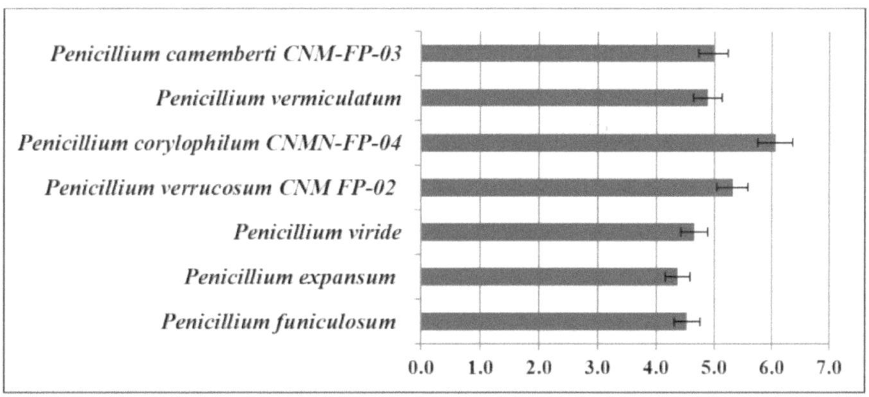

Figure 3: LDPE biodegradation in the presence of *Penicillium* spp.

Microbial consortia isolated from landfill soil demonstrated superior performance compared to individual strains. The consortia cultivated in anaerobic conditions exhibited the highest degradation rates, with LDPE weight loss reaching up to 8.5% after 270 days. These consortia consisted of a mix of bacteria, fungi, and yeasts, which interacted synergistically to enhance biodegradation. The Aerobic consortia primarily included fungal strains, such as *Trichoderma* and *Penicillium*, and showed moderate degradation rates of 5.2%. The fungi contributed to biofilm formation, while bacteria within the consortia-initiated polymer oxidation. Anaerobic consortia: With a higher proportion of bacteria (e.g., *Pseudomonas* spp.), these consortia excelled in degrading LDPE. The anaerobic environment likely facilitated the reduction of crystalline regions of LDPE, enabling enzymatic attack on the polymer chains.

In terms of mechanical characteristics, the tensile tests revealed significant reductions in LDPE tensile strength and elongation at break after microbial treatment, as shown in Figure 4.

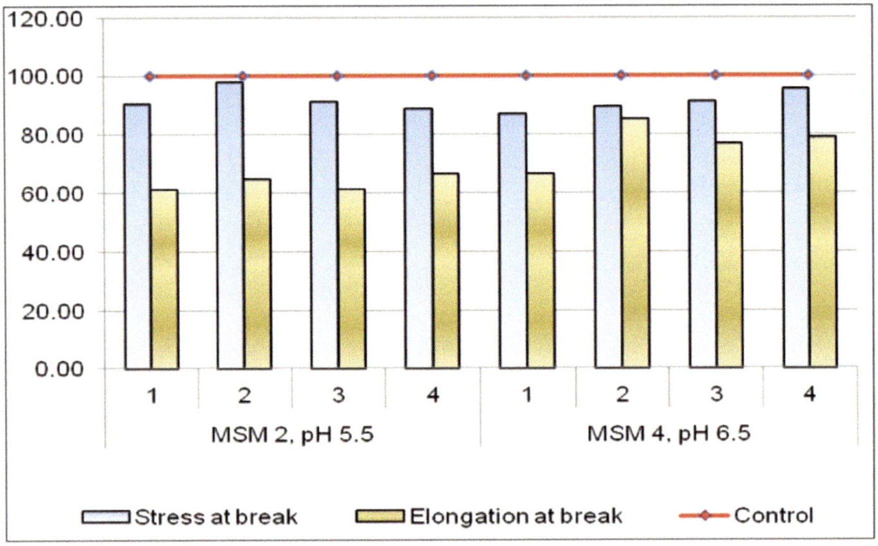

Figure 4: Tensile test of LDPE films, cultivated with microbial consortia (Rastimesina, 2023).

LDPE samples exposed to anaerobic consortia experienced the most pronounced changes, with a decrease in tensile strength by 67% and elongation by 54%. Samples treated with individual fungal strains, such as *Penicillium verrucosum*, showed a moderate reduction in elongation (30–40%). Microscopic observations confirmed biofilm formation and the presence of surface erosion, cracks, and pits on LDPE samples treated with microbial consortia. FTIR analysis revealed chemical modifications, such as the appearance of carbonyl groups and shifts in the hydroxyl group peaks, as shown in Figure 5. As shown, the characteristic peaks are at wave numbers 2915, 2848 cm^{-1}, and corresponding to CH stretching, can be observed, at around 1456 and 1377 cm^{-1}, corresponding to symmetric deformations. These and other peaks indicate changes due to oxidative degradation and hydrolysis of polymer chains.

The key factors influencing biodegradation include nutritional conditions, viz., the mineral salt media composition significantly affected biodegradation efficiency. MSM 4, with its high nitrogen content and optimal N:P ratio (4.3:1),

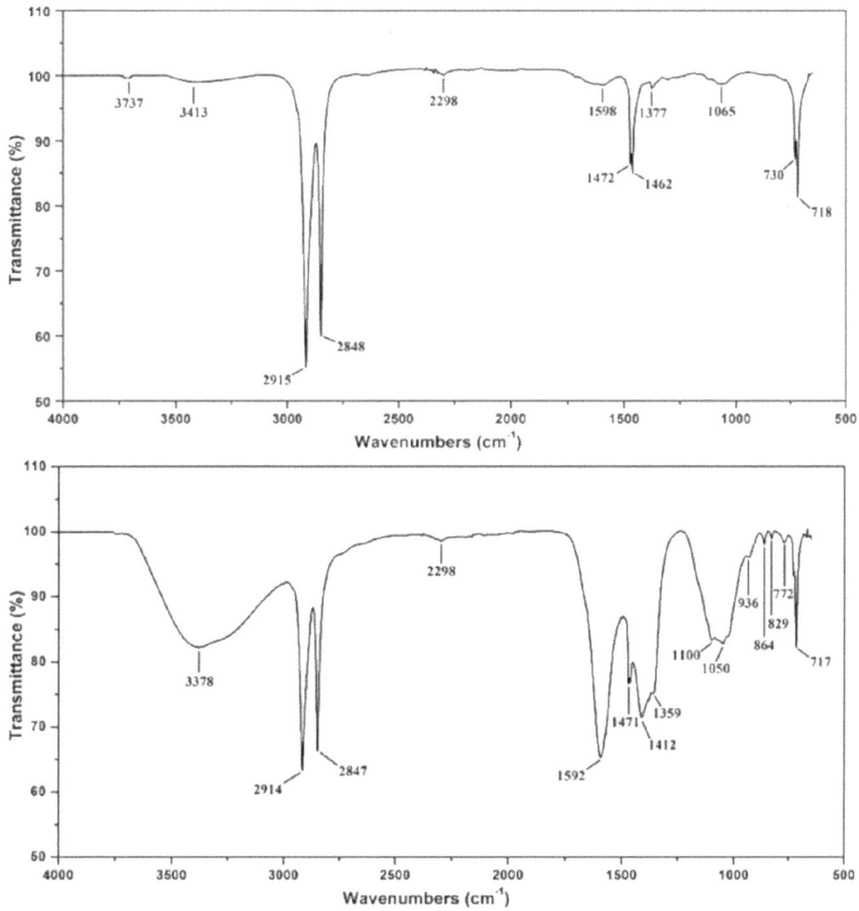

Figure 5: FT-IR chromatogram of LDPE films: (A) control LDPE, (B) LDPE film, cultivated on MSM 2 with microbial consortia (Rastimesina, 2023).

supported the highest microbial activity. Conversely, media with unbalanced nutrient ratios, such as MSM 1 (N:P 1:4.8), resulted in lower degradation rates. The addition of glucose as a growth inducer during initial cultivation stages enhanced microbial adaptation and activity. Environmental conditions, such as anaerobic conditions, were found to enhance LDPE biodegradation. These conditions likely reduced the crystallinity of LDPE, increasing its susceptibility to enzymatic attack. In contrast, aerobic conditions favored fungal growth, leading to moderate degradation efficiency but with slower biofilm development.

The presence of diverse microbial species within consortia played a critical role in enhancing LDPE degradation. For example, bacteria-initiated oxidation processes of alkane compounds C8 to C22 (hexadecane and octadecane) by introducing a hydroxyl (–OH) group to alkane compounds. The final step is the assimilation of the carboxylic acid compounds into the cells, which can then be utilized in cellular respiration via the TCA cycle, hence facilitating energy generation for bacterial cells. Therefore, AS3 was found to induce the release of these intermediate chemicals during the degradation process of biodegradation and utilize these compounds to generate energy through cellular respiration (Khampratueng et al., 2024) while fungi and yeasts utilized the resulting fragments as carbon sources, facilitating a complete degradation pathway (Dinakarkumar et al., 2024). Bacteria initiated oxidation processes, while fungi and yeasts utilized the resulting fragments as carbon sources, facilitating a complete degradation pathway.

These findings further highlight the potential of microbial consortia as a scalable solution for LDPE waste management. The synergistic interactions within consortia enhance degradation rates and promote the breakdown of even recalcitrant LDPE structures. However, challenges such as variability in degradation rates and the need for pretreatment of LDPE remain to be addressed. Future studies could focus on optimizing microbial consortia composition, exploring advanced pretreatment methods, and integrating biodegradation technologies into existing waste management systems.

3.3. Nanofiltration

With recent advances in materials design and synthesis, it is now possible to make nanomaterials, nanofibers, and various other configurations of desired shapes and sizes (Nastase et al., 2006; Stamatin et al., 2006; Vaseashta, 2006; Vaseashta & Bölgen, 2022; Vaseashta & Khudaverdyan, 2013). Typically, the mechanisms contributing to filtration include convection and diffusion transport, along with electromigration resulting from the charge associated with the membrane charge. Due to size, nanofiltration (NF) is a pressure-driven process that removes contaminants based on size exclusion and molecular interactions. The pore size in NF membranes typically ranges from 0.1–1 nanometer, making them suitable for rejecting small particles, including MNPs, as well as dissolved salts, organic compounds, and other contaminants. Recently, with the addition of cryogels and other suitable additives in nanofibers, the authors were able to successfully remove long-chain hydrocarbons, including oils, pharmaceuticals, paints, and dyes (Bolgen et al., 2013; Demir, Bölgen, et al., 2021; Demir, Vaseashta, et al., 2021) and can be used for isolating MNPs, following characterization, as identification of MNPs in aquatic environments is particularly challenging. However, nanofiltration is widely used in the field of desalination. Thus, the removal of various inorganic ions from the solution to obtain clean and drinking water can also be accomplished by NF. The NF processes can be used as a model for membrane characterization, process modeling, optimization, membrane design, and application (Hilal et al., 2007). Among the characteristics of membranes that determine their separation properties, pore sizes, hydrophobicity, zeta potential, surface charge, and ion exchange capacity (Schäefer & Fane, 2021). The influence of specific solution conditions, such as pH, temperature and pressure, total component concentration, and ionic strength, is critical to predicting retention. These parameters, when interacting with dissolved substances, lead to membrane contamination. One of the major limitations of membrane is fouling which is an inevitable by-product of effective operation. In addition to MNPs, the typical contaminants are colloidal, biological, organic, and inorganic. Nanomembranes are typically used in water and wastewater industries (Criscuoli, 2022; Mijatović et al., 2004) and food industries such as dairy and sugar, whey concentration and demineralization, dextrose whey syrup purification and nutraceutical extraction, fruit juice concentration as well as wine industries (Cassano et al., 2019). Using recent advances in nanomembranes, the filtration can successfully integrate MNP filtration, due to the increased presence of MNPs in the aquatic environment. NF is gaining more impetus in the chemical processing industry due to minimal to zero liquid discharge requirements (Y. Hu, 2023), thus necessitating solvent-tolerant membrane modeling for organophilic applications.

4. Discussion and Path Forward

This chapter highlights the need to mitigate MNPs from aquatic environments due to significant health impacts, as outlined above. While regulations and policies help reduce the MNP burden, inherently the policies and social issues take much longer to show any significant results. Due to the increased and ubiquitous presence of plastic in our immediate environment, it is necessary to find technological solutions to mitigate plastics, to make it a zero-sum outcome. Several viable alternatives are presented in this chapter, such as nano-photocatalytic degradation, NF, and the use of microorganisms, both as individual strains and in consortia, for the biodegradation of MNP of various plastics. A case study of LDPE is presented, showing that microbial degradation shows some promising results. Fungal strains, particularly *Penicillium verrucosum* and *Aspergillus flavus*, demonstrated higher degradation efficiencies compared to bacterial strains such as *Pseudomonas fluorescens*. However, microbial consortia exhibited superior performance due to synergistic interactions between diverse microbial species, enabling comprehensive polymer breakdown. Significant alterations in LDPE properties, including reduced tensile strength and elongation at break, were observed, confirming microbial impact on polymer structure. Anaerobic conditions and optimized mineral salt media further enhanced degradation efficiency by reducing LDPE crystallinity and facilitating enzymatic activity. These findings underscore the feasibility of integrating microbial biodegradation into waste management strategies for plastic pollution mitigation. Future work should focus on optimizing microbial consortia, improving pretreatment methods, and scaling up this approach for industrial applications. Several other processes are also mentioned in the chapter, which are used to pre-treat plastics to accelerate microbial degradation. It is also proposed that one or more of these processes can be applied simultaneously to obtain the MNS mitigation. We further propose strategies to mitigate MNP pollution by enforcing restrictions and bans of plastic production and the use of plastic additives in products, establishing standards for plastic production that prioritize environmentally friendly and biodegradable materials. Improved waste management, increased the efficiency of plastic waste recycling, and implemented advanced purification technologies, including modern filtration systems in treatment plants to capture MNPs before being discharged in the aquatic environment.

References

Adawi, H.I., Newbold, M.A., Reed, J.M., Vance, M.E., Feitshans, I.L., Bickford, L.R., & Lewinski, N.A. (2018). Nano-enabled personal care products: Current developments in consumer safety. *NanoImpact*, 11, 170–179. https://doi.org/10.1016/j.impact.2018.08.002.

Adegoke, K.A., Adu, F.A., Oyebamiji, A.K., Bamisaye, A., Adigun, R.A., Olasoji, S.O., & Ogunjinmi, O.E. (2023). Microplastics toxicity, detection, and removal from water/wastewater. *Marine Pollution Bulletin*, 187, 114546. https://doi.org/10.1016/j.marpolbul.2022.114546.

Ali, N., Katsouli, J., Marczylo, E.L., Gant, T.W., Wright, S., & Bernardino De La Serna, J. (2024). The potential impacts of micro- and nano-plastics on various organ systems in humans. *eBioMedicine*, 99, 104901. https://doi.org/10.1016/j.ebiom.2023.104901.

Anderson, A.G., Grose, J., Pahl, S., Thompson, R.C., & Wyles, K.J. (2016). Microplastics in personal care products: Exploring perceptions of environmentalists, beauticians and students. *Marine Pollution Bulletin*, 113(1–2), 454–460. https://doi.org/10.1016/j.marpolbul.2016.10.048.

Andler, R., Tiso, T., Blank, L., Andreeßen, C., Zampolli, J., D'Afonseca, V. et al. (2022). Current progress on the biodegradation of synthetic plastics: From fundamentals to biotechnological applications. *Reviews in Environmental Science and Bio/Technology*, 21(4), 829–850. https://doi.org/10.1007/s11157-022-09631-2

Anush, K., Shushanik, K., Susanna, T., & Ashkhen, H. (2019). Antibacterial effect of silver and iron oxide nanoparticles in combination with antibiotics on *E. coli* K12. *BioNanoScience*, 9(3), 587–596. https://doi.org/10.1007/s12668-019-00640-0.

Ariza-Tarazona, M.C., Villarreal-Chiu, J.F., Barbieri, V., Siligardi, C., & Cedillo-González, E.I. (2019). New strategy for microplastic degradation: Green photocatalysis using a protein-based porous N-TiO_2 semiconductor. *Ceramics International*, 45(7), 9618–9624. https://doi.org/10.1016/j.ceramint.2018.10.208.

Au, S.Y., Lee, C.M., Weinstein, J.E., Van Den Hurk, P., & Klaine, S.J. (2017). Trophic transfer of microplastics in aquatic ecosystems: Identifying critical research needs. *Integrated Environmental Assessment and Management*, 13(3), 505–509. https://doi.org/10.1002/ieam.1907.

Auta, H.S., Emenike, C.U., & Fauziah, S.H. (2017). Distribution and importance of microplastics in the marine environment: A review of the sources, fate, effects, and potential solutions. *Environment International*, 102, 165–176. https://doi.org/10.1016/j.envint.2017.02.013.

Avio, C.G., Gorbi, S., & Regoli, F. (2017). Plastics and microplastics in the oceans: From emerging pollutants to emerged threat. *Marine Environmental Research*, 128, 2–11. https://doi.org/10.1016/j.marenvres.2016.05.012.

Bamford, N.C., MacPhee, C.E., & Stanley-Wall, N.R. (2023). Microbial primer: An introduction to biofilms – What they are, why they form and their impact on built and natural environments – This article is part of the Microbial Primer collection. *Microbiology*, 169(8). https://doi.org/10.1099/mic.0.001338.

Beladi-Mousavi, S.M., Hermanová, S., Ying, Y., Plutnar, J., & Pumera, M. (2021). A maze in plastic wastes: Autonomous motile photocatalytic microrobots against microplastics. *ACS Applied Materials & Interfaces*, 13(21), 25102–25110. https://doi.org/10.1021/acsami.1c04559.

Benson, N.U., Agboola, O.D., Fred-Ahmadu, O.H., De-la-Torre, G.E., Oluwalana, A., & Williams, A.B. (2022). Micro(nano)plastics prevalence, food web interactions, and toxicity assessment in aquatic organisms: A review. *Frontiers in Marine Science*, 9, 851281. https://doi.org/10.3389/fmars.2022.851281.

Bergmann, M., Gutow, L., & Klages, M. (Eds.). (2015). *Marine Anthropogenic Litter*. Springer International Publishing. https://doi.org/10.1007/978-3-319-16510-3.

Bigalke, M., Filella, M., Fischer, D., Muntwyler, A., Scheurer, M., & Watts, B. (2018). Micro- and nanoplastic analysis in soils: Highlights of analytical sciences in Switzerland. *CHIMIA*, 72(12), 901. https://doi.org/10.2533/chimia.2018.901.

Bolgen, N., Demir, D., & Vaseashta, A. (2013). Development of polymeric cryogels as potential matrices for removing antibiotics from wastewater. *In:* A. Vaseashta & S. Khudaverdyan (Eds.), *Advanced Sensors for Safety and Security* (pp. 243–249). Springer Netherlands. https://doi.org/10.1007/978-94-007-7003-4_21.

Bora, S.S., Gogoi, R., Sharma, M.R., Anshu, Borah, M.P., Deka, P., Bora, J. et al. (2024). Microplastics and human health: Unveiling the gut microbiome disruption and chronic disease risks. *Frontiers in Cellular and Infection Microbiology*, 14, 1492759. https://doi.org/10.3389/fcimb.2024.1492759.

Brachner, A., Fragouli, D., Duarte, I.F., Farias, P.M.A., Dembski, S., Ghosh, M. et al. (2020). Assessment of human health risks posed by nano- and microplastics is currently not feasible. *International Journal of Environmental Research and Public Health*, 17(23), 8832. https://doi.org/10.3390/ijerph17238832.

Cao, Z., Yan, W., Ding, M., & Yuan, Y. (2022). Construction of microbial consortia for microbial degradation of complex compounds. *Frontiers in Bioengineering and Biotechnology*, 10, 1051233. https://doi.org/10.3389/fbioe.2022.1051233.

Caruso, G. (2015). Plastic degrading microorganisms as a tool for bioremediation of plastic contamination in aquatic environments. *Journal of Pollution Effects & Control*, 03(03). https://doi.org/10.4172/2375-4397.1000e112.

Cassano, A., Conidi, C., & Castro-Muñoz, R. (2019). Current and future applications of nanofiltration in food processing. *In: Separation of Functional Molecules in Food by Membrane Technology* (pp. 305–348). Elsevier. https://doi.org/10.1016/B978-0-12-815056-6.00009-7.

Chen, X., Mo, L., Zhang, L., Huang, L., Gao, Z., Peng, J. et al. (2024). Taxonomic diversity, predicted metabolic pathway, and interaction pattern of bacterial community in sea urchin *Anthocidaris crassispina*. *Microorganisms*, 12(10), 2094. https://doi.org/10.3390/microorganisms12102094.

Cheung, P.K., & Fok, L. (2016). Evidence of microbeads from personal care product contaminating the sea. *Marine Pollution Bulletin*, 109(1), 582–585. https://doi.org/10.1016/j.marpolbul.2016.05.046.

Cook, C.R., & Halden, R.U. (2020). Ecological and health issues of plastic waste. *In: Plastic Waste and Recycling* (pp. 513–527). Elsevier. https://doi.org/10.1016/B978-0-12-817880-5.00020-7.

Corcimaru, S., Gutul, T., Mereniuc, L., & Sitnic, F. (2024). *A Method of Destruction of the Low-pressure Polyethylene Film – LDPE* (Patent 12147).

Criscuoli, A. (2022). Membrane distillation for aqueous feeds treatment: *In:* A. Vaseashta, G. Duca, & S. Travin (Eds.), *Advances in Environmental Engineering and Green Technologies* (pp. 150–166). IGI Global. https://doi.org/10.4018/978-1-7998-7356-3.ch007.

Danso, I.K., Woo, J.-H., & Lee, K. (2022). Pulmonary toxicity of polystyrene, polypropylene, and polyvinyl chloride microplastics in mice. *Molecules*, 27(22), 7926. https://doi.org/10.3390/molecules27227926.

Das, A. (2023). The emerging role of microplastics in systemic toxicity: Involvement of reactive oxygen species (ROS). *Science of The Total Environment*, 895, 165076. https://doi.org/10.1016/j.scitotenv.2023.165076

Dehghani, S., Moore, F., & Akhbarizadeh, R. (2017). Microplastic pollution in deposited urban dust, Tehran metropolis, Iran. *Environmental Science and Pollution Research*, 24(25), 20360–20371. https://doi.org/10.1007/s11356-017-9674-1.

Demir, D., Bölgen, N., & Vaseashta, A. (2021). Chitosan–gelatin cryogels as bio-sorbents for removal of dyes from aqueous solutions. *In:* A. Vaseashta & C. Maftei (Eds.), *Water Safety, Security, and Sustainability* (pp. 263–274). Springer International Publishing. https://doi.org/10.1007/978-3-030-76008-3_11.

Demir, D., Vaseashta, A., & Bölgen, N. (2021). Macroporous cryogels for water purification. *In:* A. Vaseashta & C. Maftei (Eds.), *Water Safety, Security, and Sustainability* (pp. 275–290). Springer International Publishing. https://doi.org/10.1007/978-3-030-76008-3_12.

Dhali, S.L., Parida, D., Kumar, B., & Bala, K. (2024). Recent trends in microbial and enzymatic plastic degradation: A solution for plastic pollution predicaments. *Biotechnology for Sustainable Materials*, 1(1), 11. https://doi.org/10.1186/s44316-024-00011-0.

Di Pippo, F., Bocci, V., Amalfitano, S., Crognale, S., Levantesi, C., Pietrelli, L. et al. (2023). Microbial colonization patterns and biodegradation of petrochemical and biodegradable plastics in lake waters: Insights from a field experiment. *Frontiers in Microbiology*, 14, 1290441. https://doi.org/10.3389/fmicb.2023.1290441.

Dimarogona, M., Nikolaivits, E., Kanelli, M., Christakopoulos, P., Sandgren, M., & Topakas, E. (2015). Structural and functional studies of a *Fusarium oxysporum* cutinase with polyethylene terephthalate modification potential. *Biochimica et Biophysica Acta (BBA) – General Subjects*, 1850(11), 2308–2317. https://doi.org/10.1016/j.bbagen.2015.08.009.

Dinakarkumar, Y., Ramakrishnan, G., Gujjula, K.R., Vasu, V., Balamurugan, P., & Murali, G. (2024). Fungal bioremediation: An overview of the mechanisms, applications and future perspectives. *Environmental Chemistry and Ecotoxicology*, 6, 293–302. https://doi.org/10.1016/j.enceco.2024.07.002.

Dissanayake, P.D., Kim, S., Sarkar, B., Oleszczuk, P., Sang, M.K., Haque, M.N. et al. (2022). Effects of microplastics on the terrestrial environment: A critical review. *Environmental Research*, 209, 112734. https://doi.org/10.1016/j.envres.2022.112734.

Du, H., Xie, Y., & Wang, J. (2021). Microplastic degradation methods and corresponding degradation mechanism: Research status and future perspectives. *Journal of Hazardous Materials*, 418, 126377. https://doi.org/10.1016/j.jhazmat.2021.126377.

Elizalde-Velázquez, A., Crago, J., Zhao, X., Green, M.J., & Cañas-Carrell, J.E. (2020). *In vivo* effects on the immune function of fathead minnow (*Pimephales promelas*) following ingestion and intraperitoneal injection of polystyrene nanoplastics. *Science of The Total Environment*, 735, 139461. https://doi.org/10.1016/j.scitotenv.2020.139461.

Feng, K., Gong, J., Qu, J., & Niu, R. (2022). Dual-mode-driven micromotor based on foam-like carbon nitride and Fe_3O_4 with improved manipulation and photocatalytic performance. *ACS Applied Materials & Interfaces*, 14(39), 44271–44281. https://doi.org/10.1021/acsami.2c10590.

Feng, Y., Tu, C., Li, R., Wu, D., Yang, J., Xia, Y. et al. (2023). A systematic review of the impacts of exposure to micro- and nano-plastics on human tissue accumulation and health. *Eco-Environment & Health*, 2(4), 195–207. https://doi.org/10.1016/j.eehl.2023.08.002.

Gangadoo, S., Owen, S., Rajapaksha, P., Plaisted, K., Cheeseman, S., Haddara, H. et al. (2020). Nano-plastics and their analytical characterisation and fate in the marine environment: From source to sea. *Science of The Total Environment*, 732, 138792. https://doi.org/10.1016/j.scitotenv.2020.138792.

Gao, G., Zhao, X., Jin, P., Gao, K., & Beardall, J. (2021). Current understanding and challenges for aquatic primary producers in a world with rising micro- and nano-plastic levels. *Journal of Hazardous Materials*, 406, 124685. https://doi.org/10.1016/j.jhazmat.2020.124685.

Gao, L., Zhuang, J., Nie, L., Zhang, J., Zhang, Y., Gu, N. et al. (2007). Intrinsic peroxidase-like activity of ferromagnetic nanoparticles. *Nature Nanotechnology*, 2(9), 577–583. https://doi.org/10.1038/nnano.2007.260.

Ge, J., Zhang, Z., Ouyang, Z., Shang, M., Liu, P., Li, H., & Guo, X. (2022). Photocatalytic degradation of (micro)plastics using TiO_2-based and other catalysts: Properties, influencing factor, and mechanism. *Environmental Research*, 209, 112729. https://doi.org/10.1016/j.envres.2022.112729.

Geyer. (2023). *Study on Plastic Waste in the Republic of Moldova/Studiu privind deşeurile din Plastic în Republica Moldova*. https://e-circular.org/wp-content/uploads/2023/02/STUDIU-PRIVIND-DESEURILE-DIN-PLASTIC-IN-REPUBLICA-MOLDOVA.pdf.

Gholamhosseini, A., Banaee, M., Sureda, A., Timar, N., Zeidi, A., & Faggio, C. (2023). Physiological response of freshwater crayfish, *Astacus leptodactylus* exposed to polyethylene microplastics at different temperature. *Comparative Biochemistry and Physiology Part C: Toxicology & Pharmacology*, 267, 109581. https://doi.org/10.1016/j.cbpc.2023.109581.

Godoy, V., Martín-Lara, M.A., Calero, M., & Blázquez, G. (2019). Physical-chemical characterization of microplastics present in some exfoliating products from Spain. *Marine Pollution Bulletin*, 139, 91–99. https://doi.org/10.1016/j.marpolbul.2018.12.026.

Gomes, T., Bour, A., Coutris, C., Almeida, A.C., Bråte, I.L., Wolf, R. et al. (2022). Ecotoxicological impacts of micro- and nanoplastics in terrestrial and aquatic environments. *In:* M.S. Bank (Ed.), *Microplastic in the Environment: Pattern and Process* (pp. 199–260). Springer International Publishing. https://doi.org/10.1007/978-3-030-78627-4_7.

Gou, X., Fu, Y., Li, J., Xiang, J., Yang, M., & Zhang, Y. (2024). Impact of nanoplastics on Alzheimer's disease: Enhanced amyloid-β peptide aggregation and augmented neurotoxicity. *Journal of Hazardous Materials*, 465, 133518. https://doi.org/10.1016/j.jhazmat.2024.133518.

Gray, A.D., Wertz, H., Leads, R.R., & Weinstein, J.E. (2018). Microplastic in two South Carolina Estuaries: Occurrence, distribution, and composition. *Marine Pollution Bulletin*, 128, 223–233. https://doi.org/10.1016/j.marpolbul.2018.01.030.

Gricajeva, A., Nadda, A.K., & Gudiukaite, R. (2022). Insights into polyester plastic biodegradation by carboxyl ester hydrolases. *Journal of Chemical Technology & Biotechnology*, 97(2), 359–380. https://doi.org/10.1002/jctb.6745

Habumugisha, T., Zhang, Z., Ndayishimiye, J.C., Nkinahamira, F., Kayiranga, A., Cyubahiro, E. et al. (2022). Evaluation and optimization of the influence of silver cluster ions on the MALDI-TOF-MS analysis of polystyrene nanoplastic polymers. *Analytical Methods*, 14(7), 763–772. https://doi.org/10.1039/D1AY02219A.

Hamd, W., Daher, E.A., Tofa, T.S., & Dutta, J. (2022). Recent advances in photocatalytic removal of microplastics: Mechanisms, kinetic degradation, and reactor design. *Frontiers in Marine Science*, 9, 885614. https://doi.org/10.3389/fmars.2022.885614.

He, Y., Rehman, A.U., Xu, M., Not, C.A., Ng, A.M.C., & Djurišić, A.B. (2023). Photocatalytic degradation of different types of microplastics by TiOx/ZnO tetrapod photocatalysts. *Heliyon*, 9(11), e22562. https://doi.org/10.1016/j.heliyon.2023.e22562.

Hebner, T.S., & Maurer-Jones, M.A. (2020). Characterizing microplastic size and morphology of photodegraded polymers placed in simulated moving water conditions. *Environmental Science: Processes & Impacts*, 22(2), 398–407. https://doi.org/10.1039/C9EM00475K.

Hilal, N., Al-Zoubi, H., Darwish, N.A., & Mohammad, A.W. (2007). Performance of nanofiltration membranes in the treatment of synthetic and real seawater. *Separation Science and Technology*, 42(3), 493–515. https://doi.org/10.1080/01496390601120789.

Horvatits, T., Tamminga, M., Liu, B., Sebode, M., Carambia, A., Fischer, L. et al. (2022). Microplastics detected in cirrhotic liver tissue. *eBioMedicine*, 82, 104147. https://doi.org/10.1016/j.ebiom.2022.104147.

Hsiao, H.-Y., Nien, C.-Y., Shiu, R.-F., Chin, W.-C., & Yen, T.-H. (2025). Microplastic and nanoplastic exposure and risk of diabetes mellitus. *World Journal of Clinical Cases*, 13(3). https://doi.org/10.12998/wjcc.v13.i3.98110.

Hu, X., Liu, H., Wang, X., Zhang, X., Shan, Z., Zheng, W. et al. (2018). Wavelength selective photodetectors integrated on a single composition-graded semiconductor nanowire. *Advanced Optical Materials*, 6(12), 1800293. https://doi.org/10.1002/adom.201800293.

Hu, Y. (2023). Grand challenge in membrane applications: Liquid. *Frontiers in Membrane Science and Technology*, 2, 1177528. https://doi.org/10.3389/frmst.2023.1177528.

Huang, H., Lei, P., Yu, H., Du, J., Wu, B., Wang, H. et al. (2024). Micro/nano plastics in the urinary system: Pathways, mechanisms, and health risks. *Environment International*, 193, 109109. https://doi.org/10.1016/j.envint.2024.109109.

Huang, S., Deng, G., Jin, X., Lu, Y., Song, G., Huang, H. et al. (2020). The dark current suppression of black silicon photodetector by a lateral heterojunction. *Optical Materials*, 110, 110474. https://doi.org/10.1016/j.optmat.2020.110474.

Jiang, B., Kauffman, A.E., Li, L., McFee, W., Cai, B., Weinstein, J. et al. (2020). Health impacts of environmental contamination of micro- and nanoplastics: A review.

Environmental Health and Preventive Medicine, 25(1), 29. https://doi.org/10.1186/s12199-020-00870-9

Kahn, L.G., Philippat, C., Nakayama, S.F., Slama, R., & Trasande, L. (2020). Endocrine-disrupting chemicals: Implications for human health. *The Lancet Diabetes & Endocrinology*, 8(8), 703–718. https://doi.org/10.1016/S2213-8587(20)30129-7.

Kandpal, H., & Srivastava, R.K. (2024). Strategies to mitigate microplastic (MP) and nanoplastic (NP) pollution in soil: *In:* N. Gaur, E. Sharma, T.A. Nguyen, M. Bilal, & N.P. Melkania (Eds.), *Handbook of Research on Advancements in Environmental Engineering* (*Advances in Environmental Engineering and Green Technologies*) (pp. 115–138). IGI Global. https://doi.org/10.4018/979-8-3693-9163-1.ch006.

Karthik, R., Robin, R.S., Purvaja, R., Ganguly, D., Anandavelu, I., Raghuraman, R. et al. (2018). Microplastics along the beaches of southeast coast of India. *Science of The Total Environment*, 645, 1388–1399. https://doi.org/10.1016/j.scitotenv.2018.07.242.

Kazaryan, S., Farsiyan, L., Tumoyan, J., Kirakosyan, G., Ayvazyan, N., Gasparyan, H. et al. (2022). Oxidative stress and histopathological changes in several organs of mice injected with biogenic silver nanoparticles. *Artificial Cells, Nanomedicine, and Biotechnology*, 50(1), 331–342. https://doi.org/10.1080/21691401.2022.2149931.

Khalid, N., Aqeel, M., Noman, A., Hashem, M., Mostafa, Y.S., Al-hathloul, H.A.S., & Alghanem, S.M. (2021). Linking effects of microplastics to ecological impacts in marine environments. *Chemosphere*, 264, 128541. https://doi.org/10.1016/j.chemosphere.2020.128541.

Khampratueng, P., Rice, D., & Anal, A.K. (2024). Biodegradation of low-density polyethylene by the bacterial strains isolated from the dumping site community. *Discover Applied Sciences*, 6(7), 348. https://doi.org/10.1007/s42452-024-06052-4.

Kokilathasan, N., & Dittrich, M. (2022). Nanoplastics: Detection and impacts in aquatic environments – A review. *Science of The Total Environment*, 849, 157852. https://doi.org/10.1016/j.scitotenv.2022.157852

La Porta, E., Exacoustos, O., Lugani, F., Angeletti, A., Chiarenza, D.S., Bigatti, C. et al. (2023). Microplastics and kidneys: An update on the evidence for deposition of plastic microparticles in human organs, tissues, and fluids and renal toxicity concern. *International Journal of Molecular Sciences*, 24(18), 14391. https://doi.org/10.3390/ijms241814391.

Lambert, S., & Wagner, M. (2016). Characterisation of nanoplastics during the degradation of polystyrene. *Chemosphere*, 145, 265–268. https://doi.org/10.1016/j.chemosphere.2015.11.078.

Le, V.-G., Nguyen, M.-K., Lin, C., Nguyen, H.-L., Nguyen, T.Q.H., Hue, N.K. et al. (2024). Review on personal protective equipment: Emerging concerns in micro(nano)plastic pollution and strategies for addressing environmental challenges. *Environmental Research*, 257, 119345. https://doi.org/10.1016/j.envres.2024.119345.

Li, W., Zhao, W., Zhu, H., Li, Z.-J., & Wang, W. (2023). State of the art in the photochemical degradation of (micro)plastics: From fundamental principles to catalysts and applications. *Journal of Materials Chemistry A*, 11(6), 2503–2527. https://doi.org/10.1039/D2TA09523H.

Liang, B., Huang, Y., Zhong, Y., Li, Z., Ye, R., Wang, B. et al. (2022). Brain single-nucleus transcriptomics highlights that polystyrene nanoplastics potentially induce Parkinson's disease-like neurodegeneration by causing energy metabolism disorders in mice. *Journal of Hazardous Materials*, 430, 128459. https://doi.org/10.1016/j.jhazmat.2022.128459.

Liu, M., Zhang, T., Long, L., Zhang, R., & Ding, S. (2019). Efficient enzymatic degradation of poly (ε-caprolactone) by an engineered bifunctional lipase-cutinase. *Polymer Degradation and Stability*, 160, 120–125. https://doi.org/10.1016/j.polymdegradstab.2018.12.020.

Liu, S., Wang, C., Yang, Y., Du, Z., Li, L., Zhang, M. et al. (2024). Microplastics in three types of human arteries detected by pyrolysis-gas chromatography/mass spectrometry (Py-GC/MS). *Journal of Hazardous Materials*, 469, 133855. https://doi.org/10.1016/j.jhazmat.2024.133855.

Madni, A., Rehman, S., Sultan, H., Khan, M.M., Ahmad, F., Raza, M.R. et al. (2021). Mechanistic approaches of internalization, subcellular trafficking, and cytotoxicity of nanoparticles for targeting the small intestine. *AAPS PharmSciTech*, 22(1), 3. https://doi.org/10.1208/s12249-020-01873-z.

Mijatović, I., Matošić, M., Hajduk Černeha, B., & Bratulić, D. (2004). Removal of natural organic matter by ultrafiltration and nanofiltration for drinking water production. *Desalination*, 169(3), 223–230. https://doi.org/10.1016/j.desal.2004.01.002.

Mintenig, S.M., Löder, M.G.J., Primpke, S., & Gerdts, G. (2019). Low numbers of microplastics detected in drinking water from ground water sources. *Science of The Total Environment*, 648, 631–635. https://doi.org/10.1016/j.scitotenv.2018.08.178.

Mofijur, M., Ahmed, S.F., Rahman, S.M.A., Arafat Siddiki, Sk. Y., Islam, A.B.M.S., Shahabuddin, M. et al. (2021). Source, distribution and emerging threat of micro- and nanoplastics to marine organism and human health: Socio-economic impact and management strategies. *Environmental Research*, 195, 110857. https://doi.org/10.1016/j.envres.2021.110857.

Mohammadi, A., Baba, M., & Hirayama, H. (2009). *Simulation of the Carrier Trapping Effect in a Schottky CdTe Detector*, 46(11), 1032–1037.

Mohammadi, S., Moussavi, G., & Rezaei, M. (2022). Enhanced peroxidase-mediated biodegradation of polyethylene using the bacterial consortia under H_2O_2-biostimulation. *Polymer*, 240, 124508. https://doi.org/10.1016/j.polymer.2021.124508.

Mohanan, N., Montazer, Z., Sharma, P.K., & Levin, D.B. (2020). Microbial and enzymatic degradation of synthetic plastics. *Frontiers in Microbiology*, 11, 580709. https://doi.org/10.3389/fmicb.2020.580709

Montazer, Z., Habibi Najafi, M.B., & Levin, D.B. (2020). Challenges with verifying microbial degradation of polyethylene. *Polymers*, 12(1), 123. https://doi.org/10.3390/polym12010123.

Morales Ramos, J.G., Fernández Tarrillo, L.M., Guevara Bravo, A.X., Sánchez-Purihuamán, M., Carreño Farfán, C.R., Loayza Estrada, C.S. et al. (2024). Efficiency of microorganisms and effectiveness of biodegradation techniques on LDPE plastics: A systematic review. *F1000Research*, 13, 745. https://doi.org/10.12688/f1000research.151338.1.

Moraskie, M., Roshid, M.H.O., O'Connor, G., Dikici, E., Zingg, J.-M., Deo, S., & Daunert, S. (2021). Microbial whole-cell biosensors: Current applications, challenges, and future perspectives. *Biosensors and Bioelectronics*, 191, 113359. https://doi.org/10.1016/j.bios.2021.113359.

Nastase, C., Nastase, F., Vaseashta, A., & Stamatin, I. (2006). Nanocomposites based on functionalized nanotubes in polyaniline matrix by plasma polymerization. *Progress in Solid State Chemistry*, 34(2–4), 181–189. https://doi.org/10.1016/j.progsolidstchem.2005.11.021.

Nelis, J.L.D., Schacht, V.J., Dawson, A.L., Bose, U., Tsagkaris, A.S., Dvorakova, D. et al.

(2023). The measurement of food safety and security risks associated with micro- and nanoplastic pollution. *TrAC Trends in Analytical Chemistry*, 161, 116993. https://doi.org/10.1016/j.trac.2023.116993.

Pan, Y., Gao, S.-H., Ge, C., Gao, Q., Huang, S., Kang, Y. et al. (2023). Removing microplastics from aquatic environments: A critical review. *Environmental Science and Ecotechnology*, 13, 100222. https://doi.org/10.1016/j.ese.2022.100222.

Parida, D., Sangtani, R., & Bala, K. (2023). Microplastics: The stemming environmental challenge and the quest for the missing mitigation strategies. *International Biodeterioration & Biodegradation*, 179, 105581. https://doi.org/10.1016/j.ibiod.2023.105581.

Park, S.Y., & Kim, C.G. (2019). Biodegradation of micro-polyethylene particles by bacterial colonization of a mixed microbial consortium isolated from a landfill site. *Chemosphere*, 222, 527–533. https://doi.org/10.1016/j.chemosphere.2019.01.159.

Peeken, I., Primpke, S., Beyer, B., Gütermann, J., Katlein, C., Krumpen, T. et al. (2018). Arctic sea ice is an important temporal sink and means of transport for microplastic. *Nature Communications*, 9(1), 1505. https://doi.org/10.1038/s41467-018-03825-5.

Peng, G., Zhu, B., Yang, D., Su, L., Shi, H., & Li, D. (2017). Microplastics in sediments of the Changjiang Estuary, China. *Environmental Pollution*, 225, 283–290. https://doi.org/10.1016/j.envpol.2016.12.064

Peng, G., Xu, P., Zhu, B., Bai, M., & Li, D. (2018). Microplastics in freshwater river sediments in Shanghai, China: A case study of risk assessment in mega-cities. *Environmental Pollution*, 234, 448–456. https://doi.org/10.1016/j.envpol.2017.11.034.

Pivokonsky, M., Cermakova, L., Novotna, K., Peer, P., Cajthaml, T., & Janda, V. (2018). Occurrence of microplastics in raw and treated drinking water. *Science of The Total Environment*, 643, 1644–1651. https://doi.org/10.1016/j.scitotenv.2018.08.102.

Postolachi, O., Rastimesina, I., Vorona, V., & Mamaliga, V. (2022). Screening of cultivation media for LDPE biodegradation by *Pseudomonas fluorescens*. *Scientific Bulletin. Series F. Biotechnologies*, XXVI(2), 151–157.

Prattichizzo, F., Ceriello, A., Pellegrini, V., La Grotta, R., Graciotti, L., Olivieri, F. et al. (2024). Micro-nanoplastics and cardiovascular diseases: Evidence and perspectives. *European Heart Journal*, 45(38), 4099–4110. https://doi.org/10.1093/eurheartj/ehae552.

Prüst, M., Meijer, J., & Westerink, R.H.S. (2020). The plastic brain: Neurotoxicity of micro- and nanoplastics. *Particle and Fibre Toxicology*, 17(1), 24. https://doi.org/10.1186/s12989-020-00358-y.

Ramsperger, A.F.R.M., Bergamaschi, E., Panizzolo, M., Fenoglio, I., Barbero, F., Peters, R. et al. (2023). Nano- and microplastics: A comprehensive review on their exposure routes, translocation, and fate in humans. *NanoImpact*, 29, 100441. https://doi.org/10.1016/j.impact.2022.100441.

Rastimesina, I. (2023). Isolation of microbial consortia from polyethylene films surface. *Book of Abstracts SIMI 2023*, 84–85. https://doi.org/10.21698/simi.2023.ab31.

Revel, M., Châtel, A., & Mouneyrac, C. (2018). Micro(nano)plastics: A threat to human health? *Current Opinion in Environmental Science & Health*, 1, 17–23. https://doi.org/10.1016/j.coesh.2017.10.003.

Roohi, Null, Bano, K., Kuddus, M., Zaheer, M.R., Zia, Q., Khan, M.F., et al. (2017). Microbial enzymatic degradation of biodegradable plastics. *Current Pharmaceutical Biotechnology*, 18(5), 429–440. https://doi.org/10.2174/1389201018666170523165742.

Rubio, L., Marcos, R., & Hernández, A. (2020). Potential adverse health effects of ingested micro- and nanoplastics on humans. Lessons learned from *in vivo* and *in vitro* mammalian models. *Journal of Toxicology and Environmental Health, Part B*, 23(2), 51–68. https://doi.org/10.1080/10937404.2019.1700598.

Schäefer, A.I., & Fane, A.G. (Eds.). (2021). *Nanofiltration: Principles, Applications, and New Materials* (1st Edn.). Wiley. https://doi.org/10.1002/9783527824984.

Schymanski, D., Goldbeck, C., Humpf, H.-U., & Fürst, P. (2018). Analysis of microplastics in water by micro-Raman spectroscopy: Release of plastic particles from different packaging into mineral water. *Water Research*, 129, 154–162. https://doi.org/10.1016/j.watres.2017.11.011.

Sharma, S., Mohler, J., Mahajan, S.D., Schwartz, S.A., Bruggemann, L., & Aalinkeel, R. (2023). Microbial biofilm: A review on formation, infection, antibiotic resistance, control measures, and innovative treatment. *Microorganisms*, 11(6), 1614. https://doi.org/10.3390/microorganisms11061614.

Shi, X., Chen, Z., Wei, W., Chen, J., & Ni, B.-J. (2023). Toxicity of micro/nanoplastics in the environment: Roles of plastisphere and eco-corona. *Soil & Environmental Health*, 1(1), 100002. https://doi.org/10.1016/j.seh.2023.100002.

Shiwakoti, S., Ko, J.-Y., Gong, D., Dhakal, B., Lee, J.-H., Adhikari, R. et al. (2022). Effects of polystyrene nanoplastics on endothelium senescence and its underlying mechanism. *Environment International*, 164, 107248. https://doi.org/10.1016/j.envint.2022.107248.

Sidorenko, L., Diaz-Ramirez, I., Vovc, V., & Baumann, G. (2018). *New Approach to Heart Rate Variability Analysis Based on Cardio Physiological Biomarkers*. https://doi.org/10.5281/ZENODO.1465926.

Singh, S., Kumar Naik, T.S.S., Anil, A.G., Dhiman, J., Kumar, V., Dhanjal, D.S. et al. (2022). Micro (nano)plastics in wastewater: A critical review on toxicity risk assessment, behaviour, environmental impact, and challenges. *Chemosphere*, 290, 133169. https://doi.org/10.1016/j.chemosphere.2021.133169.

Smolders, R., Schramm, K.-W., Stenius, U., Grellier, J., Kahn, A., Trnovec, T. et al. (2009). A review on the practical application of human biomonitoring in integrated environmental health impact assessment. *Journal of Toxicology and Environmental Health, Part B*, 12(2), 107–123. https://doi.org/10.1080/15287390802706397.

Stabnikova, O., Stabnikov, V., Marinin, A., Klavins, M., & Vaseashta, A. (2022). The role of microplastics biofilm in accumulation of trace metals in aquatic environments. *World Journal of Microbiology and Biotechnology*, 38(7), 117. https://doi.org/10.1007/s11274-022-03293-6.

Stamatin, I., Berlic, C., & Vaseashta, A. (2006). On the computer-aided modelling of analyte–receptor interactions for an efficient sensor design. *Thin Solid Films*, 495(1–2), 312–315. https://doi.org/10.1016/j.tsf.2005.08.299.

Sundar, K.P., & Kanmani, S. (2020). Progression of photocatalytic reactors and its comparison: A review. *Chemical Engineering Research and Design*, 154, 135–150. https://doi.org/10.1016/j.cherd.2019.11.035.

Suresh, S.S., Mohanty, S., & Nayak, S.K. (2017). Composition analysis and characterization of waste polyvinyl chloride (PVC) recovered from data cables. *Waste Management*, 60, 100–111. https://doi.org/10.1016/j.wasman.2016.08.033.

Thomas, R.T., & Sandhyarani, N. (2013). Enhancement in the photocatalytic degradation of low density polyethylene–TiO_2 nanocomposite films under solar irradiation. *RSC Advances*, 3(33), 14080. https://doi.org/10.1039/c3ra42226g.

Tumoyan, J., Kazaryan, S., & Hovhannisyan, A. (2024). The impact of biogenic silver nanoparticles on the enzymatic antioxidant system of Wistar Rats' kidney. *In:* V. Sontea,

I. Tiginyanu, & S. Railean (Eds.), *6th International Conference on Nanotechnologies and Biomedical Engineering* (Vol. 91, pp. 546–553). Springer Nature Switzerland. https://doi.org/10.1007/978-3-031-42775-6_58.

Tuteja, J., Vyas, A., & Sand, A. (2024). Introductory Chapter: Polyethylene – Its properties and application in industry and in households. *In:* A. Sand & J. Tuteja (Eds.), *Polyethylene: New Developments and Applications*. IntechOpen. https://doi.org/10.5772/intechopen.1003974.

Urso, M., & Pumera, M. (2022). Nano/Microplastics capture and degradation by autonomous nano/microrobots: A perspective. *Advanced Functional Materials*, 32(20), 2112120. https://doi.org/10.1002/adfm.202112120.

Vaseashta, A. (2006). Carbon nanotubes based devices and sensors. *Materials and Manufacturing Processes*, 21(7), 710–716. https://doi.org/10.1080/10426910600613595.

Vaseashta, A., & Khudaverdyan, S. (Eds.). (2013). *Advanced Sensors for Safety and Security*. Springer Netherlands. https://doi.org/10.1007/978-94-007-7003-4.

Vaseashta, A., Gevorgyan, G., Kavaz, D., Ivanov, O., Jawaid, M., & Vasović, D. (2021). Exposome, biomonitoring, assessment, and data analytics to quantify universal water quality. *In:* A. Vaseashta & C. Maftei (Eds.), *Water Safety, Security and Sustainability* (pp. 67–114). Springer International Publishing. https://doi.org/10.1007/978-3-030-76008-3_4.

Vaseashta, A., Ivanov, V., Stabnikov, V., & Marinin, A. (2021). Environmental safety and security investigations of neustonic microplastic aggregates nearwater-air interphase. *Polish Journal of Environmental Studies*, 30(4), 3457–3469. https://doi.org/10.15244/pjoes/131947.

Vaseashta, A., & Bölgen, N. (Eds.). (2022). *Electrospun Nanofibers: Principles, Technology, and Novel Applications*. Springer International Publishing. https://doi.org/10.1007/978-3-030-99958-2.

Vaseashta, A., Duca, G., & Travin, S.O. (Eds.). (2022). *Handbook of Research on Water Sciences and Society*. Engineering Science Reference.

Verma, R., Vinoda, K.S., Papireddy, M., & Gowda, A.N.S. (2016). Toxic pollutants from plastic waste – A review. *Procedia Environmental Sciences*, 35, 701–708. https://doi.org/10.1016/j.proenv.2016.07.069

Vethaak, A.D., & Legler, J. (2021). Microplastics and human health. *Science*, 371(6530), 672–674. https://doi.org/10.1126/science.abe5041.

Vitali, C., Peters, R.J.B., Janssen, H.-G., & Nielen, M.W.F. (2023). Microplastics and nanoplastics in food, water, and beverages; Part I: Occurrence. *TrAC Trends in Analytical Chemistry*, 159, 116670. https://doi.org/10.1016/j.trac.2022.116670.

Wang, B., Liang, B., Huang, Y., Li, Z., Zhang, B., Du, J. et al. (2023). Long-chain acyl carnitines aggravate polystyrene nanoplastics-induced atherosclerosis by upregulating MARCO. *Advanced Science*, 10(19), 2205876. https://doi.org/10.1002/advs.202205876.

Wang, K., Li, S., Petersen, M., Wang, S., & Lu, X. (2018). Detection and characterization of antibiotic-resistant bacteria using surface-enhanced Raman spectroscopy. *Nanomaterials*, 8(10), 762. https://doi.org/10.3390/nano8100762.

Wang, M., Zhang, S., Xu, Y., He, Y., Zhang, Y., Zhang, Z., & Liu, Y. (2021). Frequency response measurement of high-speed photodiodes based on a photonic sampling of an envelope-modulated microwave subcarrier. *Optics Express*, 29(7), 9836. https://doi.org/10.1364/OE.420662.

Weinstein, J.E., Crocker, B.K., & Gray, A.D. (2016). From macroplastic to microplastic: Degradation of high-density polyethylene, polypropylene, and polystyrene in a salt marsh habitat. *Environmental Toxicology and Chemistry*, 35(7), 1632–1640. https://doi.org/10.1002/etc.3432.

Windheim, J., Colombo, L., Battajni, N.C., Russo, L., Cagnotto, A., Diomede, L. et al. (2022). Micro- and nanoplastics' effects on protein folding and amyloidosis. *International Journal of Molecular Sciences*, 23(18), 10329. https://doi.org/10.3390/ijms231810329.

Xu, X., Goros, R.A., Dong, Z., Meng, X., Li, G., Chen, W. et al. (2023). Microplastics and nanoplastics impair the biophysical function of pulmonary surfactant by forming hetero aggregates at the alveolar–capillary interface. *Environmental Science & Technology*, 57(50), 21050–21060. https://doi.org/10.1021/acs.est.3c06668.

Yang, D., Shi, H., Li, L., Li, J., Jabeen, K., & Kolandhasamy, P. (2015). Microplastic pollution in table salts from China. *Environmental Science & Technology*, 49(22), 13622–13627. https://doi.org/10.1021/acs.est.5b03163.

Yang, Y., Xie, E., Du, Z., Peng, Z., Han, Z., Li, L. et al. (2023). Detection of various microplastics in patients undergoing cardiac surgery. *Environmental Science & Technology*, 57(30), 10911–10918. https://doi.org/10.1021/acs.est.2c07179.

You, C., Han, C., Wang, X., Zheng, Y., Li, Q., Hu, X., & Sun, H. (2012). The progress of silver nanoparticles in the antibacterial mechanism, clinical application, and cytotoxicity. *Molecular Biology Reports*, 39(9), 9193–9201. https://doi.org/10.1007/s11033-012-1792-8.

You, X., You, M., Lyu, Y., Peng, G., & Sun, W. (2022). Single and combined exposure to micro(nano)plastics and azithromycin disturbing the photosynthetic carbon fixation of *Synechocystis* sp. *Environmental Science: Nano*, 9(12), 4354–4366. https://doi.org/10.1039/D2EN00204C.

Yu, X., Ladewig, S., Bao, S., Toline, C.A., Whitmire, S., & Chow, A.T. (2018). Occurrence and distribution of microplastics at selected coastal sites along the southeastern United States. *Science of The Total Environment*, 613–614, 298–305. https://doi.org/10.1016/j.scitotenv.2017.09.100.

Yuan, Z., Nag, R., & Cummins, E. (2022). Human health concerns regarding microplastics in the aquatic environment: From marine to food systems. *Science of The Total Environment*, 823, 153730. https://doi.org/10.1016/j.scitotenv.2022.153730.

Zaki, M.R.M., & Aris, A.Z. (2022). An overview of the effects of nanoplastics on marine organisms. *Science of The Total Environment*, 831, 154757. https://doi.org/10.1016/j.scitotenv.2022.154757.

Zampolli, J., Mangiagalli, M., Vezzini, D., Lasagni, M., Ami, D., Natalello, A. et al. (2023). Oxidative degradation of polyethylene by two novel laccase-like multicopper oxidases from *Rhodococcus opacus* R7. *Environmental Technology & Innovation*, 32, 103273. https://doi.org/10.1016/j.eti.2023.103273.

Zeenat, Elahi, A., Bukhari, D.A., Shamim, S., & Rehman, A. (2021). Plastics degradation by microbes: A sustainable approach. *Journal of King Saud University – Science*, 33(6), 101538. https://doi.org/10.1016/j.jksus.2021.101538.

Zhang, S., Gumpper, R.H., Huang, X.-P., Liu, Y., Krumm, B.E., Cao, C. et al. (2022). Molecular basis for selective activation of DREADD-based chemogenetics. *Nature*, 612(7939), 354–362. https://doi.org/10.1038/s41586-022-05489-0.

Zhang, X., Chen, X., Liu, Z., Pan, X., Zheng, X., Li, Y. et al. (2024). Microbial colonization of microplastic (MP) in aquatic environments: MP toxicity, microbial degradation potential, and their interactions. *TrAC Trends in Analytical Chemistry*, 181, 118028. https://doi.org/10.1016/j.trac.2024.118028

Zhao, M., Huang, L., Arulmani, S.R.B., Yan, J., Wu, L., Wu, T. et al. (2022). Adsorption of different pollutants by using microplastic with different influencing factors and mechanisms in wastewater: A review. *Nanomaterials*, 12(13), 2256. https://doi.org/10.3390/nano12132256.

Zheng, H., Vidili, G., Casu, G., Navarese, E.P., Sechi, L.A., & Chen, Y. (2024). Microplastics and nanoplastics in cardiovascular disease: A narrative review with worrying links. *Frontiers in Toxicology*, 6, 1479292. https://doi.org/10.3389/ftox.2024.1479292.

Zhong, Y., Feng, Y., Huang, Y., Wang, B., Shi, W., Liang, B. et al. (2024). Polystyrene nanoplastics accelerate atherosclerosis: Unraveling the impact on smooth muscle cells through KIF15-mediated migration. *Ecotoxicology and Environmental Safety*, 284, 116983. https://doi.org/10.1016/j.ecoenv.2024.116983.

Zhu, X., Wang, C., Duan, X., Liang, B., Genbo Xu, E., & Huang, Z. (2023). Micro- and nanoplastics: A new cardiovascular risk factor? *Environment International*, 171, 107662. https://doi.org/10.1016/j.envint.2022.107662.

Index